GSM/EDGE

GSM/EDGE
EVOLUTION AND PERFORMANCE

Edited by

Mikko Säily, *Nokia Siemens Networks, Finland*

Guillaume Sébire, *Nokia, Finland*

Dr. Eddie Riddington, *Nokia Siemens Networks, UK*

A John Wiley and Sons, Ltd., Publication

Library of Congress Cataloging-in-Publication Data

Säily, Mikko.
  GSM/EDGE : evolution and performance / Mikko Säily, Guillaume Sébire, Eddie Riddington.
    p. cm.
  Includes index.
  ISBN 978-0-470-74685-1 (cloth)
  1. Global system for mobile communications.   I. Sébire, Guillaume.   II. Riddington, Eddie.   III. Title.
  TK5103.483.S24 2010
  621.3845′6–dc22

                                                                                    2010019375

A catalogue record for this book is available from the British Library.

Print ISBN: 9780470746851 (HB)
ePDF ISBN: 9780470669617
oBook ISBN: 9780470669624

Typeset in 10/12pt Times by Aptara Inc., New Delhi, India
Printed and Bound in Singapore by Markono Print Media Pte Ltd

Contents

Jürgen Hofmann, Vlora Rexhepi-van der Pol, Guillaume Sébire
and Sergio Parolari

Mikko Säily, Jari Hulkkonen, Kent Pedersen, Carsten Juncker,
Rafael Paiva, Renato Iida, Olli Piirainen, Seelan Sundaralingam,
Alexandre Loureiro, Jon Helt-Hansen, Robson Domingos and
Fernando Tavares

Robert Müllner, Carsten Ball, Kolio Ivanov, Markus Mummert,
Hubert Winkler and Kurt Kremnitzer

Acknowledgements

The editors would like to acknowledge the priceless work of contributors from Nokia Siemens Networks, Nokia and Instituto Nokia de Tecnologia (INdT): Carsten Ball, Alessandra Celin, Robson Domingos, David Gallegos, Cristina Gangai, Marcin Grygiel, Piotr Grzybowski, Khairul Hasan, Jon Helt-Hansen, Jürgen Hofmann, Michal Hronec, Jari Hulkkonen, Renato Iida, Kolio Ivanov, Rauli Järvelä, Carsten Juncker, Juha Karvinen, Kurt Kremnitzer, Krystian Krysmalski, Sebastian Lasek, Grzegorz Lehmann, Alexandre Loureiro, Sebastian Lysiak, Andrzej Maciolek, Krystian Majchrowicz, Robert Müllner, Markus Mummert, David Navrátil, Rafael Paiva, Sergio Parolari, Kent Pedersen, Olli Piirainen, Vlora Rexhepi-van der Pol, Sathiaseelan Sundaralingam, Fernando Tavares, Dariusz Tomeczko, Hubert Winkler, Eduardo Zacarías and Fabrizio Zizza.

We would also like to express our gratitude to our colleagues for fruitful discussions during the manuscript preparation and their support for the project, in particular: Riku Pirhonen, Timo Halonen, Peter Merz, Kari Niemelä, Eric Jones, Harry Kuosa, Jarmo Mäkinen and Kari Pehkonen.

We are indebted to Instituto Nokia de Tecnologia for joining the project with valuable contribution. We would like to thank Leonardo Aguayo and Daniel Barboni for their support.

Great thanks and a big hand go to the publishing team at Wiley: Mark Hammond, Sarah Tilley, Katharine Unwin, Sophia Travis and Alexandra King.

Our loving thanks go to our families and the families of all contributors for their patience, support and understanding during the late evenings and weekends devoted to the writing of this book.

Kiitos kaikille! Merci à tous! Thank you all!

The editors and contributors welcome comments, suggestions and improvements that could be implemented in possible forthcoming editions of this book. Please send your feedback to editors' e-mail addresses:

mikko.saily@nsn.com, guillaume.sebire@nokia.com and eddie.riddington@nsn.com, or use the following generic address: gsm_evo_book@ovi.com.

Acronyms

16QAM	Sixteen Quadrature Amplitude Modulation
2G	Second Generation
32QAM	Thirty-two Quadrature Amplitude Modulation
3G	Third Generation
3GPP	Third Generation Partnership Project
8-PSK	Octal Phase Shift Keying
α-QPSK	Alpha-QPSK
Abis	Interface between BTS and BSC
ABQP	Aggregate BSS QoS Profile
AC	Admission Control
ACCH	Associated Control Channel(s)
A-CELP	Algebraic CELP
Ack, ACK	Acknowledgement (positive)
ACR	Absolute Category Rating
AD, A/D	Analogue to Digital
ADC	AD Converter
AGC	Automatic Gain Control
AGI	Antenna Gain Imbalance
A-GNSS	Assisted GNSS
A-GPS	Assisted GPS
AM	Acknowledged Mode
AM	Amplitude Modulation
AMR	Adaptive Multi-Rate (speech)
AMR-NB	AMR Narrowband
AMR-WB	AMR Wideband
AoIP, AOIP	A interface over IP
APN	Access Point Name
AQPSK, α-QPSK	Adaptive Quadrature Phase Shift Keying
ARFCN	Absolute Radio Frequency Channel Number
ARIB	Association of Radio Industries and Businesses
ARNS	Aeronautical Radio Navigation Service
ARP	Allocation Retention Priority
ARQ	Automated Repeat reQuest
ASCI	Advanced Speech Call Items

ATIS	Alliance for Telecommunications Industry Solutions
BA	BCCH Allocation
BC	Band Category
BCC	Base Station Colour Code
BCCH	Broadcast Control Channel
BCR	Blocked Call Ratio
BEM	Block Edge Mask
BEP	Bit Error Probability
BER	Bit Error Rate
BLER	Block Error Ratio
BQC	Bad Quality Call
BQP	Bad Quality Probability
BS	Base Station
BSC	Base Station Controller
BSIC	Base Station Identity Code
BSS	Base Station Sub-system
BSSAP	Base Station Subsystem Application Part
BSSGP	Base Station Sub-system GPRS Protocol
BSSMAP	Base Station Subsystem Management Application Part
BTS	Base Transceiver Station
BTTI	Basic TTI (20 ms)
BVC	BSSGP Virtual Connection
C/I	Carrier to Interferer power ratio
CA	Certificate Authority
CAMEL	Customised Applications for Mobile networks Enhanced Logic
CAPEX	Capital Expenditures
CBC	Cell Broadcast Center
CBS	Cell Broadcast Service
CC	Convolutional Coding
CCN	Cell Change Notification
CCSA	China Communications Standards Association
CDF	Cumulative Distribution Function
CDMA	Code Division Multiple Access
CDR	Call Drop Rate
CELP	Code Excited Linear Prediction
CEPT	Conférence Européenne des Postes et Télécommunications (European conference of postal and telecommunications administrations)
CHD	Channel Decoder
CHE	Channel Encoder
CIC	Circuit Identification Code
CIR	Carrier to Interferer Ratio
CMC	Codec Mode Command
CMI	Codec Mode Indicator
CMR	Codec Mode Request
CN	Core Network

CRC	Cyclic Redundancy Check
CS	Circuit Switched
CS	Coding Scheme
CSD	Circuit-Switched Data
CSG	Closed Subscriber Group
CV_BEP	Coefficient of Variation of the Bit Error Probability
DA, D/A	Digital to Analogue
DAC	DA Converter
DAPD	Digital Adaptive Predistortion
DARP	Downlink Advanced Receiver Performance
DAS	EGPRS2-A Downlink modulation and coding scheme
DBS	EGPRS2-B Downlink modulation and coding scheme
DCA	Dynamic Channel Allocation
DCDL/DLDC	Dual Carrier Downlink
DCR	Dropped Call Ratio/Rate
DCS (DCS1800)	Digital Cellular System (in the 1800 MHz band)
DECT	Digital Enhanced Cordless Telecommunications
DFR	Double Full Rate (OSC)
DHR	Double Half Rate (OSC)
DiffServ	Differentiated Services
DIR	Dominant to Rest Interference Ratio
DL	Downlink (network to mobile)
DMA	Dynamic MAIO Allocation
DME	Distance Measurement Equipment
DSP	Digital Signal Processor
DTAP	Direct Transfer Application Part
DTM	Dual Transfer Mode
DTS	DARP Test Scenario
DTX	Discontinuous Transmission
EAP	Extensible Authentication Protocol
ECC PT	Electronic Communications Committee Project Team
ECSD	Enhanced CSD
EDA	Extended Dynamic Allocation
EDGE	Enhanced Data rates for Global Evolution
EFL	Effective Frequency Load
EFR	Enhanced Full-Rate (speech)
EGPRS	Enhanced GPRS
EGPRS2	Enhanced General Packet Radio Service 2
EGPRS2-A	EGPRS2 level A
EGPRS2-B	EGPRS2 level B
E-GSM	Extended GSM band
EMC	Electromagnetic Compatibility
EMR	Enhanced Measurement Reporting
E-OTD	Enhanced-Observed Time Difference
EPC	Evolved Packet Core (network)

EPS	Evolved Packet System
ETSI	European Telecommunications Standards Institute
E-UTRAN	Evolved UTRAN
FACCH	Fast Associated Control Channel
FANR	Fast Ack/Nack Reporting
FCC	Federal Communications Commission
FDD	Frequency Division Duplex
FEP	Frame Error Probability
FER	Frame Erasure Rate
FH	Frequency Hopping
FL	Fractional Load
FLO	Flexible Layer One
FN	Frame Number
FR	Full-Rate (speech)
FTP	File Transfer Protocol
GAN	Generic Access Network
GANC	GAN controller
GEA	GPRS Encryption Algorithm
GERAN	GSM/EDGE Radio Access Network
GGSN	Gateway GPRS Support Node
GMM	GPRS Mobility Management
GMSK	Gaussian Minimum Shift Keying
GNSS	Generic Navigational Satellite System
GPRS	General Packet Radio Service
GPS	Global Positioning System
GSM	Global System for Mobile communications
GSM	Groupe Spécial Mobile (Special Mobile Group)
GSM-R	GSM Railway
HeNB	Home eNode B
HLR	Home Location Register
HNB	Home Node B
HO	Handover
HR	Half-Rate (speech)
HS	Harmonized Standard
HSCSD	High-Speed CSD
HSN	Hopping Sequence Number
HSPA	High-Speed Packet Access
HW	Hardware
ICI	Inter-Channel Interference
IE	Information Element
IETF	Internet Engineering Task Force
IF	Intermediate Frequency
IM	Intermodulation
IMD	Intermodulation Distortion
IMS	IP Multimedia Subsystem

IntServ	Integrated Services
IP	Internet Protocol
IRC	Interference Rejection Combining
ISDN	Integrated Services Data Network
ITU-T	International Telecommunication Union – Telecommunications Standardization Sector
JD	Joint Detection
KPI	Key Performance Indicator
L1	Layer 1 (Physical Layer)
L2	Layer 2 (Data Link Layer)
L2S	Link-to-System Mapping
L3	Layer 3 (Network Layer)
LAN	Local Area Network
LAPDm	Link Access Protocol on the Dm channel
LGMSK	Linearized Gaussian Minimum Shift Keying
LLC	Logical Link Control (protocol)
LNA	Low Noise Amplifier
LO	Local Oscillator
LPC	Linear Predictive Coding
LTE	Long Term Evolution
LTP	Long Term Predictor/Prediction
M3UA	MTP Level 3 (MTP3) User Adaptation Layer (SIGTRAN)
MA	Mobile Allocation
MAC	Medium Access Control (protocol)
MAIO	Mobile Allocation Index Offset
MBMS	Multimedia Broadcast and Multicast Service
MCBTS	Multicarrier BTS
MCPA	Multicarrier Power Amplifier
MCS	Modulation and Coding Scheme
MEAN_BEP	Mean of Bit Error Probability
MGW	Media Gateway
MIMO	Multiple Input Multiple Output
MMI	Man-Machine Interface
MOS	Mean Opinion Score
MPLS	Multi-protocol Label Switching
MS	Mobile Station
MSC	Mobile Switching Center
MSR	Multi-Standard Radio
MSRD	MS Receiver Diversity
MTP	Message Transfer Part
MTS	MUROS Test Scenario
MUROS	Multi-User Reusing One Slot
MUSHRA	MUlti Stimulus test with Hidden Reference and Anchors
NACC	Network Assisted Cell Change
Nack, NACK	Negative Acknowledgement

NB	Narrowband
NCCR	Network Controlled Cell Re-selection
NPM	Non-Persistent Mode
NRT	Non-Real-Time
OPEX	Operating expenditures
OSC	Orthogonal Sub-channels
O-TCH/WFS	Octal Traffic Channel Wideband Full-Rate Speech
O-TCH/WHS	Octal Traffic Channel Wideband Half-Rate Speech
OTD	Observed Time Difference
PA	Power Amplifier
PACCH	Packet Associated Control Channel
PAMR	Private Access Mobile Radio
PAN	Piggy-backed Ack/Nack
PAR	Peak-to-Average power Ratio
PC	Power Control
PCM	Pulse Code Modulation
PCS	Personal Communications Service
PCU	Packet Control Unit
PDAN	Packet Downlink Ack/Nack
PDCH	Packet Data Channel
PDH	Plesiochronous Digital Hierarchy
PDP	Packet Data Protocol
PDTCH	Packet Data Traffic Channel
PDU	Packet Data Unit / Protocol Data Unit
PESQ	Perceptual Evaluation of Speech Quality
PFC	Packet Flow Context
PFI	Packet Flow Identifier
PFM	Packet Flow Management
P-GSM	Primary GSM band
PIN	Personal Identification Number
PLMN	Public Land Mobile Network
PM	Phase Modulation
POTS	Plain Old Telephony Service
PPC	Progressive Power Control
PS	Packet-Switch(ed)
PS	Packet Switched
PSI	Packet System Information
PSK	Phase Shift Keying
PSN	Packet Switched Network
PSTN	Public-Switched Telephone Network
PTCCH	Packet Timing Control Channel
PUAN	Packet Uplink Ack/Nack
QAM	Quadrature Amplitude Modulation
QC	Quality Control
QoS	Quality of Service

QPSK	Quadrature Phase Shift Keying
RA	Routing Area
RAT	Radio Access Technology
RB	Radio Block
RF	Radio Frequency
RFC	Request For Comment
RLC	Radio Link Control (protocol)
RLF	Radio Link Failure
RLT	Radio Link Timeout
RPE	Regular Pulse Excitation
RPE-LTP	Regular Pulse Excited Long Term Prediction
RR	Radio Resource
RRC	Radio Resource Control
RRC	Root Raised Cosine
RRH	Remote Radio Head
RRM	Radio Resource Management
RT	Real Time
RTT	Round Trip Time
RTTI	Reduced TTI (10 ms)
Rx, RX	Receive
RXLEV	Received Power level
RXQUAL	Received Quality
SABM	Set Asynchronous Balanced Mode
SACCH	Slow Associated Control Channel
SAIC	Single Antenna Interference Cancellation
SCBTS	Single carrier BTS
SCCP	Signaling Connection Control Part
SCH	Synchronization Channel
SCPIR	Sub-channel Power Imbalance Ratio
SCTP	S Common Transport Protocol
SDCA	Service Dependent Channel Allocation
SDCCH	Stand-alone Dedicated Control Channel
SDH	Synchronous Digital Hierarchy
SDU	Service Data Unit
SE	Spectrum Efficiency
SEGW	Security Gateway
SGSN	Serving GPRS Support Node
SI	System Information
SIC	Successive Interference Cancellation
SID	Silence Indicator Description
SIGTRAN	Signaling Transport protocol stack for PSTN signaling over IP
SIM	Subscriber Identity Module
SIM ATK	SIM Application Toolkit
SINR	Signal to Interference and Noise Ratio
SMA	Static Mobile Allocation

S-MIC	Successive Mono Interference Cancellation
SMLC	Serving Mobile Location Center
SMS	Short Message Service
SNDCP	Sub Network Dependent Convergence Protocol
SNR	Signal to Noise Ratio
SPD	Speech Decoder
SPE	Speech Encoder
SSID	Service Set Identifier
SSN	Starting Sequence Number
STIRC	Space Time Interference Rejection Combining
SW	Software
TAPS	TETRA Advanced Packet Service
TBF	Temporary Block Flow
TC	Transcoder
TC	Turbo Coding
TC MSG	Technical Committee Mobile Standards Group
TCH	Traffic Channel
TCH/AFS	Traffic Channel using Adaptive Multi-Rate Full Rate Speech
TCH/AHS	Traffic Channel using Adaptive Multi-Rate Half Rate Speech
TCH/EFS	Traffic Channel using Enhanced Full Rate Speech
TCH/FS	Traffic Channel using Full Rate Speech
TCH/HS	Traffic Channel using Half Rate Speech
TCH/WFS	Traffic Channel using Wideband Adaptive Multi-Rate Full Rate Speech and GMSK
TCO	Total Cost of Ownership
TCP	Transmission Control Protocol
TDD	Time Division Duplex
TDM	Time Division
TDMA	Time Division Multiple Access
TETRA	Terrestrial Trunked Radio
TFES	Task Force for ETSI ERM and MSG for Harmonized Standards for IMT-2000
TFO	Tandem Free Operation
T-GSM	Trunked GSM
THP	Traffic Handling Priority
THR	Threshold
TIA/IS	Telecommunications Industry Association / International Standard
TOA	Time Of Arrival
TOM	Tunneling of Messages
TOP	Temporary Overpower
TP	Throughput
TR	Technical Report
TRAU	Transcoder and Rate Adaptation Unit
TrFO	Transcoder Free Operation
TRX	Transceiver

TS	Technical Specification
TS	Time slot
TSC	Training Sequence Code
TSG	Technical Specification Group
TTA	Telecommunications Technology Association
TTC	Telecommunication Technology Committee
TTI	Transfer Time Interval
TUxx	Typical Urban environment with MS speed of xx km/hr
Tx; TX	Transmit
UA	Unnumbered Acknowledgement
UAS	EGPRS2-A Uplink modulation and coding scheme
UBS	EGPRS2-B Uplink modulation and coding scheme
UDP	User Datagram Protocol
UE	User Equipment
UEM	Unwanted Emission Mask
UL	Uplink (mobile to network)
UM	Unacknowledged Mode
UMA	Unlicensed Mobile Access
UMTS	Universal Mobile Telecommunication System
URL	Uniform Resource Locator
USF	Uplink State Flag
U-TDOA	Uplink Time Difference Of Arrival
UTRAN	UMTS Terrestrial Radio Access Network
VAMOS	Voice over Adaptive Multi-user channels on One Slot
VGCS	Voice Group Call Service
VoIP	Voice over IP
V-SELP	Vector Sum Excited Linear Prediction
WAPECS	Wireless Access Policy for Electronic Communications Services
WB	Wideband
WCDMA	Wideband Code Division Multiple Access
WG	Working Group
WLAN	Wireless LAN
WWW	World Wide Web

Part I
GSM/EDGE Standardization

1

GSM Standardization History

Guillaume Sébire

1.1 Introduction

GSM, the Global System for Mobile communications, owes its worldwide success to the continued progressive and backward-compatible evolution of its open industry standard and to visionary yet simple ideas such as global roaming – enabling, thanks to a harmonized spectrum, the use of a device with the same number outside its home network, multivendor environment – enabling different vendors to implement with sufficient freedom compatible products based on the same standard, SMS[1] –enabling people to text each other, etc.

First aimed at providing mobile voice communications, GSM developed early on into a rich system offering supplementary services and other data communications, well ahead of the analogue systems then sporadically deployed in several regions of the world and which were incompatible. GSM has been and is by far the most widely used and most successful communications system of all time, enabling, at the time of writing, over four billion subscribers [1] to communicate in just about every single country of the world, just about everywhere (including airplanes) and with virtually everyone. The success of GSM is, simply put, staggering. Of all the active digital mobile subscriptions worldwide, more than 80% are GSM [2].

This chapter relates the history of GSM standardization from the early 1980s to the late 2000s and lists the main features and functionalities that have gradually been introduced in GSM specifications.

---

[1] SMS: Short Message Service.

GSM/EDGE: Evolution and Performance   Edited by Mikko Säily, Guillaume Sébire and Eddie Riddington
© 2011 John Wiley & Sons, Ltd

1.2   History

Initially launched as a European initiative in 1982 by CEPT,[2] the Groupe Spécial Mobile (Special Mobile Group) was tasked to develop a standard for mobile telephony across Europe in the 900 MHz band. Five years later in 1987, the signature by thirteen countries[3] of a Memorandum of Understanding to develop a pan-European common cellular telephony system in the 900 MHz band marked the official birth of GSM, set for service launch in 1991. ETSI, the European Telecommunications Standards Institute, created in 1988 by CEPT to handle all telecommunication standardization activities, became in 1989 the sole entity responsible for the GSM standard.

By 1990, the first set of specifications, GSM Phase 1, was frozen and published. By 1995, GSM Phase 2 was available, followed a couple of years later by GSM Phase 2+ which also introduced the concept of "yearly" release. The publication of the specifications into backward-compatible phases/releases has been a cornerstone of the evolution of the GSM standard and a model for future standards. It has ensured the availability in the specifications of a same phase/release of a consistent set of services, functionalities and features on both network and terminal sides and the inherent compatibility between equipment of different phases/releases. Since the first "release" in Phase 2+, known as Release 96 (or R96), nine others have been published (or are being developed): R97, R98, R99, Release 2000 later renamed Rel-4, Rel-5, Rel-6, Rel-7, Rel-8 and Rel-9. Release 9 is still in the making at the time of writing while stage 1 requirements for Release 10 are being laid out. Release 4 marked the transfer of GSM specifications within the Third Generation Partnership Project or 3GPP in the year 2000.

3GPP was established in December 1998 as a collaboration project between ETSI (Europe), ARIB[4] (Japan), TTC[5] (Japan), ATIS[6] (North America), TTA[7] (South Korea) and CCSA[8] (China) to develop a global third generation mobile phone system specification, that is UMTS commonly referred to as "3G". Though originating from GSM concepts and seen as part of the GSM family, UMTS is not as such an evolution of GSM. It was developed as a new system using GSM as a model. UMTS requires a new radio interface and the deployment of brand new radio networks, and thus is not backward-compatible with GSM – a UMTS phone cannot work in a GSM system nor can a GSM phone work in a UMTS system. The UMTS core network and architecture, however, though requiring new IP-based interfaces, were largely based on the GSM core network and architecture.

In the following sub-sections, non-exhaustive lists of services, features and functionalities characterizing each GSM phase/release are provided.

---

[2] CEPT: Conférence Européenne des Postes et Télécommunications (the European Conference of Postal and Telecommunications Administrations).

[3] Belgium, Denmark, Finland, France, Germany, Ireland, Italy, the Netherlands, Norway, Portugal, Spain, Sweden and the United Kingdom.

[4] ARIB: Association of Radio Industries and Businesses.

[5] TTC: Telecommunication Technology Committee.

[6] ATIS: Alliance for Telecommunications Industry Solutions.

[7] TTA: Telecommunications Technology Association.

[8] CCSA: China Communications Standards Association.

1.3   Phase 1

GSM Phase 1 contains the following items:

- Basic telephony using full-rate speech codec (FR) at 13 kbit/s [3] with a speech quality comparable to that of POTS[9] wireline.
- Emergency calls using a single number ("112") even if the SIM[10] is not present or the PIN[11] is not entered. The growing deployment of GSM outside Europe led to the introduction in Phase 2+ (R96) of additional numbers (in the SIM) to be regarded as emergency numbers.
- Support for multiple data services (up to 9.6 kbit/s) allowing, for example interconnection with ISDN,[12] modem connection through PSTN.[13]
- Security through authentication and confidentiality in order to protect operators and subscribers against malicious actions by third parties. Authentication to verify and confirm a subscriber's identity. Confidentiality to preserve the privacy of a given piece of information [4]. See Chapter 4 for more details on the evolution of GSM security.
- Short message service (SMS) either point-to-point or using cell broadcast [5, 6].
- Supplementary services pertaining to call barring and call forwarding such as barring of all incoming calls, barring of incoming calls when roaming outside the home network, call forwarding on no reply, call forwarding on mobile subscriber busy, etc.
- Support for facsimile (fax) communications (Group 3: the most widely used) [7, 8].

1.4   Phase 2

GSM Phase 2 contains the following items:

- Half-rate speech codec (HR) at 5.6 kbit/s allowing a higher maximum number of voice users compared to FR speech, at the expense of speech quality [9].
- Enhanced Full-Rate speech codec (EFR) at 12.2 kbit/s. EFR provides a considerable speech quality improvement over FR [10].
- Half-rate data services allowing a higher maximum number of data users.
- SMS enhancements such as SMS concatenation, replacement.
- Supplementary services such as enhancements to call barring and forwarding, calling line identification presentation and restriction, multiparty calls, etc.
- Fax enhancements.
- Support of GSM in the 1800 MHz band that is DCS[14] 1800 as well as interworking between GSM 900 and DCS 1800, and multi-band operation by a single operator.

---

[9] POTS: Plain Old Telephone Service.
[10] SIM: Subscriber Identity Module.
[11] PIN: Personal Identification Number.
[12] ISDN: Integrated Services Data Network.
[13] PSTN: Public-Switched Telephone Network.
[14] DCS: Digital Cellular System.

1.5   Phase 2+

1.5.1   Phase 2+, R96

Release 96 contains the following items:

- Data services at 14.4 kbit/s.
- High-Speed Circuit-Switched Data (HSCSD) allowing the use of multiple 9.6 kbit/s or 14.4 kbit/s channels in one direction for considerably faster data transfers. HSCSD offers data rates up to 38.4 kbit/s (four times 9.6 kbit/s) or 57.6 kbit/s (four times 14.4 kbit/s) for Type 1 mobile stations that is mobile stations not required to transmit and receive at the same time [11].
- ASCI (Advanced Speech Call Items) Phase 1 for GSM railway systems (GSM-R) containing for example Voice Broadcast Service (VBS) calls supporting one talker and several listeners and Voice Group Call Service (VGCS) allowing calls supporting several talkers and listeners [12, 13].
- CAMEL (Customized Applications for Mobile networks Enhanced Logic) Phase 1. CAMEL enables the definition of services on top of existing GSM services such as allowing using the same phone number when roaming outside one's home network. CAMEL Phase 1 offers call control related functionalities.
- SIM Application Toolkit (SIM ATK) which provides standardized means for applications (e.g. banking, weather) residing on the SIM to interact proactively with the mobile station.
- Support of additional call set-up MMI procedures allowing emergency calls to be placed with emergency numbers stored in the SIM thus catering for the expansion of GSM in countries using other numbers than "112" for emergencies.

1.5.2   Phase 2+, R97

Release 97 contains the following items:

- GPRS (General Packet Radio Service) allowing packet-switched data connections down to the GSM radio interface thus providing a more efficient use of network and radio resources compared to circuit-switched data. Resources are assigned when data are transmitted, and released otherwise thus creating packet transmissions. Four coding schemes, CS-1 to CS-4 using GMSK[15] modulation, and link adaptation allow adaption of the channel coding to the channel conditions thus enabling an efficient use of radio resources. Data rates up to 20 kbit/s per time slot per direction are possible [13].
- GPRS encryption using the GPRS-A5 algorithm (GEA[16]). See Chapter 4 for more details on the evolution of GSM security.
- Security mechanisms for SIM ATK.
- ASCI Phase 2.
- CAMEL Phase 2.

---

[15] GMSK: Gaussian Minimum Shift Keying.
[16] GEA: GPRS Encryption Algorithm.

1.5.3 Phase 2+, R98

Release 98 contains the following items:

- AMR (Adaptive Multi-Rate speech codec): Definition of mechanisms to support AMR speech (narrow-band or AMR-NB) in GSM enabling the adaptation of the speech codec to the link quality and/or capacity requirements by means of an optimized link and codec adaptation. To this end, eight codec modes are defined: 4.75, 5.15, 5.9, 6.7, 7.4, 7.95, 10.2 and 12.2 kbit/s.[17] The higher the bitrate, the higher the source coding and the weaker the channel coding. In GSM, AMR channel coding is defined for GMSK full-rate channels (TCH/AFS) for all codec modes and for GMSK half-rate channels (TCH/AHS) for 4.75 kbit/s to 7.95 kbit/s codec modes [15]. AMR has also been defined as the default speech codec for UMTS.
- Location Services (LCS) in CS Domain: Definition of mechanisms to support location technologies in GSM based on cell identity (or Cell ID), E-OTD,[18] TOA[19] and A-GPS.[20] LCS supports both mobile station-assisted positioning (the terminal takes the measurements while the network calculates the position) and mobile station-based positioning (the mobile station both takes the measurements and calculates the position).
- Support of GSM/GPRS in the 1900 MHz band (PCS 1900).

1.5.4 Phase 2+, R99

Release 99 contains the following items:

- EDGE (Enhanced Data rates for Global Evolution): EDGE was specified as a global evolution path for GSM operators, and, in the US, TDMA operators. Through the introduction of the 8-PSK[21] modulation on the GSM air interface for both packet-switched data (EGPRS – Enhanced GPRS) and circuit-switched data (ECSD – Enhanced CSD), EDGE boosts peak and average data rates as well as network capacity. EGPRS provides data rates up to 59.2 kbit/s per time slot per direction (i.e. up to 473.6 kbit/s per 200kHz carrier), supports incremental redundancy (Hybrid Type II ARQ[22]), introduces a wide range of modulation and coding schemes using GMSK and 8-PSK modulations, as well as new link quality measurement quantities, the combination of which allows accurate link adaptation in varying channel conditions [16]. ECSD provides data rates up to 43.2 kbit/s per time slot, hence significantly reducing the need to allocate multiple time slots to increase data rates for CS data and thus addressing a main issue inherent to HSCSD: capacity. ECSD also allows fast power control.
- DTM (Dual Transfer Mode): Definition of mechanisms to support parallel CS and PS connections on the same carrier for a given mobile station, hence significantly reducing the

---

[17] AMR 12.2 speech codec is compatible with EFR.
[18] E-OTD: Enhanced-Observed Time Difference.
[19] TOA: Time Of Arrival.
[20] A-GPS: Assisted GPS.
[21] 8-PSK: Octal Phase Shift Keying.
[22] ARQ: Automated Repeat reQuest.

complexity otherwise implied for class A mobile stations (where the CS and PS connections are independent) [17]. A DTM mobile station is hence referred to as a simple Class A mobile station. DTM multislot classes 1, 5 and 9 are supported [18].

- Enhanced Measurement Reporting: Definition of mechanisms allowing the reporting of more than six neighbor cells for mobility purpose thus providing benefits for multiband and/or multisystem scenarios (e.g. GSM and UMTS).
- GSM/UTRAN[23] Interworking: Definition of mechanisms to support interworking (mobility) between GSM and UMTS. Handover from GSM to UMTS is specified in the CS domain, and cell re-selection otherwise.
- Support of A5/3 and GEA3: Definition of the support in GSM of the KASUMI f8 algorithm introduced in UMTS, using a ciphering key of 64 bits. See Chapter 4 for more details on the evolution of GSM security.
- GSM 450 band: Uplink 450.4–457.6 MHz and Downlink 460.4–467.6 MHz (GSM400).
- GSM 480 band: Uplink 478.8–486 MHz and Downlink 488.8–496 MHz (GSM400).
- GSM 850 band: Uplink 824–849 MHz and Downlink 869–894 MHz (GSM850).

1.5.5   Phase 2+, Rel-4

Release 4 contains the following items:

- NACC (Network Assisted Cell Change): definition of mechanisms allowing the mobile station to notify the cell to which it will reselect (Cell Change Notification), and the network to provide in turn system information pertaining to this cell thereby reducing the access time in this cell.
- Delayed Downlink TBF Release/Extended Uplink TBF mode: Definition of mechanisms maintaining a layer 2 link (TBF[24]) in a given direction between a mobile station and the network when no data is exchanged between the mobile station and the network on this TBF, hence avoiding TBF re-establishment when new data is incoming, thus improving latency and reducing signaling.
- DTM Multislot Class 11 [19].
- Extended DTM Multislot Class: allowing the support of half-rate PDCH (Packet Data Channel) together with full-rate PDCH(s) in DTM.
- Gb over IP: Definition of IP transport over the Gb interface between the BSC[25] and the SGSN,[26] as an alternative to frame relay.
- GERAN/UTRAN Interworking additions: Definition of interworking between GERAN and UTRAN low chip-rate TDD.[27]

---

[23] UTRAN: UMTS Terrestrial Radio Access Network.
[24] TBF: Temporary Block Flow.
[25] BSC: Base Station Controller.
[26] SGSN: Serving GPRS Support Node.
[27] TDD: Time Division Duplex.

- Dynamic ARFCN Mapping: Definition of mechanisms to allow dynamic mapping of ARFCN (Absolute Radio Frequency Channel Number) to different bands hence overriding the fixed ARFCN numbering. An ARFCN identifies a GSM carrier (200 kHz).
- GSM 750 band: Downlink 747–762 MHz and Uplink 777–792 MHz (GSM700).

1.5.6   Phase 2+, Rel-5

Release 5 contains the following items:

- Support of High Multislot Classes for (E)GPRS.
- GERAN Iu mode: Definition of architecture and mechanisms to connect the GSM/EDGE Radio Access Network to the same Core Network as UTRAN via the Iu interface (Iu-ps and Iu-cs). The Iur-g interface was also defined allowing the connection and exchange of control-plane information between two BSCs or between a BSC and an RNC, similar to the Iu-r interface in UTRAN [20].
- Wideband AMR (AMR-WB): Definition of mechanisms to support AMR-WB speech codec [21] in GSM. AMR-WB yields a major improvement of speech quality over other speech codecs and exceeds POTS wireline voice quality. Five codec modes (6.60, 8.85, 12.65, 15.85 and 23.85 kbit/s)[28] are supported through the use of 8-PSK full-rate channels (O-TCH/WFS). Codec modes up to and including 12.65 kbit/s are supported through the use of GMSK full-rate channels (TCH/WFS) and 8-PSK half-rate channels (O-TCH/WHS). There is no support for GMSK half-rate channels.
- AMR 8-PSK HR: Definition of layer 1 and layer 3 (RR[29]) support for AMR (4.75–12.2 kbit/s) through the use of 8-PSK half-rate channels (O-TCH/AHS).
- EPC (Enhanced Power Control): Definition of Enhanced Power Control allowing faster Power Control for GMSK and 8-PSK channels, through the use of power control (and reporting) on every SACCH[30] block (occurring every 120 ms) instead of every SACCH frame. A SACCH frame consists of four SACCH blocks thus occurs every 480 ms.
- eNACC (External NACC): Definition of mechanisms to support external NACC, that is NACC between two BSCs, through the introduction of RIM (RAN Information Management) procedures allowing the exchange of information (e.g. system information) between two BSCs [22].
- Flow Control over the Gb interface: Definition of mechanisms to allow the SGSN to adapt its scheduling of data over the Gb interface, according to the scheduling (leak rate) of the PFCs[31] on the radio interface [23, 24].
- Connection of a BSC to multiple Core Network nodes: Definition of mechanisms allowing a BSC to connect to multiple SGSNs, MSCs[32].

---

[28] It should be noted that a total of nine codec modes are defined for AMR-WB: 6.60, 8.85, 12.65, 14.25, 15.85, 18.25, 19.85, 23.05 and 23.85 kbit/s.

[29] RR(C): Radio Resource (Control) protocol.

[30] SACCH: Slow Associated Control Channel.

[31] PFC: Packet Flow Context.

[32] MSC: Mobile Switching Centre.

- Improvements to GSM/UTRAN interworking, for example compressed inter RAT handover information.
- Location Services in the PS Domain.

1.5.7   Phase 2+, Rel-6

Release 6 contains the following items:

- PS Handover: Definition of mechanisms allowing the assignment of PS resources to a mobile station in a target cell prior to the mobile station being handed over to that cell [25].
- Multiple TBFs: Definition of MAC mechanisms allowing parallel TBFs in downlink and/or uplink between a mobile station and the network to enable better multiplexing between data flows of different quality of service [26].
- MBMS (Multimedia Broadcast and Multicast Service): Definition of mechanisms to support Multimedia Broadcast and Multicast Service in GERAN, allowing the network to send (MBMS) data to a plurality of mobile stations on the same radio resources [27].
- GAN (Generic Access Network): Definition of mechanisms and architecture allowing access to GSM services (via the A and Gb interfaces) through an internet access (using e.g. Wireless LAN, BlueTooth) by tunneling non-access stratum protocols between the network and the mobile station. This allows for example access to GSM outside GSM radio coverage [28].
- DARP Phase 1 (Downlink Advance Receiver Performance Phase 1): Improvements of the reception performance of the mobile station through the use of single antenna interference cancellation (SAIC).
- ACCH Enhancements: Definition of mechanisms to increase the robustness of FACCH[33] and SACCH by means of repetition and, when supported, chase combining as the robustness of the traffic channel increases. ACCH enhancements are supported with legacy terminals.
- FLO (Flexible Layer One): Definition of mechanisms allowing the configuration of the layer 1 at call set-up, for PS domain, thus allowing optimized support of IMS[34] services in GERAN Iu mode only [29].
- DTM Enhancements: Definition of mechanisms to allow direct transition between packet transfer mode and dual transfer mode, without releasing the PS resources (i.e. without transit through packet idle mode) [30].
- Support of High Multislot Classes for DTM (E)GPRS.
- Definition of A5/4 and GEA4: Definition of the support in GSM of the KASUMI f8 algorithm using a 128-bit encryption key, thus aligning the security level of all 3GPP radio access technologies. However, the signaling support allowing the use of A5/4 and GEA4 was completed in Release 9. See Chapter 4 for more details on the evolution of GSM security.
- TETRA[35] (TAPS) – T-GSM 380 band: Uplink 380.2–389.8 MHz and Downlink 390.2–399.8 MHz (GSM400).

---

[33] FACCH: Fast Associated Control Channel.

[34] IMS: IP Multimedia Sub-system.

[35] TETRA: Terrestrial Trunked Radio. TETRA specified by ETSI is meant for professional usage (PMR: Professional Mobile Radio) by for example transportation, public safety or other military organizations. TAPS, TETRA Advanced Packet Service, is an adaptation of GPRS/EGPRS for data communications only.

- TETRA (TAPS) – T-GSM 410 band: Uplink 410.2–419.8 MHz and Downlink 420.2–29.8 MHz (GSM400).
- TETRA (TAPS) – T-GSM 900 band: Uplink 870.4–915.4 MHz and Downlink 915.4–921 MHz (GSM900).
- U-TDOA (Uplink time difference of arrival): Definition of mechanisms to support location service for both GSM and GPRS using the time difference between the received signals from a mobile station to determine its position. The support for U-TDOA was driven by FCC E911 requirements in the US [31].

1.5.8 Phase 2+, Rel-7

Release 7 contains the following items:

- GSM Onboard Aircrafts: Not part of 3GPP specifications the work to define GSMOBA,[36] initiated by a mandate of the European Commission, spanned 3GPP Releases 6 and 7 timeframes. To ensure compliance with 3GPP requirements and essentially compatibility with legacy GSM terminals, 3GPP involvement was necessary. It consisted of 3GPP reviewing and guiding the design of the GSMOBA system which was under the responsibility of the ETSI GSMOBA and CEPT ECC PT SE7 groups. The airborne GSM system provides GSM/GPRS connectivity in an aircraft cabin during cruise flight (above 3000 meters) enabling phone calls, SMS, and other data exchange, for example e-mail. It operates in the 1800 MHz band and ensures, by means of a Network Control Unit (NCU) installed in the cabin, that any harmful interference to a terrestrial mobile network is prevented. To this end, the NCU transmits on at least the GSM400, GSM900, DCS1800 and UMTS bands, a wideband noise signal of which the power can be adjusted as a function of the altitude of the aircraft. GSM coverage in the cabin is provided by an onboard GSM BTS (OBTS) which is further connected by means of a satellite link to the terrestrial mobile network. The OBTS makes use of uplink power control to limit the transmit power of terminals to its lowest specified level that is 0dBm [34, 35].
- EGPRS2: Definition of mechanisms to support (combinations of) higher order modulations (16QAM, 32QAM), turbo coding and higher modulation symbol rate to boost data rates up to twice those of EGPRS. Two levels, EGPRS2-A and EGPRS2-B, were specified in both uplink and downlink directions.
- LATRED (Latency Reduction): Latency Reduction features, that is RTTI and FANR.
- RTTI (Reduced TTI[37]): 10 ms over two time slots, instead of 20 ms over one time slot.
- FANR (Fast Ack/Nack reporting): Definition of a mechanism which consists of piggy-backing RLC[38] acknowledgement information within RLC/MAC blocks for data transfer. It is also known as PAN (Piggy-backed Ack/Nack).
- RLC non-persistent mode: Definition of RLC protocol behavior where RLC retransmissions are allowed for a limited amount of time. This can be seen as a hybrid RLC mode between RLC acknowledged mode and RLC unacknowledged mode where the RLC performance

---

[36] GSMOBA: GSM OnBoard Aircrafts. Also known as MCA (Mobile Communications onboard Aircrafts).
[37] TTI: Transfer Time Interval.
[38] RLC: Radio Link Control.

is significantly increased compared to unacknowledged mode while maintaining a delay budget unlike the RLC acknowledged mode. It was first introduced in Release 6 for MBMS, and expanded in Release 7 to other applications.

- DCDL or DLDC (Downlink Dual Carrier): Definition of mechanisms for the transmission of data to a mobile station over two simultaneous independent downlink carriers (200 kHz), hence enabling higher data rates compared to EGPRS.
- DARP Phase 2 (Downlink Advanced Receiver Performance Phase 2): Also known as mobile station receiver diversity (MSRD), DARP Phase 2 improves the reception of a transmitted signal by using two antennas to enable diversity techniques between the two received signals.
- PS Handover between GAN and GERAN and GAN and UTRAN.
- DTM Handover: Definition of mechanisms to support concurrent handovers of CS and PS resources in DTM [36, 37].
- SIGTRAN[39] support over A, Lb, Lp interfaces: Definition of IP transport for control-plane traffic on the A interface (between the BSC and the MSC), the Lb interface (between the BSC and the SMLC[40]) and the Lp interface (between the MSC and the SMLC).
- A-GNSS: Definition of a generic signaling method to support navigational satellite systems other than GPS, for example support of Galileo [31] (see Release 7 version).
- GSM 710 band: Downlink 698–716 MHz and Uplink 728–746 MHz (GSM700).
- T-GSM 810 band: Uplink 806–821 MHz and Downlink 851–866 MHz.
- A-GPS minimum performance requirements, aligned with UTRAN.
- Mobile Station Antenna Performance Evaluation Method and Requirements.
- VGCS[41] Enhancements.

1.5.9   Phase 2+, Rel-8

Release 8 contains the following items:

- GERAN/E-UTRAN[42] Interworking: Definition of mechanisms allowing interworking (mobility) between GERAN and E-UTRAN, in the direction from GERAN to E-UTRAN.
- A interface over IP: Definition of IP transport for user-plane traffic over the A interface (specifically, between the BSC and the MGW) [32].
- Gigabit Gb interface: Definition of a 1000-fold increase of the data rates supported over the Gb interface, up to 6 Gbit/s, hence allowing more mobile stations with faster data transfers and thereby preventing the radio access from being limited by the capacity of the Gb interface.
- GAN Iu: Generic Access to the Iu interface, reusing the same principles as introduced for GAN, in Release 6 [33].

---

[39] SIGTRAN: Signaling Transport protocol stack for PSTN signaling over IP.
[40] SMLC: Serving Mobile Location Center.
[41] VGCS: Voice Group Call Service.
[42] E-UTRAN: Evolved UTRAN. E-UTRAN refers to the radio access network part of the EPS (Evolved Packet System), the core network part being referred to as EPC (Evolved Packet Core). E-UTRAN and EPC are commonly known as LTE (Long-Term Evolution).

- MUROS (Multiple User Re-using One Slot): Feasibility study to select a technique allowing several voice users sharing the same radio resources, hence yielding voice capacity improvement.
- WIDER (Wideband pulse shape for RED HOT level B): Feasibility study to select an optimized pulse shape for EGPRS2-B in the downlink in order to exploit fully the benefits of the higher modulation symbol rate used in EGPRS2-B.

1.5.10 Phase 2+, Rel-9

- GERAN aspects of Home (e)Node B Enhancements: Definition of mechanisms allowing interworking (mobility) between GERAN and (E-)UTRAN Home (e)Node Bs, in connected mode.
- Local-Call Local Switch: Definition of mechanisms allowing the two parties of a call served by the same BSS to be locally switched by the BSS (with involvement of the MSC).
- Support of A5/4 and GEA4: Definition of the signaling means in GSM to allow the use of the KASUMI f8 algorithm using a 128-bit encryption introduced earlier in Release 6. See Chapter 4 for more details on the evolution of GSM security.
- CBC-BSC Interface: Definition of the interface and related protocol between the Cell Broadcast Center and the BSC.
- VAMOS (Voice services over Adaptive Multi-user channels on One Slot): Definition of mechanisms to support concurrent voice users sharing the same radio resources at the same time, thus allowing up to two full-rate speech users or up to four half-rate speech users on the same timeslot.

References

[1]   GSA, The Global Suppliers Association, http://www.gsacom.com.
[2]   Subscriptions by Technology, World Cellular Information Service, Informa Telecoms and Media, http://www.wcisdata.com/newt/l/wcis/research/subscriptions_by_technology.html.
[3]   GSM 06.10 v3.2.0, "Full Rate Speech Transcoding", January 1995.
[4]   GSM 03.20 v3.3.2, "Security-related Network Functions", January 1995.
[5]   GSM 03.40 v3.7.0, "Technical Realization of the Short Message Service (SMS)", January 1995.
[6]   GSM 03.41 v3.4.0, "Technical Realization of Short Message Service Cell Broadcast (SMSCB)", January 1995.
[7]   GSM 03.45 v3.3.0, "Technical Realization of Facsimile Group 3 Service – Transparent", January 1995.
[8]   GSM 03.46 v3.2.1, "Technical Realization of Facsimile Group 3 Service – Non-transparent", January 1995.
[9]   GSM 06.20 v4.3.1, "Half Rate Speech Transcoding", May 1998.
[10] GSM 06.60 v4.1.1, "Enhanced Full Rate Speech Transcoding", August 2000.
[11] GSM 03.34 v5.2.0, "High Speed Circuit Switched Data (HSCSD); Stage 2", February 1999.
[12] GSM 03.69 v5.6.0, "Voice Broadcast Service (VBS); Stage 2", October 2003.
[13] GSM 03.68 v5.6.0, "Voice Group Call Service (VGCS); Stage 2", October 2003.
[14] GSM 03.64 v6.4.0, "General Packet Radio Service (GPRS); Overall Description of the GPRS Radio Interface; Stage 2", November 1999.
[15] GSM 06.90 v7.2.0, "Adaptive Multi-Rate (AMR) Speech Transcoding", December 1999.
[16] GSM 03.64 v8.12.0, "General Packet Radio Service (GPRS); Overall Description of the GPRS Radio Interface; Stage 2", May 2004.
[17] 3GPP TS 05.02 v8.11.0, "Multiplexing and Multiple Access on the Radio Path", July 2003.
[18] 3GPP TS 03.55 v8.4.0, "Dual Transfer Mode (DTM); Stage 2", February 2005.
[19] 3GPP TS 45.002 v4.8.0, "Multiplexing and Multiple Access on the Radio Path", July 2003.

[20] 3GPP TS 43.051 v5.10.0, "GSM/EDGE Radio Access Network (GERAN) Overall Description; Stage 2", September 2003.
[21] 3GPP TS 26.190 v5.1.0, "AMR Wideband Speech Codec; Transcoding Functions", December 2001.
[22] 3GPP TR 44.901 v5.1.0, "External Network Assisted Cell Change (NACC)", May 2002.
[23] 3GPP TS 23.060 v5.13.0, "General Packet Radio Service (GPRS); Service Description; Stage 2", December 2006.
[24] 3GPP TS 48.018 v5.14.0, "General Packet Radio Service (GPRS); Base Station System (BSS) – Serving GPRS Support Node (SGSN); BSS GPRS Protocol (BSSGP)", December 2006.
[25] 3GPP TS 43.129 v6.12.0, "Packed-switched Handover for GERAN A/Gb Mode; Stage 2", June 2007.
[26] 3GPP TS 43.064 v6.11.0, "General Packet Radio Service (GPRS); Overall Description of the GPRS Radio Interface; Stage 2", July 2006.
[27] 3GPP TS 43.246 v6.10.0, "Multimedia Broadcast/Multicast Service (MBMS) in the GERAN; Stage 2", December 2006.
[28] 3GPP TS 43.318 v6.12.0, "Generic Access Network (GAN); Stage 2", June 2008.
[29] 3GPP TR 45.902 v6.8.0, "Flexible Layer One (FLO)", February 2005.
[30] 3GPP TS 43.055 v6.15.0, "Dual Transfer Mode (DTM); Stage 2", March 2007.
[31] 3GPP TS 43.059 v6.6.0, "Functional Stage 2 Description of Location Services (LCS) in GERAN", May 2006.
[32] 3GPP TR 43.903 v8.3.0, "A Interface over IP Study (AINTIP)", December 2008.
[33] 3GPP TS 43.318 v8.4.0, "Generic Access Network (GAN); Stage 2", March 2009.
[34] EN 302 480, "Harmonized EN for the GSM Onboard Aircraft System Covering Essential Requirements of Article 3.2 of the R&TTE Directive".
[35] ETSI TS 102 576, "GSM Onboard Aircraft; Technical and Operational Requirements of the GSM Onboard Aircraft System".
[36] 3GPP TS 43.129 v8.1.0, "Packet-switched Handover for GERAN A/Gb Mode; Stage 2", March 2009.
[37] 3GPP TS 43.055 v8.1.0, "Dual Transfer Mode (DTM); Stage 2", March 2009.
[38] 3GPP TS 43.059 v8.1.0, "Functional Stage 2 Description of Location Services (LCS) in GERAN", September 2008.

2

3GPP Release 7

Eddie Riddington, David Navrátil, Jürgen Hofmann, Kent Pedersen and
Guillaume Sébire

2.1 Introduction

GSM, the world's most successful and widely deployed communications technology of all
times recording over four billion subscribers, reached another evolution milestone with the
completion of the 3GPP Release 7 specifications.

Making the internet truly mobile. While mobility has revolutionized the internet and enabled
access where the internet had never been before, it has, at the same time, brought about
constraints of coverage, data rates, latency and spectrum efficiency.[1] When 3GPP laid the
foundation for evolving EGPRS further, key requirements were thus defined [1]:

- at least 50% improvement in coverage and spectrum efficiency;
- at least double the data rates in both uplink (terminal to network) and downlink (network to
  terminal) reaching one Mbit/s and above;
- at most halve the round-trip time between a terminal and the network (~150 ms).

These requirements paved the way for carefully studying, selecting and specifying techniques,
such as higher order modulations, turbo-coding, antenna diversity, and associated features
which individually or combined achieve the goals above: EGPRS2, Downlink Dual Carrier
and Mobile Station Receiver Diversity(DARP Phase 2) and Latency Reductions.

Leveraging operators' investments, enabling new technologies. Massive investments have
been made to deploy EGPRS in GSM networks across the globe, thereby making it the most
widely used cellular packet data technology. In addition, besides a large base of installed
GSM (radio network) infrastructure worldwide, green-field deployments, network expansions

---

[1] How efficient the usage of the spectrum is: the more the amount of bits that can be sent in a given time in a given
frequency allocation, the better.

GSM/EDGE: Evolution and Performance   Edited by Mikko Säily, Guillaume Sébire and Eddie Riddington
© 2011 John Wiley & Sons, Ltd

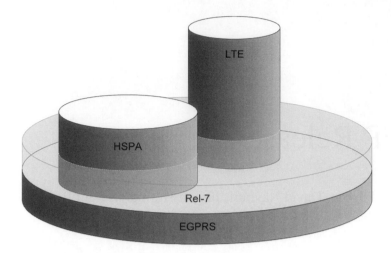

Figure 2.1   Minimizing technology gaps and raising the minimum user experience.

and modernization are also continuously bringing the latest generation hardware to the field. Underwriting these investments is the full exploitation of these pieces of equipment and their capabilities.

Omnipresent in most countries (and getting there rapidly in for example India, China or Africa), GSM is the technology providing ubiquitous access to voice and data services. Being able to use mobile services anywhere is taken for granted; it is the result of an evolution that is continuously raising the minimum user experience. Moreover with the deployment of, for example HSPA and with the forthcoming LTE (Rel-8), GSM consistently acts as the underlying technology; GSM coverage typically reaches far beyond any other overlay technologies. GSM hence provides a nurturing base for other technologies to grow and interworking with GSM becomes thus a de facto requirement for such technologies to commercially take off. Interworking ensures service continuity and availability.

Minimizing through the EGPRS evolution the technology gap between GSM and other technologies, as illustrated in Figure 2.1,[2] thus serves not only the competitiveness of GSM but also that of other technologies that it works with, while further improving the overall user experience.

This chapter describes the features making up the EGPRS Evolution as specified in 3GPP Release 7: EGPRS2, Downlink Dual Carrier, Mobile Station Receiver Diversity and Latency Reductions.

2.2   EGPRS2

2.2.1   Introduction

Specified in 3GPP Release 99 and designed to boost GPRS peak data rates, throughput, coverage and capacity, EGPRS features two key elements: 8-PSK modulation tripling peak

---

[2] This figure is only illustrative.

data rates compared to GMSK modulation, and incremental redundancy (Type II Hybrid ARQ) yielding significantly better RLC (layer 2) performance than GPRS' Type I ARQ by exploiting erroneous transmissions otherwise lost in the case of GPRS.

Though very similar, EGPRS and GPRS RLC protocols are, due to incremental redundancy, incompatible. EGPRS consists of nine modulation and coding schemes, MCS-1 to MCS-9, covering a wide range of channel conditions and arranged in four families defined by payload; A, B, C and A with padding. EGPRS MCSs offers data rates ranging from 8.8 kbit/s to 59.2 kbit/s per time slot (or up to 473.6 kbit/s per carrier). MCS-1 to MCS-4 use GMSK modulation and are the peers of GPRS CS1 to CS4 but, as opposed to these, support incremental redundancy. MCS-5 to MCS-9 use 8-PSK modulation.

EGPRS2, specified in 3GPP Release 7, complements EGPRS and is compatible with it. Higher order modulations, turbo codes and a higher modulating symbol rate are the key additional building blocks that characterize EGPRS2 compared to EGPRS. EGPRS2 consists of two separate levels both in uplink and in downlink: EGPRS2-A and EGPRS2-B. EGPRS2 data rates range from 22.4 kbit/s up to 118.4 kbit/s per time slot.

In Sections 2.2.2 and 2.2.3, an overview of the two levels of EGPRS2 is given, with each section split between downlink and uplink and with the characteristics of the new modulation and coding schemes summarized in Tables 2.1, 2.2, 2.3 and 2.4. Then in Section 2.2.4, more detail is provided about the new modulations and the new symbol rate, including a description of the operations that define a modulation scheme such as the bit to symbol mapping, symbol rotation and transmit pulse shaping. In Section 2.2.5, the error correction coding in EGPRS2 is described, that is the turbo coding and the convolutional coding. This section also covers the puncturing schemes used, the interleaving and the bit swapping operations. Finally, Section 2.2.6 addresses link adaptation, including a description of the link adaptation families supported by EGPRS2.

2.2.2 EGPRS2-A

2.2.2.1 Introduction

EGPRS2-A is defined differently for downlink and uplink. In downlink, 8-PSK, 16QAM, 32QAM modulations and turbo coding are used while in uplink, only 16QAM modulation is used alongside convolutional coding. Its data rates range from 22.4 kbit/s to 96.4 kbit/s per time slot in downlink and from 44.8 to 76.8 kbit/s per time slot in uplink.

2.2.2.2 EGPRS2-A Downlink

In EGPRS2-A Downlink, eight modulation and coding schemes (DAS-5 to DAS-12) are defined in addition to the four GMSK modulated coding schemes of EGPRS (MCS-1 to MCS-4). The main attributes of DAS-5 to DAS-12 are summarized in Table 2.1.

2.2.2.3 EGPRS2-A Uplink

In EGPRS2-A Uplink, five modulation and coding schemes (UAS-7 to UAS-11) are defined in addition to the four GMSK modulated coding schemes and two of the 8-PSK modulation

Table 2.1   Modulation and coding schemes in EGPRS2-A downlink

Scheme	Mod. and symbol rate (ksymb/s)	Mother code	No. punc. schemes	Initial code rate	RLC data block size (octets)	No. RLC data blocks	Peak data rate (kbit/s)
DAS-5	8-PSK 270.8	1/3 TC	2	0.37	56	1	22.4
DAS-6	8-PSK 270.8	1/3 TC	2	0.45	68	1	27.2
DAS-7	8-PSK 270.8	1/3 TC	2	0.54	82	1	32.8
DAS-8	16QAM 270.8	1/3 TC	2	0.56	56	2	44.8
DAS-9	16QAM 270.8	1/3 TC	3	0.68	68	2	54.4
DAS-10	32QAM 270.8	1/3 TC	2	0.64	82	2	65.6
DAS-11	32QAM 270.8	1/3 TC	3	0.80	68	3	81.6
DAS-12	32QAM 270.8	1/3 TC	3	0.96	82	3	98.4

schemes of EGPRS (MCS-1 to MCS-6). The main attributes of UAS-7 to UAS-11 are summarized in Table 2.2.

2.2.3   EGPRS2-B

2.2.3.1   Introduction

EGPRS2-B uses modulations QPSK, 16QAM and 32QAM at the higher modulating symbol rate of 325 ksymb/s. In the downlink, turbo coding is used while in the uplink convolutional coding is used. EGPRS2-B data rates range from 22.4 to 118.4 kbit/s per time slot.

2.2.3.2   EGPRS2-B Downlink

In EGPRS2-B Downlink, eight modulation and coding schemes (DBS-5 to DBS-12) are defined in addition to the four GMSK modulated coding schemes of EGPRS (MCS-1 to MCS-4). The main attributes of DBS-5 to DBS-12 are summarized in Table 2.3.

Table 2.2   Modulation and coding schemes in EGPRS2-A uplink

Scheme	Mod. and symbol rate (ksymb/s)	Mother code	No. puncturing schemes	Initial code rate	RLC data block size (octets)	No. RLC data blocks	Peak data rate (kbit/s)
UAS-7	16QAM 270.8	1/3 CC	2	0.55	56	2	44.8
UAS-8	16QAM 270.8	1/3 CC	2	0.62	64	2	51.2
UAS-9	16QAM 270.8	1/3 CC	3	0.71	74	2	59.2
UAS-10	16QAM 270.8	1/3 CC	3	0.84	56	3	67.2
UAS-11	16QAM 270.8	1/3 CC	3	0.95	64	3	76.8

Table 2.3 Modulation and coding schemes in EGPRS2-B downlink

Scheme	Mod. and symbol rate (ksymb/s)	Mother code	No. puncturing Schemes	Initial code rate	RLC data block size (octets)	No. RLC data blocks	Peak data rate (kbit/s)
DBS-5	QPSK 325	1/3 TC	2	0.49	56	1	22.4
DBS-6	QPSK 325	1/3 TC	2	0.63	74	1	29.6
DBS-7	16QAM 325	1/3 TC	2	0.47	56	2	44.8
DBS-8	16QAM 325	1/3 TC	2	0.60	74	2	59.2
DBS-9	16QAM 325	1/3 TC	3	0.71	56	3	67.2
DBS-10	32QAM 325	1/3 TC	3	0.72	74	3	88.8
DBS-11	32QAM 325	1/3 TC	3	0.91	68	4	108.8
DBS-12	32QAM 325	1/3 TC	3	0.98	74	4	118.4

2.2.3.3   EGPRS2-B Uplink

In EGPRS2-B Uplink, eight modulation and coding schemes (UBS-5 to UBS-12) are defined in addition to the four GMSK modulated coding schemes of EGPRS (MCS-1 to MCS-4). The main attributes of UBS-5 to UBS-12 are summarized in Table 2.4.

2.2.4   Modulation and Pulse Shaping

2.2.4.1   Introduction

EGPRS2 introduces a number of additions compared to EGPRS, especially in the number of modulations schemes supported.[3] Five new modulation schemes are defined: 16QAM and 32QAM at the symbol rate of 270.8 ksymb/s and QPSK, 16QAM and 32QAM at the higher symbol rate of 325 ksymb/s [12]. This is in addition to GMSK and 8-PSK at the legacy symbol rate of 270.8 ksymb/s.

Each modulation scheme represents a different trade-off between bandwidth efficiency and robustness to noise and collectively they provide a high degree of flexibility to the physical layer of EGPRS2 to adapt to a wide range of channel conditions. For example, in areas such as indoors or at the cell border, the data rates might benefit more from the support of the QPSK modulation scheme, while benefit might be expected in areas close to the base station with the support of 32QAM modulation. The raw bit rate of a modulation scheme is the bit rate at the air interface. It provides an indication of the modulation scheme's bandwidth efficiency and is depicted in Table 2.5 for each of the modulation schemes supported by EGPRS and EGPRS2. In the case of EGPRS2, a wide range of raw bit rates are supported which extend up to twice the maximum of EGPRS.

To facilitate the early adoption of the EGPRS2 feature by the telecommunications industry, the modulation schemes have been divided into two levels: EGPRS2-A, which corresponds to the schemes at the legacy symbol rate of 270.8 ksymb/s (the same symbol rate as EGPRS), and EGPRS2-B, which corresponds to the schemes at the higher symbol rate of 325 ksymb/s.

---

[3] A modulation scheme in this context is defined as the combination of a modulation and its symbol rate.

Table 2.4 Modulation and coding schemes in EGPRS2-B uplink

Scheme	Mod. and symbol rate (ksymb/s)	Mother code	No. punc. schemes	Initial code rate	RLC block size (octets)	No. RLC data blocks	Peak data rate (kbit/s)
UBS-5	QPSK 325	1/3 CC	2	0.47	56	1	22.4
UBS-6	QPSK 325	1/3 CC	2	0.62	74 & 68	1	29.6
UBS-7	16QAM 325	1/3 CC	2	0.46	56	2	44.8
UBS-8	16QAM 325	1/3 CC	2	0.60	74 & 68	2	59.2
UBS-9	16QAM 325	1/3 CC	3	0.70	56	3	67.2
UBS-10	32QAM 325	1/3 CC	3	0.71	74 & 68	3	88.8
UBS-11	32QAM 325	1/3 CC	3	0.89	68	4	108.8
UBS-12	32QAM 325	1/3 CC	3	0.96	74	4	118.4

Each has been specified in 3GPP so that neither level is dependent on the other. This is to allow a mobile vendor to implement EGPRS2-A in a first phase and EGPRS2-B in a later second phase. A mobile indicates its level of support during the establishment phase of a data connection with the help of two indicator bits in the MS Radio Access Capability Information Element [3]. These bits are used to signal to the network the support of either EGPRS2-A, both EGPRS2-A and EGPRS2-B, or neither EGPRS2-A nor EGPRS2-B. Separate indicator bits are defined for uplink and downlink.

Each modulation scheme in EGPRS2 can be described by the three operations which are depicted in Figure 2.2 [12].

2.2.4.2 Bit to Symbol Mapping

Bits entering the modulator are mapped into modulating symbols (or vectors) using the constellations in Figures 2.3, 2.4 and 2.5. The "square" constellations in Figure 2.3 and Figure

Table 2.5 Modulation schemes in EGPRS and EGPRS2

Level	Modulation	No. bits per symbol	Symbol rate	Raw bit rate
EGPRS2-B	32QAM	5	325 ksymb/s	1625 kbit/s
	16QAM	4	325 ksymb/s	1300 kbit/s
	QPSK	2	325 ksymb/s	650 kbit/s
	GMSK	1	270.8 ksymb/s	270.8 kbit/s
EGPRS2-A	32QAM[1]	5	270.8 ksymb/s	1354.2 kbit/s
	16QAM	4	270.8 ksymb/s	1083.3 kbit/s
	8-PSK	3	270.8 ksymb/s	812.5 kbit/s
	GMSK	1	270.8 ksymb/s	270.8 kbit/s
EGPRS	8-PSK	3	270.8 ksymb/s	812.5 kbit/s
	GMSK	1	270.8 ksymb/s	270.8 kbit/s

Note: [1]Downlink only.

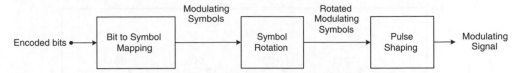

Figure 2.2   Modulator in EGPRS2.

2.4 are for QPSK and 16QAM modulations respectively and the "cross" constellation in Figure 2.5 is for 32QAM modulation. All constellation points are Gray coded with the exception of the 32QAM constellation, where a perfect Gray coding was not possible. In this case, the coding was optimized by simulation, with the non-Gray pairs being indicated in Figure 2.5.

2.2.4.3   Symbol Rotation

After the bit to symbol mapping, the modulating symbols are rotated to avoid transitions through the origin. This minimizes the variations in the modulating signal which in turn minimizes the linearity requirements of the amplifier (resulting in a more efficient amplifier) as well as maximizing the power capability of the respective modulation and hence its coverage. The impact of rotation can be seen in Figures 2.6 and 2.7, where transitions between the constellation points of 16QAM modulation are shown first without rotation (Figure 2.6) and then with rotation applied (Figure 2.7). For the QPSK and QAM modulation schemes, the

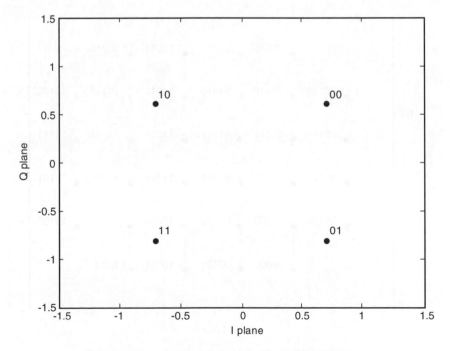

Figure 2.3   Constellation diagram for QPSK modulation.

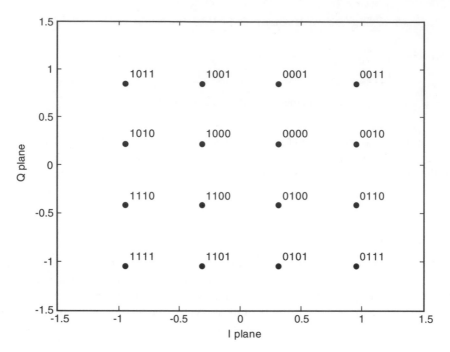

Figure 2.4   Constellation diagram for 16QAM modulation.

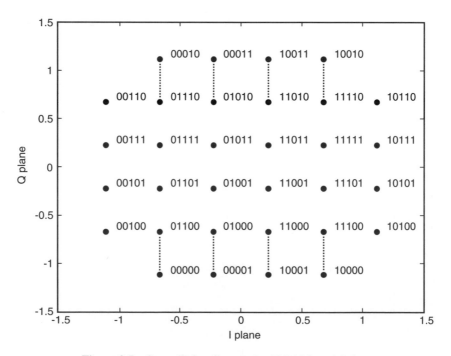

Figure 2.5   Constellation diagram for 32QAM modulation.

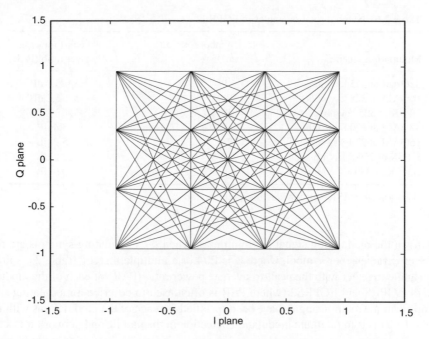

Figure 2.6 Transitions between symbols using 16QAM modulation with no rotation applied (prior to pulse shaping).

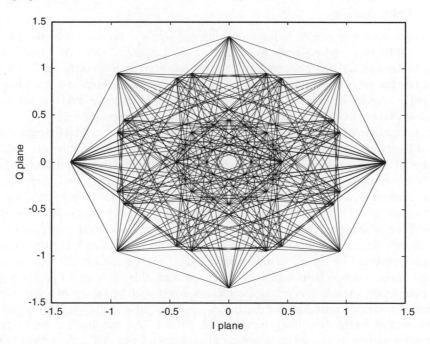

Figure 2.7 Transitions between symbols using 16QAM modulation with PI/4 rotation applied (prior to pulse shaping).

Table 2.6 Symbol rotation angles in EGPRS2

Modulation scheme	Symbol rotation angle (φ)	Peak-to-average power ratio (PAR)
32QAM at 325 ksymb/s	$-$PI/4	5.9/5.3 dB*
16QAM at 325 ksymb/s	PI/4	5.8/5.2 dB*
QPSK at 325 ksymb/s	3PI/4	3.4/3.1 dB*
32QAM at 270.8 ksymb/s	$-$PI/4	5.4 dB
16QAM at 270.8 ksymb/s	PI/4	5.2 dB
8-PSK at 270.8 ksymb/s	3PI/8	3.2 dB
GMSK at 270.8 ksymb/s	n/a	0.0 dB

*Linearized GMSK pulse/Wide pulse, see Section 2.2.4.4.

variations in the modulating signal are minimized when the modulating symbols are rotated at a rate of φ radians per symbol, where φ is PI/4 or a multiple thereof. Table 2.6 shows the rotation angle together with the peak-to-average power ratio (PAR) of each of the modulating signals of EGPRS and EGPRS2 (where PAR is often used to describe a modulating signal's variation). Unique rotation angles have been specified for each of the modulations within each symbol rate. This is to facilitate modulation detection in the mobile and network transceivers.

2.2.4.4   Pulse Shaping

Transmit pulse shaping is used to limit a modulating signal's spectral bandwidth and in the case of EGPRS2 two pulse shapes are supported: the Linearized Gaussian Minimum Shift Keying (linearized GMSK) pulse and a pulse that has been optimized for the 325 ksymb/s symbol rate for use in the uplink direction. This optimized pulse, referred to hereafter as the "wide pulse", has a wider bandwidth than the linearized GMSK pulse and results in a higher throughput performance thanks to its reduced inter-symbol interference.

The spectral shape of both pulses is shown in Figure 2.8. One of the consequences of a widened bandwidth is a lowered adjacent channel protection. For example, the wide pulse in Figure 2.8 exhibits a 5.5 dB lower adjacent channel protection than the linearized GMSK pulse. However, the level of interference introduced by a pulse shape not only depends on the adjacent channel protection provided, but also on factors such as the proportion of time slots supporting the EGPRS2 service or the pulse shape's time slot occupancy. Interestingly, this latter aspect can be expected to be lowered as a result of the higher throughput performance. Moreover, any interference that might be introduced by the use of a wide pulse in uplink can easily be suppressed by the use of interference cancellation algorithms that are commonly deployed in dual antenna base stations. For the downlink, this latter aspect is less certain in the mobile station where the suppression capability is restricted by the use of a single antenna and limited baseband capability. For this reason, no wide pulse shape has been specified yet for the downlink (at the time of writing). Instead, a feasibility study into the performance of a wide pulse and on its impact on legacy mobiles is ongoing in 3GPP, the results of which are being collated in [2]. More information about the pulse shapes being considered in this feasibility study can be found in Chapter 4 on Release 9 and beyond.

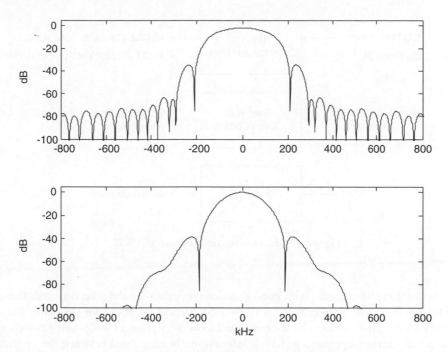

Figure 2.8 Transmit pulse shaping filters in EGPRS2: the upper figure depicts the wide pulse shape and the lower figure the linearized GMSK pulse shape.

2.2.5   Coding and Puncturing

2.2.5.1   Introduction

EGPRS2 provides a wide range of modulation and coding schemes (MCSs) to be used by the radio resource controller to adapt to the constantly changing channel. Their main attributes are shown in Tables 2.1, 2.2, 2.3 and 2.4 (corresponding to EGPRS2-A Downlink and EGPRS2-A Uplink and EGPRS2-B Downlink and EGPRS2-B Uplink respectively). In these tables, the robustness of an MCS to noise is indicated by its modulation and symbol rate and its initial code rate (the ratio of information bits to coded bits). The given peak bit rate is given at the RLC layer, exclusive of all overheads up to and including the RLC layer.

Figure 2.9 depicts the error correction codes that are used in EGPRS2. In addition to the convolutional codes used for the RLC/MAC header and for each of the RLC data blocks in uplink, EGPRS2 introduces turbo codes for the RLC data blocks on the downlink and new block codes for the Uplink State Flag (USF) [10].

2.2.5.2   Turbo Code

Turbo codes are well suited to the long block lengths associated with the new modulation schemes of EGPRS2. The turbo encoder, a structure consisting of constituent encoders which in parallel are separated with an interleaver, has been found to be particularly efficient at

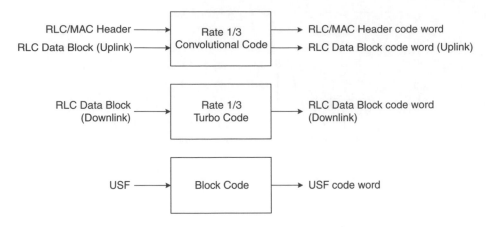

Figure 2.9   Error correction codes in EGPRS2.

constructing long block codes with good distance properties. At the receiving end, the turbo decoder has a similar arrangement, with two constituent decoders for decoding the constituent codes. Turbo decoders have been found to be very efficient at decoding turbo codes, thanks to an iterative operation in which information is exchanged between the constituent decoders.

The turbo encoder in EGPRS2 is depicted in Figure 2.10. It is the same turbo encoder as in UTRAN: a parallel concatenated convolutional encoder, consisting of two recursive convolutional encoders arranged in parallel and separated by a pseudo random interleaver. Using the same encoder in this way allows a mobile vendor to exploit the already available hardware within their dual-mode GSM/EDGE and UTRAN mobile platforms. Hence the turbo code is only specified on the downlink. Furthermore, as turbo codes are not very optimal for short block lengths, they have been specified only for the RLC data blocks.

2.2.5.3   Convolutional Code

The encoder for the RLC/MAC header and the RLC data blocks in uplink is depicted in Figure 2.11. It is the same encoder as in EGPRS, a 7-state feed-forward convolutional encoder, operating at a 1/3 rate.

2.2.5.4   Puncturing and Incremental Redundancy

To achieve the desired code rate for each modulation and coding scheme (the initial code rate in Tables 2.1 to 2.4), puncturing is applied to the convolutional or turbo encoded RLC data blocks using one of two (or three[4]) puncturing schemes. The first of these puncturing schemes applies to the initial transmission of the block and the second (and third) scheme

---

[4] Two puncturing schemes are needed when the initial code rate is less than or equal to 2/3 and three puncturing schemes are needed when the initial code rate is greater than 2/3.

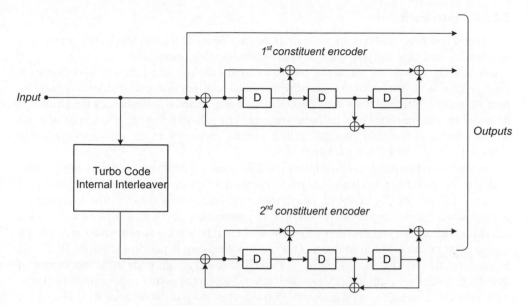

Figure 2.10   Turbo encoder used in EGPRS2.

applies when the block is re-transmitted (assuming RLC acknowledged mode of operation). Specifying the puncturing schemes in this way allows the initial transmission of an RLC block to be combined with the first (and/or second) retransmission of the block so that additional redundancy is added leading to a reduction in the code rate. To achieve this, the puncturing scheme optimization used the following criteria:

- decoding performance of the initial transmission;
- joint decoding performance between the initial transmission and the first and second retransmission;
- self-decodability of the retransmissions (this criterion is needed for the case of a lost initial transmission for example due to an RLC header decode failure).

In the uplink, these criteria were met by a computer search, while in the downlink a modified UTRAN Rate Matching algorithm has been used [11] (where the modifications relate to improvements to the criteria on joint decoding performance and self-decodability).

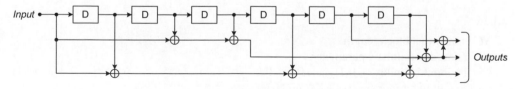

Figure 2.11   Convolutional encoder used in EGPRS2.

2.2.5.5 Interleaving

The coded and punctured bits are mapped to the four bursts of a radio block after performing interleaving and after carrying out a number of bit swapping operations.

Interleaving spreads the bits across the bursts so that the impact of channel disturbances is averaged across all four bursts that comprise the radio block. For example, interleaving provides time diversity when the bursts are subjected to a fading channel, or enhances the frequency diversity benefits provided by a frequency hopping channel. Interleaving also spreads the bits within a burst so that their strength relative to their proximity to the training sequence is averaged over the duration of the burst.

Channel decoding performance improves with interleaving depth except for coding schemes with little or no coding (very high code rates). For these schemes, the interleaving depth should be as short as possible. In EGPRS2, two formulae for interleaving are specified: Equation 2.1 is used when interleaving is performed over an interleaving depth of 4 bursts and Equation 2.2 is used for the very high code rate schemes where interleaving is performed over a depth of one burst or one and a third bursts. One burst interleaving is performed for the RLC data blocks of UBS-11, DBS-11 and DBS-12 (which have very high code rates and consist of four RLC blocks per radio block), one and a third burst interleaving is performed for UAS-11 and DAS-12 (which also have very high code rates and consist of three RLC blocks per radio block) and four burst interleaving is performed for the RLC data in all the remaining modulation and coding schemes and for the RLC header in all the modulation and coding schemes.

The interleaving formulae are applied by first reserving the bit positions in the burst for the RLC/MAC header, the stealing bits and the USF (if applicable). These bits are placed in the symbol positions closest to the training sequence where the symbol error probability is lowest. The rectangular interleaving over four bursts (Equation 2.1) is then performed as a single operation after concatenating all RLC blocks together. When interleaving over a single burst or over one and a third bursts, each RLC data block is interleaved independently (Equation 2.2).

Both interleaving formulae include one non-deterministic parameter, a. This parameter has been optimized by simulation to maximize the interleaving performance for each modulation and coding scheme.

$$j = \overbrace{\frac{NC \cdot B}{4}}^{\substack{\text{spreads the} \\ \text{bits within} \\ \text{the burst}}} + \overbrace{\left(((k \text{ div } 4) + (NC \text{ div } 16) \cdot B) \cdot a \mod \frac{NC}{4} \right)}^{\substack{\text{spreads the bits} \\ \text{between the bursts}}}$$

where

NC is the input block length
a is a non-deterministic parameter
k is the input bit index $= 0,1,\ldots,NC - 1$ (2.1)
j is the output bit index $= 0,1,\ldots,NC-1$
$B = 2 \cdot (k \mod 2) + (k \mod 4) \text{ div } 2$

$$j = k \cdot a \bmod NC$$

where

NC is the input block length
a is a non-deterministic parameter
k is the input bit index $= 0, 1, \ldots, NC - 1$ (2.2)
j is the output bit index $= 0, 1, \ldots, NC - 1$

2.2.5.6 Bit Swapping

The constellations for 16QAM and 32QAM have the property that not all the bit positions in the symbol have equal bit error probability. A similar phenomenon exists also for 8-PSK of EGPRS, where the most reliable bits are in the first two positions (of three), while in the case of 16QAM, the most reliable bits are in the first two positions (of four) and in the case of 32QAM, the most reliable bits are in the first and fourth positions (of five).

Already the RLC/MAC header, the stealing bits and the USF (downlink only) are mapped to the symbol positions that are closest to the training sequence. This is where the symbol error probability is lowest. This is further refined with bit swapping where the RLC/MAC header bits are swapped with data bits so that the RLC/MAC header is mapped to the most reliable bits closest to the training sequence. This is depicted in Figure 2.12 for the 32QAM modulated schemes. Bit swapping brings benefits to incremental redundancy performance where a robust RLC/MAC header is essential.

2.2.6 Link Adaptation

Link adaptation when in RLC acknowledged mode is supported by EGPRS and EGPRS2 through a concept of coding scheme families, whereby each family of coding schemes support a common RLC data block size. Different code rates or bit rates within a family are then obtained by transmitting different numbers of RLC data blocks within each radio block. For example, Figure 2.13 depicts the families supported by EGPRS2-A Downlink. Four families have been defined: C, B padding2, B and A padding6, which are based on common RLC data block sizes: 22, 82, 28 and 68 octets respectively. When performing link adaptation within Family A, an RLC data block (be it an initial transmission or a retransmission) may be sent

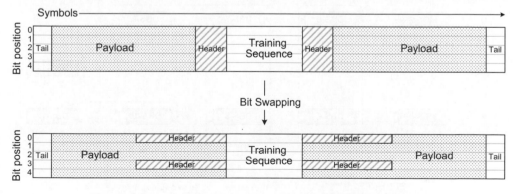

Figure 2.12   Bit swapping operation for a 32QAM modulated burst.

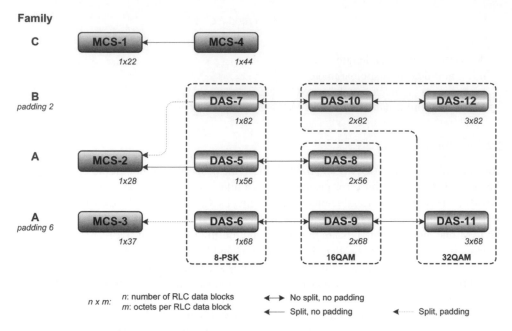

Figure 2.13   Link adaptation families in EGPRS2-A downlink.

using DAS-8, DAS-5 or MCS-2 (the latter requiring the block to be transmitted in two radio blocks), where the quality of service and the prevailing channel conditions will dictate which scheme to use. Figures 2.14, 2.15 and 2.16 depict the coding scheme families for EGPRS2-B Downlink, EGPRS2-A Uplink and EGPRS2-B Uplink respectively.

In many cases, a modulation and coding scheme in one family is re-used in another family through the use of padding, whereby the modulation and coding scheme's native RLC data

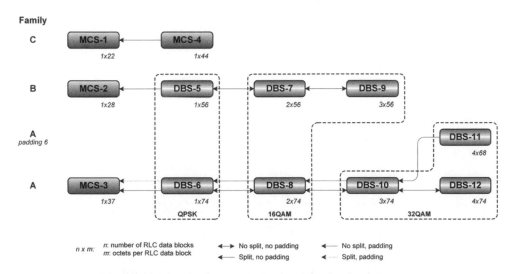

Figure 2.14   Link adaptation families in EGPRS2-B downlink.

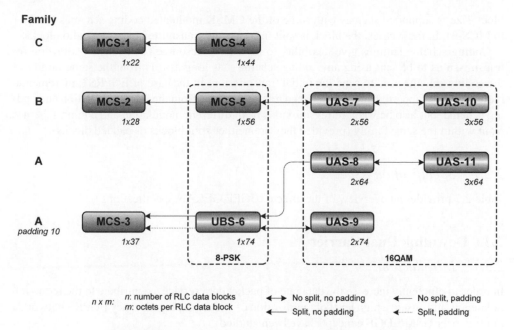

Figure 2.15 Link adaptation families in EGPRS2-A uplink.

block size is effectively shortened by appending padding octets prior to encoding. For example, in EGPRS2-B Downlink (Figure 2.14), DBS-6, DBS-8 and DBS-10 are native to Family A (74 octets) but become members of Family A padding6 (68 octets) when 6 padding octets are appended to each RLC data block. While inefficient, this has the benefit of reducing the number of modulation and coding schemes that need to be implemented.

The families of EGPRS2 have been designed to be compatible with EGPRS so that many of the coding schemes of EGPRS can be adopted by EGPRS2. In some cases, a half RLC data

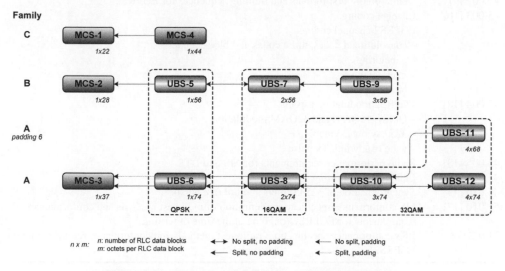

Figure 2.16 Link adaptation families in EGPRS2-B uplink.

block size is supported such as with some of the GMSK modulated coding schemes (MCS-1 to MCS-3). In these cases, the block is split into two and transmitted over two radio blocks.

Coding scheme families give flexibility to the Radio Resource Controller by allowing retransmissions to be sent using any coding scheme – as long as it is from the same family as the coding scheme used to send the initial transmission. In the case of EGPRS2, incremental redundancy is supported not only between retransmissions from the same modulation and coding scheme, but also between retransmissions from different modulation and coding schemes from within the same family (provided they are neither split blocks or padded blocks).

2.2.7   Specification Aspects

Table 2.7 provides an overview of the affected 3GPP GERAN specifications.

2.3   Downlink Dual Carrier

2.3.1   Introduction

In order to efficiently increase the data rate of packet data services, contribute to the reduction of latency, and ensure a high level of reutilization of widely deployed legacy EGPRS networks, multi-carrier GSM/EDGE concepts have been studied.

Table 2.7   3GPP specifications relevant to EGPRS2

Specification	Aspects
24.008 [3]	Signaling of the mobile station's EGPRS2 capabilities
43.064 [4]	Functional description of EGPRS2
44.060 [7]	RLC/MAC protocol support for EGPRS2, e.g. assignments, incremental redundancy
45.001 [8]	Overview of EGPRS2 physical layer
45.002 [9]	Definition of burst formats and training sequences for EGPRS2
45.003 [10]	Channel coding EGPRS2 channel coding: – convolutional codes, turbo codes and blocks codes – puncturing – interleaving – bit swapping
45.004 [12]	EGPRS2 modulation aspects: – QPSK, 16QAM and 32QAM modulation – 325 ksymb/s symbol rate – wide bandwidth Tx pulse
45.005 [13]	Radio performance requirements for MS and BTS
45.008 [14]	Link quality reporting
45.010 [15]	Synchronization of 325 ksymb/s symbol rate bursts
51.010 [17]	MS conformance specification: definition of radio test methods and conformance requirements for EGPRS2 (ongoing at the time of writing)
51.021[18]	BSS conformance specification: radio test methods and conformance requirements for EGPRS2

In the uplink direction, from the mobile station to the network, two dual carrier allocation schemes were considered where the two carriers were either independent or within a certain frequency range. However, for neither of them could feasibility be shown. RF impacts such as inter-modulation distortion, high peak to average power ratio and low power amplifier efficiency resulting in power consumption issues implied a complex and non-cost efficient design of the RF module at the transmitter side. Both carriers could indeed be spread over the entire frequency band. As a result, the use of two simultaneous uplink carriers is not specified.

In addition, the asymmetrical downlink-biased characteristic of popular packet data services such as web browsing, download of emails or videos, leads to focusing on the downlink direction. Further, to minimize complexity and costs, while complying with initial requirements (see Section 2.1), multi-carriers eventually converged towards Downlink Dual Carrier for EGPRS.

EGPRS providing a peak data rate of 59.2 kbit/s per time slot and 473.6 kbit/s per carrier (i.e. 8 time slots), Downlink Dual Carrier can yield up to 947.2 kbit/s. In real network conditions, an average throughput performance ranging from 100 kbit/s to 200 kbit/s is feasible using four EGPRS time slots. With Downlink Dual Carrier, this figure can thus double up to 400 kbit/s. Combined with EGPRS2-B, the Downlink Dual Carrier can reach a peak data rate of 1.9 Mbit/s, and in real network conditions, up to about 600 kbit/s average throughput is expected [19]. See Chapter 6 for further evaluation.

Downlink Dual Carrier allows the reuse of legacy network implementations since the network is only required to distribute the user data on two legacy carriers. Hence the impact on the network is limited to software upgrade. New mobile stations, however, are required that support the reception of two independent carriers at the same time.

2.3.2 Functional Description

In this section a summary of the functional description of the Downlink Dual Carrier is provided.

2.3.2.1 Basic Concept

Downlink Dual Carrier is based on the simultaneous reception, per TDMA frame, of two independent GSM carriers belonging to the same serving cell, and is compatible with slow frequency hopping. Note the term carrier is used hereafter such that for a network not using slow frequency hopping, the carrier is fixed and defined by the absolute radio frequency channel number (ARFCN). In the case of a network employing slow frequency hopping, the term carrier refers to the temporary GSM carrier in each TDMA frame which is derived according to the parameters Mobile Allocation (MA), Mobile Allocation Offset (MAIO) and Hopping Sequence Number (HSN). By means of this simultaneous reception it is possible for the mobile to receive on more downlink time slots in a shorter period of time in the TDMA frame thus allowing transmitting on a higher number of uplink time slots per TDMA frame. Consequently the uplink data rate can also be increased for a mobile supporting Downlink Dual Carrier. Aside from reception and transmission, the mobile is required to use idle times in each TDMA frame for neighbor cell monitoring. This is depicted in Figure 2.17.

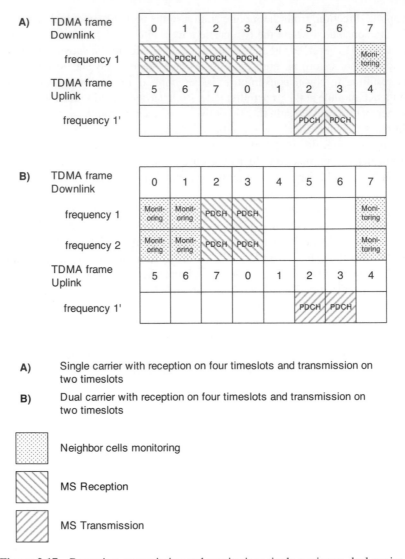

A)   Single carrier with reception on four timeslots and transmission on two timeslots

B)   Dual carrier with reception on four timeslots and transmission on two timeslots

        Neighbor cells monitoring

        MS Reception

        MS Transmission

Figure 2.17   Reception, transmission and monitoring: single carrier vs. dual carrier.

As can be seen in Figure 2.17, the time for neighbor cell monitoring can be increased remarkably if the same number of downlink and uplink time slots is supported as in single carrier reception. Alternatively time slots 0 and 1 can be used for downlink data reception, providing doubled throughput with neighbor cell monitoring on two carriers in time slot 7.[5]

---

[5] Each radio block on a Packet Data Channel (PDCH) in downlink and in uplink is mapped to the same time slot and same carrier in four consecutive TDMA frames, resulting in a radio block period of 20 ms conforming to the Basic Transmit Time Interval (BTTI) of 20 ms.

The performance of a mobile station, as determined by its RF and Baseband processing capability, is indicated by its multislot class. The multislot class depicts the capability of the mobile to receive and transmit on a given number of time slots including switching times between reception and transmission and vice versa. This capability was originally defined for single carrier reception in Release 1997, extended in Rel-5 and is now mapped to dual carrier operation.

Downlink Dual Carrier is also applicable to Dual Transfer Mode (DTM), that is the simultaneous support of a circuit switched (CS) connection carrying a voice traffic channel and of a packet switched (PS) connection carrying constrained or unconstrained delay data. Thus, with DTM, Downlink Dual Carrier improves the use of multimedia applications (e.g. web browsing, email reception, file sharing, etc.) also during a voice call. If the mobile supports DTM and Downlink Dual Carrier, both CS voice and PS data connections are mapped onto both carriers in downlink, while the uplink carrier is assigned to both the CS voice and the PS data connections. Note, the specification prescribes that all uplink PDCHs need to be assigned on the same carrier as the uplink CS time slot. An example of a Downlink Dual Carrier channel configuration in DTM with one time slot for voice, four downlink and three uplink time slots for packet data is depicted in Figure 2.18 for case B.

Although fast frequency synthesizers are assumed, the monitoring slot is a bit shorter than for single carrier reception shown in case A, to allow for tuning from the Tx frequency to the monitoring frequency and from the monitoring frequency to the Rx frequency. This allocation of a total of nine time slots for transmit and receive gives a gain of 80% in the overall throughput compared with a state-of-the-art single carrier mobile supporting a sum of five transmit and receive slots as depicted in Figure 2.18.

Downlink Dual Carrier multislot capabilities have been defined using single carrier capabilities as a basis. Due to the fact that the monitoring task can be shifted to the second receiver tuned to carrier frequency 2 in Figure 2.18, the first transceiver on carrier frequency 1 employed both for reception and transmission can be used in a more efficient manner and thus the peak throughput in Downlink can be more than doubled. This is shown in Table 2.8, depicting the transmit and receive multislot capabilities for the single carrier and Downlink Dual Carrier operation.

2.3.2.2   Multiplexing of a Downlink Dual Carrier Mobile with Legacy Mobiles

On the uplink, the Downlink Dual Carrier mobile cannot serve two carriers at the same time, however, it has the flexibility to alternate between the uplink carriers that correspond to the allocated downlink carriers according to the dynamic allocation. This is illustrated in Figure 2.19, which shows a Downlink Dual Carrier mobile A, supporting up to eight receive time slots on carrier frequency 1 or carrier frequency 2. When multiplexed with two legacy single carrier mobiles B and C, supporting up to two receive slots and one transmit slot, the Downlink Dual Carrier mobile in this case uses in two consecutive radio block periods either carrier frequency 1 or carrier frequency 2 on uplink, as commanded by the network and hence the uplink multiplexing with legacy single carrier mobiles is considerably improved.

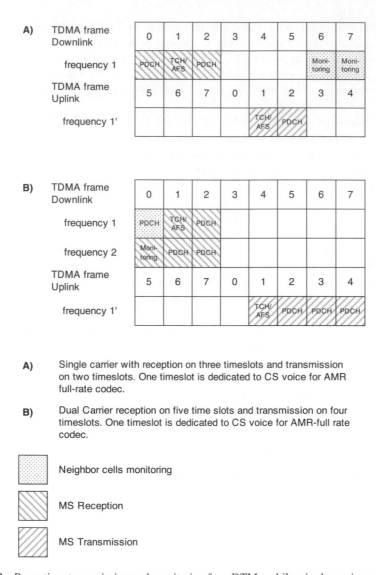

A) Single carrier with reception on three timeslots and transmission on two timeslots. One timeslot is dedicated to CS voice for AMR full-rate codec.

B) Dual Carrier reception on five time slots and transmission on four timeslots. One timeslot is dedicated to CS voice for AMR-full rate codec.

Neighbor cells monitoring

MS Reception

MS Transmission

Figure 2.18 Reception, transmission and monitoring for a DTM mobile: single carrier vs. dual carrier.

2.3.2.3   Feedback Control Information

Feedback control information sent by the mobile to the network includes ACK/NACK reports and channel quality reports related to the reception of downlink data. This information is sent on PACCH after the mobile has been polled by the network. In case of simultaneous ongoing CS voice, measurement reports are transmitted via SACCH. If a paired uplink timeslot is not available (e.g. see Figure 2.18 where there is a paired uplink timeslot only for frequency 1 on frequency 1' but not for frequency 2), the PACCH message has to include the ACK/NACK report and Channel Quality report for both carriers.

Table 2.8 Transmit and receive multislot capabilities for the mobile in single carrier and Downlink Dual Carrier operation

	Single carrier operation			Downlink Dual Carrier operation			
Multislot class	max RX slots	max TX slots	sum:= TX+ RX slots	max RX slots	max TX slotx	sum:= TX+ RX slots	Gain in DL peak throughput over single carrier operation (%)
8	4	1	5	5–10 (*)	1	6–11 (*)	25–150
10	4	2	5	5–10 (*)	2	6–11 (*)	25–150
11	4	3	5	5–10 (*)	3	6–11 (*)	25–150
12	4	4	5	5–10 (*)	4	6–11 (*)	25–150
30	5	1	6	6–10 (*)	1	7–11 (*)	20–100
31	5	2	6	6–10 (*)	2	7–11 (*)	20–100
32	5	3	6	6–10 (*)	3	7–11 (*)	20–100
33	5	4	6	6–10 (*)	4	7–11 (*)	20–100
34	5	5	6	6–10 (*)	5	7–11 (*)	20–100
40	6	1	7	7–12 (*)	1	8–13 (*)	16.7–100
41	6	2	7	7–12 (*)	2	8–13 (*)	16.7–100
42	6	3	7	7–12 (*)	3	8–13 (*)	16.7–100
43	6	4	7	7–12 (*)	4	8–13 (*)	16.7–100
44	6	5	7	7–12 (*)	5	8–13 (*)	16.7–100
45	6	6	7	7–12 (*)	6	8–13 (*)	16.7–100

Note: (*) depending on the maximum supported number of downlink receive time slots in Downlink Dual Carrier operation the MS signals to the network as part of the MS capabilities.

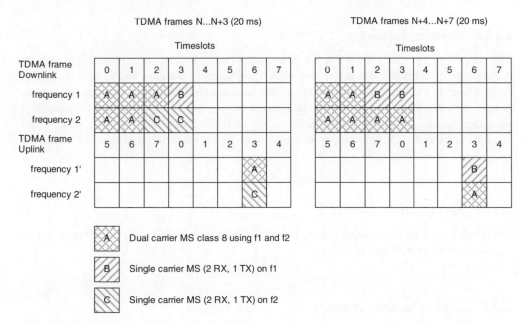

Figure 2.19 Multiplexing of a Downlink Dual Carrier MS with a legacy MS in two consecutive radio block periods.

2.3.2.4   Channel Quality Reporting

Channel Quality Reporting by the mobile station to the network includes averaged measurements such as mean and variance of the Bit Error Probability (MEAN_BEP, CV_BEP) on each assigned carrier, signal level of the serving cell and interference level on each assigned time slot. This information is used for Downlink Power Control and Link Adaptation procedures executed at the BSS. In a Downlink Dual Carrier configuration, the mobile station needs to support the channel quality measurements for each carrier independently and also to support reporting of these measurements in a message sent to the network. If the message size available for reporting is too short, then the mobile only includes channel quality measurements for the carrier on which the polling message was received.

2.3.2.5   Application to Different Frequency Bands

Several GSM operators today have allocations in different frequency bands (such as GSM 900 and DCS 1800 in Europe) and use a common BCCH in the low band (900 MHz band) and deploy traffic channels both in the low band (900 MHz) and in the high band (1800 MHz). The issue was assessed whether Downlink Dual Carrier should also support configurations with one carrier allocated in the low band and one carrier allocated in the high band. However, because frequency planning due to different propagation losses for low band and high band is different and because of the impact on the mobile receiver design (such a configuration would force a narrowband implementation), the application of Downlink Dual Carrier has been restricted in the specifications to two carriers in the same frequency band.

2.3.2.6   Signaling Support

The support of Downlink Dual Carrier is indicated by the mobile including information on the supported multislot class in the Downlink Dual Carrier operation and is sent via higher control layer to the network as part of the mobile station's capabilities. The maximum number of time slots in a TDMA frame that the MS can receive is implicitly specified by the applicable multislot class as depicted in Table 2.8, according to 3GPP TS 45.002. The MS signals to the network whether it supports the possible maximum or only a reduced number of time slots relative to this maximum.

2.3.3   Specification Aspects

Downlink Dual Carrier is one of the major Release 7 features in 3GPP. Table 2.9 provides an overview of the related specifications.

2.3.4   Implementation Aspects

In this section, implementation aspects are considered for radio access and core network and for the mobile.

Table 2.9 3GPP specifications relevant to Downlink Dual Carrier

Specification	Aspects
24.008 [3]	Signaling of the mobile station's Downlink Dual Carrier Capability, including: – Multislot Capability Reduction for Downlink Dual Carrier – Downlink Dual Carrier for DTM Capability
43.055 [5]	Overview of Downlink Dual Carrier in dual-transfer mode
43.064 [4]	Functional description of Downlink Dual Carrier: – Downlink Dual Carrier configuration – Assignment for PTCCH, PDTCH, PACCH – MS Multislot Capability – Uplink power control – Channel quality measurements – Uplink/Downlink Packet Transfer
44.018 [6]	RRC protocol support for Downlink Dual Carrier, e.g. in dual-transfer mode
44.060 [7]	RLC/MAC protocol support for Downlink Dual Carrier, e.g. assignments
45.002 [9]	Definition of: – Mapping of uplink packet traffic and control channels – Multislot configurations and multislot capabilities in Downlink – Dual Carrier configurations
45.005 [13]	MSRD requirements in Downlink Dual Carrier configurations
45.008 [14]	Channel quality reporting in packet transfer mode
51.010 [17]	MS conformance specification: definition of radio test methods and conformance requirements for Downlink Dual Carrier
51.021 [18]	BSS conformance specification: radio test methods and conformance requirements for Downlink Dual Carrier

2.3.4.1 Impacts on the Network

Impacts on the radio access network are generally limited to SW-based functionalities such as the multiplexing of Downlink Dual Carrier mobiles performed in RLC/MAC layer, the support of Downlink Dual Carrier radio resource signaling (at assignment and resource reconfiguration) and Link Adaptation. The core network needs to be made aware of the Downlink Dual Carrier capability of the mobile.

2.3.4.2 Impact on the Mobile Station

The major difference between a state-of-the-art single carrier mobile and a Downlink Dual Carrier mobile is that simultaneous reception on two carriers is required. In general, two receiver types may fulfill this requirement:

1. *Separate narrowband receiver chains*: The dual-carrier terminal exploits an architecture, where the receiver branches can be tuned to different frequencies, whereby the receiver branches can use either the same antenna or separate antennas. The latter case is depicted in Figure 2.20.

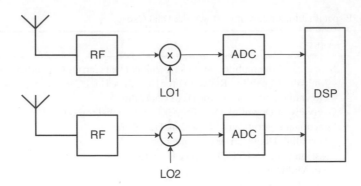

Figure 2.20 Downlink Dual Carrier receiver with separate receiver chains for each carrier using two receiver antennas.

2. *Wideband receiver*: Another option for Downlink Dual Carrier operation may be a wideband receiver. In this architecture, a single analogue to digital converter (ADC) samples the frequency range of the dual carrier assignment and baseband processing is commonly done for all carriers. Due to lower channel selectivity than for option (1), receiver blocking and receiver intermodulation requirements are tougher with this option.

2.4 Mobile Station Receiver Diversity

Mobile Station Receiver Diversity (MSRD) is a downlink feature which aims to improve the reception of the radio link in the downlink direction. This is achieved by means of diversity provided by an additional antenna in the MS. Antenna diversity is commonly used in the BTS where a significant improvement in uplink performance is achieved. MSRD aims to bring some of these benefits to the MS.

MSRD is characterized by a new set of performance requirements to the MS. These performance requirements are referred to as Downlink Advanced Receiver Performance Phase 2 (DARP Phase 2). MSRD can be considered a natural extension to the Single Antenna Interference Cancellation (SAIC) feature specified in 3GPP Release 6; denoted DARP Phase 1. The introduction of SAIC showed that significant link and system level gains can be achieved by enhancing the MS receiver for GMSK modulated signals. SAIC is described further in Section 8.3. MSRD provides a natural extension to this by adding further gains to GMSK modulated speech and data services, but also by improving the performance of channels with higher order modulation, such as EGPRS.

This means that whereas DARP Phase 1 was targeting improvements in the spectral efficiency of speech services, DARP Phase 2 also allows improved data services. If the market penetration of DARP Phase 2 capable MS is high, it will allow a significant increase in the network capacity. For lower MS penetration rates, the user will observe an increase in the average data rate and thereby a better user experience for data services.

2.4.1   The Concept of MSRD

The basic principle of receiver diversity is that the MS receives the downlink transmission using two antennas and combines the signals from the antennas before decoding. This mitigates the impact of the fading radio channel, also known as multi-path fading. Ideally, the two received signals are independent, that is uncorrelated and thus the combined signal will have less fades, thereby increasing the signal-to-noise level of the received transmission.

In practice, the signals received at the two antennas will be correlated, and the higher the correlation "ρ", the less diversity gain can be achieved. The amount of correlation depends on factors such as the radio propagation environment, the antenna design, the physical design of the MS as well as the presence of a user. A simple model of the environment surrounding a diversity-capable MS is shown in Figure 2.21.

The left-hand side of Figure 2.21 illustrates the transmitter at the base station sending the desired downlink signal to the MS. This signal is received by the MS, and the signals received at the two antennas are correlated with a factor, ρ. Besides the desired signal, Figure 2.21 also shows signals originating from other base station transmitters. Like the desired signal, these signals are received by the MS antennas and are correlated with factors $\rho_2 \ldots \rho_n$.

Another impairment which affects the performance of a receiver diversity MS is the antenna gain imbalance or branch power difference, G. This impairment simply stems from the fact that the received signal at the two antennas may be attenuated by different amounts. As for the correlation, the antenna gain imbalance depends on the design of the MS and the user interaction. For instance, the user may attenuate the signal to one antenna by covering its signal with the hand during reception. Figure 2.22 illustrates the impact of antenna correlation for an example MSRD implementation. The scenario is a typical urban scenario with a mobile speed of 50 km/h with a co-channel interferer. The desired user signal and the interferer signal are correlated with different correlation factors. Each point of the surface illustrates the required CIR in dB to achieve a bit error rate of 1% in the receiver. It can be observed from Figure 2.22 that there is a difference of 5–6 dB between a correlation factor of 0 and 1 (100%).

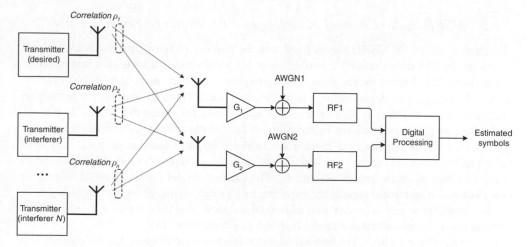

Figure 2.21   Model of the environment surrounding a receiver diversity-capable MS.

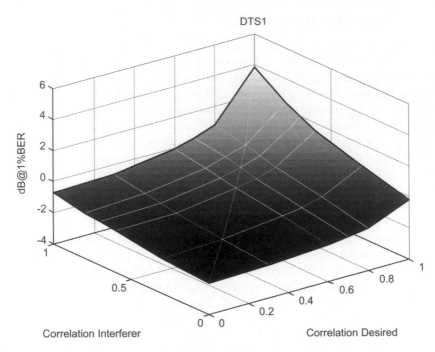

Figure 2.22   Impact of antenna correlation on diversity performance.

Note: Typical urban environment; MS speed of 50 km/h; co-channel interference.

In order to take these impairments into account when specifying the performance requirements to MSRD (DARP Phase 2), a simple channel model was adopted by 3GPP and included in 3GPP TS 45.005 [13].

2.4.2   MSRD Radio Channel Modeling and Performance Requirements

The channel model for MSRD was derived with the purpose of setting performance requirements for the MS which reflects a performance that would be achievable in a real network. Furthermore, the channel model allows setting requirements that ensure a robust MS performance. The channel model takes these two objectives into account by including antenna gain imbalance and correlation as parameters. The channel model is shown in Figure 2.23.

Antenna correlation is modeled by the parameter ρ denoting the magnitude of the complex correlation whereas antenna gain imbalance is modeled by an attenuation G applied to one of the signal branches. As can be observed from the model, $\rho = 0$ results in uncorrelated signals at the two outputs of the model, Y_1 and Y_2. The channel model reflects one signal path or radio link and if additional signals are required, for example to model interfering signals, the channel model is simply extended with additional instances, and each signal is summed at the antenna ports. A multi-interferer model is shown in Annex N of [13].

For DARP Phase 1 (SAIC), a new set of test scenarios were defined denoted DARP Test Scenarios (DTS), see Section 8.3. The basic idea behind these is to better reflect the complex

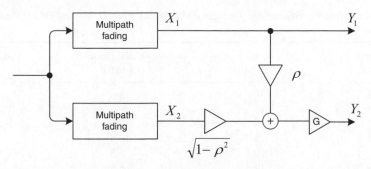

Figure 2.23 Simple channel model for MSRD. Reproduced by permission of 3GPP TS 45.005 v7.20.0 figure N.2.1. © 2009. 3GPP™ TSs and TRs are the property of ARIB, ATIS and TTC, CCSA, ETSI, TTA and TTC who jointly own the copyright in them. They are subject to further modifications and are therefore provided to you "as is" for information purposes only. Further use is strictly prohibited.

interference scenarios that a DARP-capable MS will be influenced by in a real network. For MSRD, these test scenarios are extended to receiver diversity scenarios by means of the channel model in Figure 2.23 extended to a multi-interferer scenario, see Annex N of [13].

Two sets of antenna correlation and gain imbalance have been defined reflecting the best case and worst-case impairments (Table 2.10).

2.4.3 Link Level Performance of MSRD

The link level performance of MSRD can be exemplified by comparing the performance requirements in 3GPP TS 45.005 for DARP Phase 2 with those of a legacy and a DARP Phase 1 receiver.

Table 2.11 shows the performance requirements of a legacy MS, a DARP Phase 1 MS and a DARP Phase 2 MS, for GPRS and EGPRS PDTCH using GMSK modulation and with co-channel interference. The requirements are specified as the CIR level in dB for which a BLER of at most 10% is reached.

It can be observed that the requirements for DARP Phase 1 are roughly 5 to 8 dB tighter than those of a normal MS. The DARP Phase 2 requirements are in the area of 15 to 20 dB tighter than those of the DARP Phase 1 MS, and more than 20 dB tighter than those of the

Table 2.10 DARP Phase 2 (MSRD) diversity parameters. Reproduced by permission of 3GPP TS 45.005 v7.20.0. © 2009. 3GPP™ TSs and TRs are the property of ARIB, ATIS, CCSA, ETSI, TTA and TTC who jointly own the copyright in them. They are subject to further modifications and are therefore provided to you "as is" for information purposes only. Further use is strictly prohibited.

Parameter set	Antenna correlation, ρ	Antenna gain imbalance, G
Set 1	0.0	0 dB
Set 2	0.7	−6 dB

Table 2.11   Performance requirements for legacy MS, DARP Phase 1 MS and DARP Phase 2 MS at BLER ≤ 10%

Modulation and coding scheme	Legacy MS (dB)	DARP Phase 1 (dB)	DARP Phase 2 (dB)
PDTCH CS-1	10	3	−12.5
PDTCH CS-2	14	6	−9.5
PDTCH CS-3	16	8.5	−8.0
PDTCH CS-4	24	19.5	0.0
PDTCH MCS-1	10.5	3.5	−11.5
PDTCH MCS-2	12.5	5.5	−10.0
PDTCH MCS-3	17	11	−6.5
PDTCH MCS-4	22	18	−1.0

Notes: Typical urban channel conditions. Speed 50 km/h. GSM 900 MHz. Co-channel interference (DTS-1).
Source: From 3GPP TS 45.005, Tables 2a, 2o and 2q. GMSK modulated PDTCHs (GPRS/EGPRS).

normal MS. Figure 2.24 provides a graphical representation of the gains of DARP Phase 1 and DARP Phase 2 compared to a legacy MS.

It is clear that DARP Phase 2 MS effectively cancels the impact of a co-channel interferer and thus provides very large gains in performance compared to a legacy receiver.

Table 2.12 shows the performance requirements of the 8-PSK modulated PDTCHs of EGPRS for a legacy MS and a DARP Phase 2 MS. No requirements are specified for DARP Phase 1, as SAIC only targets GMSK modulated channels, as described in Section 8.3. The

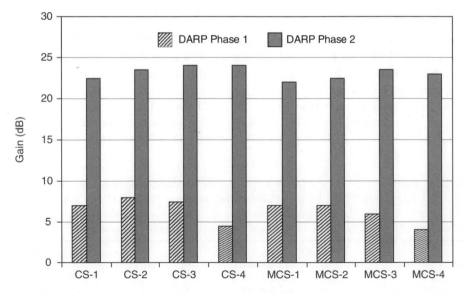

Figure 2.24   Gains of DARP Phase 1 and DARP Phase 2 compared to a legacy MS.

Table 2.12 Performance requirements for legacy MS and DARP Phase 2 MS at BLER ≤ 10%

Modulation and coding scheme	Legacy MS (dB)	DARP Phase 1 (dB)	DARP Phase 2 (dB)
PDTCH/MCS-5	15.5	–	−6.5
PDTCH/MCS-6	18	–	−4.0
PDTCH/MCS-7	25	`–`	1.5
PDTCH/MCS-8	25.5*	–	1.5*
PDTCH/MCS-9	30.5*	–	6.0*

Notes: Typical urban channel conditions. Speed 50 km/h. GSM 900 MHz. Co-channel interference.
* means the requirement is specified at the 30% BLER level.
Source: From 3GPP TS 45.005, Tables 2a and 2q. 8-PSK modulated PDTCH (EGPRS).

differences between the requirements are represented in Figure 2.25. It is clear that DARP Phase 2 also provides significant gains for 8-PSK modulated services, as the performance requirements are more than 20 dB tighter than those for the legacy receiver.

Table 2.13 shows the reference sensitivity performance requirements of a legacy MS compared to those of a DARP Phase 2 MS. The DARP Phase 2 requirements are shown for both best and worst case impairments. That is, no correlation and no gain imbalance as well as a correlation of 0.7 and a gain imbalance of −6 dB. The gains compared to legacy MS requirements are represented in Figure 2.26 where "AGI" denotes the antenna gain imbalance.

When comparing the legacy requirements to those of the DARP Phase 2 receiver, it is clear that even with worst case parameters, the DARP Phase 2 MS provides a significant improvement in reference sensitivity level. The gains are in the order of 0.5 to 7 dB, and

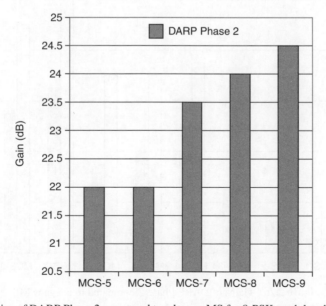

Figure 2.25 Gains of DARP Phase 2 compared to a legacy MS for 8-PSK modulated coding schemes.
Note: Co-channel interference.

Table 2.13 Performance requirements for legacy MS and DARP phase 2 MS at BLER $\leq$ 10%

Modulation and coding scheme	Legacy MS (dBm)	DARP Phase 2 $\rho = 0$, AGI = 0 dB (dBm)	DARP Phase 2 $\rho = 0.7$, AGI = -6 dB (dBm)
PDTCH MCS-1	-102.5	-105.0	-103.0
PDTCH MCS-2	-100.5	-105.0	-102.0
PDTCH MCS-3	-96.5	-102.5	-99.0
PDTCH MCS-4	-91	-98.5	-95.0
PDTCH MCS-5	-93	-100.0	-97.0
PDTCH MCS-6	-91	-98.0	-94.5
PDTCH MCS-7	-84	-94.0	-90.5
PDTCH MCS-8	-83^*	-92.0^*	-88.5^*
PDTCH MCS-9	-78.5^*	-89.0^*	-85.5^*

Notes: Typical urban channel conditions. Speed 50 km/h. GSM 900 MHz Sensitivity.
* means the requirement is specified at the 30% BLER level.
Source: From 3GPP TS 45.005, Tables 1a, 1c and 1j. GMSK and 8-PSK modulated PDTCH (EGPRS).

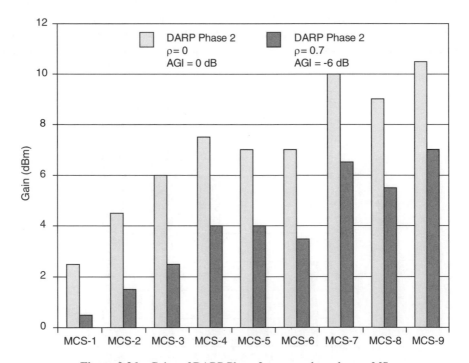

Figure 2.26 Gains of DARP Phase 2 compared to a legacy MS.

Note: Reference sensitivity level for different impairments.

Table 2.14 3GPP specifications relevant to MSRD

Specification	Aspects
24.008 [3]	Signaling of the mobile station's DARP Phase 2 capability
45.005 [13]	Performance requirements to DARP Phase 2 terminals
	Description of interferer scenarios
	Description of channel model
45.015 [16]	Release-independent DARP implementation guidelines
51.010 [17]	MS conformance specification: definition of radio test methods and conformance requirements for DARP Phase 2

increase to 2.5 to 11 dB when considering the best case parameters. Thus, it is obvious that large gains can be achieved when taking these impairments into consideration in the MS design.

2.4.4 Specification Aspects

Table 2.14 shows the 3GPP specifications relevant to MSRD.

2.4.5 Implementation Aspects

MSRD is a receiver enhancement which to a large extent only affects the MS implementation. The network implementation may, however, require minor changes to procedures such as link adaptation algorithms and radio resource management in order to take the enhanced downlink performance into account.

The impacts on the MS implementation are significant compared to a DARP Phase 1 implementation as an increase in processing power is required as well as additional hardware in order to implement the second receiver branch. Antenna design and placement also pose significant challenges to the form factor and design of the MS. Such constraints do not exist in the BTS and thus receiver diversity is commonly used in the uplink. However, cost, size and mechanics design are very important parameters in the MS. Therefore, the introduction of MSRD often requires trade-offs between these parameters and the actual performance of the feature.

2.5 Latency Reductions

2.5.1 Introduction

Latency reductions aim at decreasing the round-trip time (RTT) between the mobile station and the network's PCU,[6] the unit operating RLC and MAC protocols. A short RTT enables

---

[6] PCU: Packet Control Unit.

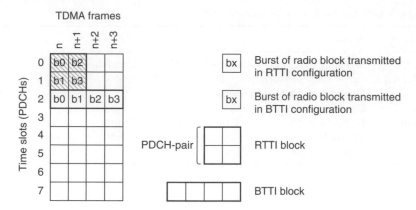

Figure 2.27   BTTI and RTTI configurations.

low-delay services, but also generally improves the performance of TCP-based services including web browsing. Fast Ack/Nack Reporting (FANR) and Reduced Transmission Time Interval (RTTI) are the two features specified in Release 7 to reduce latency. In addition, while FANR and RTTI are closely linked to the support of packet-switched conversational services in GERAN, RLC non-persistent mode (NPM) was also defined in Release 7 to increase the radio link performance of point-to-point connections used for delay-sensitive applications.

2.5.2   Reduced Transmission Time Interval

2.5.2.1   Concept

A radio block, which consists of four normal bursts, is transmitted using either a basic transmission time interval (BTTI) configuration or a reduced transmission time interval (RTTI) configuration.

The BTTI configuration corresponds to the original radio block configuration as of the first release of GPRS, R97. In this configuration the four bursts of a radio block are transmitted on the same time slot in four consecutive TDMA frames.

The RTTI configuration was introduced as a means to reduce latency. In this configuration the four bursts of a radio block are transmitted on two time slots (PDCH-pair[7]) in two consecutive TDMA frames. Thus a radio block is sent in RTTI configuration in half the time it takes with a BTTI configuration.

Figure 2.27 illustrates the transmission of a radio block in BTTI and RTTI configurations. A radio block is transmitted with four bursts (b0, b1, b2, b3). The legacy frame structure is strictly followed in RTTI configuration. The transmission of an RTTI block always occurs during the first or the second half of a BTTI period.

---

[7] A PDCH (Packet Data Channel) corresponds to a time slot.

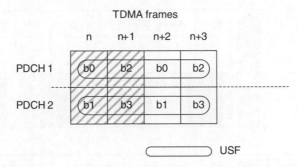

Figure 2.28 BTTI USF Mode for a TBF in RTTI configuration.

2.5.2.2 Assignment of RTTI Resources

The network assigning a TBF using an RTTI configuration specifies a set of PDCH-pairs for both the uplink and the downlink directions. In a default configuration the first PDCH-pair consists of the PDCH on time slot 0 and PDCH on time slot 1, the second PDCH-pair comprises PDCH on time slot 2 and PDCH on time slot 3 and so forth.

Each PDCH-pair in one direction has a unique corresponding PDCH-pair in the opposite direction for the purpose of reception and transmission of control signaling. In a default configuration, the corresponding PDCH-pair is allocated on the same time slot numbers as the PDCH-pair in the opposite direction. However, some configurations require the corresponding PDCH-pair(s) to use a different time slot in particular when the dual transfer mode is used, in which case the network provides an explicit indication of the assigned PDCH-pairs for both directions.

2.5.2.3 Multiplexing with Legacy Terminals

In order to avoid radio resource segregation between RTTI-capable mobile stations and legacy ones, it is required that the uplink resource be allocated to a legacy mobile station when RTTI radio blocks are transmitted in the downlink and that the uplink resource can be allocated to a mobile station in RTTI configuration when BTTI radio blocks are transmitted in the downlink.

This requirement led to the introduction of two USF[8] modes: BTTI USF mode and RTTI USF mode. In RTTI USF mode, the USF is transmitted as part of the RTTI radio block within two TDMA frames. The RTTI USF mode is used if the radio resource is shared only between mobile stations operating in RTTI configuration. In BTTI USF mode, two USF values are transmitted on a PDCH-pair. The first USF value is carried in bursts b0, b2 of the two RTTI blocks transmitted on the PDCH-pair. The second USF value is carried in bursts b1, b3 of the two RTTI blocks transmitted on the PDCH-pair (Figure 2.28).

---

[8] USF: Uplink State Flag. The USF is sent in downlink RLC/MAC blocks to allocate dynamically one or more radio blocks in uplink direction. At assignment of an uplink TBF to a mobile station, a USF is assigned to that TBF (and mobile station), the later scheduling of which in downlink will determine the scheduling of the uplink TBF. The USF granularity controls whether the USF allocates one or four radio blocks on a given time slot (USF granularity one and four, respectively).

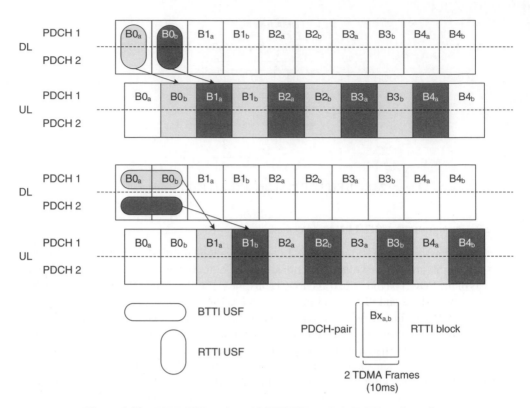

Figure 2.29   BTTI USF mode and RTTI USF mode in RTTI configuration.

If multiplexing with legacy terminals is not used and thus RTTI USF mode is used, a USF value assigned to a TBF in RTTI configuration allocates one RTTI block in the next RTTI block period. When USF granularity four is used, the USF sent during the first half of a BTTI period allocates an RTTI block in the second half of the current and of the next three BTTI periods. Similarly, the USF sent during the second half of a BTTI period allocates an RTTI block in the first half of each of the four BTTI periods immediately following the BTTI period in which the USF was sent.

If multiplexing with legacy terminals is required and thus BTTI USF mode is used, a USF on the lower-numbered PDCH (lower time slot number) of a PDCH-pair allocates an RTTI block in the first half of the next BTTI period. A USF on the higher numbered PDCH (higher time slot number) of a PDCH-pair allocates an RTTI block in the second half of the next BTTI period. If USF granularity "four" is used, the USF allocates an RTTI block in the first or the second half of the next four BTTI periods depending on which PDCH the USF was sent. The uplink allocation with USF granularity four in BTTI and RTTI USF mode is shown in Figure 2.29.

In BTTI USF mode, the network may also transmit the USF with mixed modulation. In this case, the modulation used for the transmission of the first RTTI block differs from the modulation used for the transmission of the second RTTI block. However, the network must ensure that only the modulations supported by the mobile station to which the USF is addressed are used. If the USF is sent with mixed modulation, then the coding of the USF bits conveyed

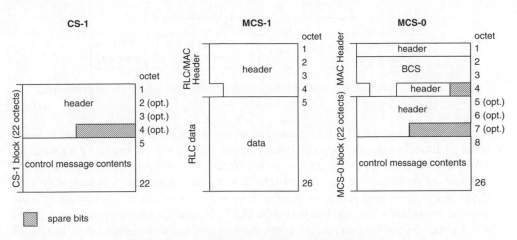

Figure 2.30   CS-1, MCS-1 and MCS-0 block structures.

by the first and the second RTTI block is performed according to the modulation and coding scheme used for these blocks.

Because CS-1 cannot be used to send RLC/MAC control blocks during a TBF in RTTI configuration when BTTI USF mode is used, a new modulation and coding scheme, MCS-0, is used instead. MCS-0 may also be used in RTTI configuration with RTTI USF mode.

MCS-0 is largely derived from MCS-1. The MCS-0 header includes the RLC/MAC header of the RLC/MAC control block as well as the 18-bit block check sequence (BCS) of the payload it carries (the RLC/MAC control message). It corresponds to the size of the RLC/MAC header in MCS-1. The encoding of the MCS-0 header follows that of the MCS-1 RLC/MAC header, while the encoding of the MCS-0 payload is the same as that of the MCS-1 payload. The interleaving and burst mapping are common to both schemes. The structure of MCS-0 compared to CS-1 and MCS-1 is depicted in Figure 2.30.

2.5.3   Fast Ack/Nack Reporting

2.5.3.1   Concept

Since GPRS was specified in R97, the transfer of RLC data blocks[9] between two peer RLC entities in RLC acknowledged mode has relied on the receiver indicating the reception status of these blocks to the transmitter, for potential retransmission(s). A (positive) acknowledgement (Ack) indicates the corresponding data block was correctly received while a negative acknowledgement (Nack) indicates otherwise, that is a retransmission is necessary. Ack/Nack information pertaining to a number of blocks is provided by the RLC receiver by means of a specific control message (Ack/Nack message) which then carries an Ack/Nack bitmap, of which each bit indicates the reception status of a given data block. Given that the RLC

---

[9] RLC data blocks are carried within RLC/MAC blocks for data transfer. An RLC/MAC block for data transfer may carry only one RLC data block in GPRS, up to two RLC data blocks in EGPRS and up to four RLC data blocks in EGPRS2.

RLC/MAC block			
RLC/MAC header	RLC data block 1	RLC data block 2 (conditional)	PAN (optional)

Figure 2.31   Structure of RLC/MAC blocks for data transfer including PAN field (example).

control-plane data and the RLC user-plane data share the same radio resources and are multiplexed in time-domain, the network usually keeps the amount of resources allocated to
control-plane signaling at a minimum so as not to negatively impact the user-plane traffic.
As a result, an Ack/Nack message is transmitted with delay after a block is sent; only once
multiple blocks have been sent is an Ack/Nack message issued. Consequently, the retransmission of erroneously received blocks, that is blocks intended for a given mobile station of
which the decoding of the data part or of both the data and header parts has failed, is delayed
proportionally to the frequency of the Ack/Nack message transmission.

Fast Ack/Nack Reporting (FANR) addresses the problems above by allowing, in addition
to Ack/Nack messages, the piggy-backing of Ack/Nack information within RLC/MAC blocks
for data transfer. Fast reporting of Ack/Nack information between peer RLC entities is thus
made possible by means of a new PAN (Piggy-backed Ack/Nack) field in those blocks, as
shown in Figure 2.31. Two ways of reporting are defined: SSN-based, described in Section
2.5.3.2 and time-based, described in Section 2.5.3.3, which differ only by how the PAN is
constructed and formatted.

The introduction of this new structure required the definition of new modulation and coding
scheme (MCS) variants for all MCSs except MCS-4 and MCS-9. The PAN field could not be
included in MCS-4 or MCS-9 because the code rate of the payload is 1.0, and because the
encoding of the header could not be changed. Tables 2.15 and 2.16 list the coding parameters
for the EGPRS and EGPRS2 coding schemes. The code rate of the payload when a PAN is
included is indicated with an asterisk (*).

Similarly to Ack/Nack messages, the network sends PAN fields pertaining to an uplink
TBF in an unsolicited manner. However, the mobile station transmits an Ack/Nack message
or a PAN pertaining to a downlink TBF upon polling from the network. In addition, the
network may enable an event-based fast Ack/Nack reporting for a downlink TBF, the aim
of which is to report an erroneous transmission as soon as it is detected by transmitting a
negative acknowledgement to the RLC transmitter. When an event is detected, the mobile
station transmits either a PAN if possible (e.g. it has other data to transmit in uplink), or if not,
an Ack/Nack message. An erroneous transmission is detected by one of the following events:

- The RLC/MAC header of an RLC/MAC block for data transfer is successfully received but
  the decoding of the RLC data block(s) included in the RLC/MAC block fails.
- An out-of-sequence reception occurs. The mobile station receives a new RLC data block
  which is not the next in-sequence data block to the most recent[10] RLC data block received.
  The in-sequence data block(s) have thus not been received.

---

[10] Recent here refers to the sequence number domain, not to the time domain.

Table 2.15 EGPRS modulation and coding scheme details, with and without PAN

Scheme	Code rate (payload)	Code rate (header)	PAN code rate (if present)	Modulation	RLC data blocks per Radio Block	Raw data per RLC data block (bit)	Family	Data rate kbit/s
MCS-9	1.0	0.36	n/a[1]		2	2×592	A	59.2
MCS-8	0.92 0.98*	0.36	0.42		2	2×544	A	54.4
MCS-7	0.76 0.81*	0.36	0.42	8-PSK	2	2×448	B	44.8
MCS-6	0.49 0.52*	1/3	0.39		1	592 48+544	A	29.6 27.2
MCS-5	0.37 0.40*	1/3	0.39		1	448	B	22.4
MCS-4	1.0	0.53	n/a		1	352	C	17.6
MCS-3	0.85 0.96*	0.53	0.63		1	296 48+248 and 296	A	14.8 13.6
MCS-2	0.66 0.75*	0.53	0.63	GMSK	1	224	B	11.2
MCS-1	0.53 0.60*	0.53	0.63		1	176	C	8.8

Note: [1] not available. *Code rate of the payload when a PAN is included.

While PAN enables the fast exchange of Ack/Nack information between peer RLC entities, reaction times are defined to minimize both the time by which the mobile station should be ready to retransmit a block indicated as erroneous in the received PAN,[11] and the time by which the mobile station should be ready to transmit a PAN or Ack/Nack message upon detection of an erroneous RLC data block.[12]

2.5.3.2 SSN-Based Fast Ack/Nack Reporting

SSN[13]-Based FANR is always used in the uplink direction. In this case it conveys acknowledgement information pertaining to a downlink data transfer. In the downlink direction, the network may use either SSN-Based FANR or Time-Based FANR. The variant used is signaled to the mobile station at establishment and reconfiguration of an uplink TBF.

---

[11] This reaction time is defined as 5 or 6 TDMA frames since the last TDMA frame of the RLC/MAC block containing the PAN.

[12] This reaction time is defined as 5 or 6 TDMA frames since the last TDMA frame of the RLC/MAC block in which the erroneous data block was detected.

[13] SSN: Starting Sequence Number. The SSN contains the sequence number of the first data block reported in a PAN or an Ack/Nack message.

Table 2.16 EGPRS2 modulation and coding schemes details

Scheme	Code rate (payload)	Code rate (header)	PAN code rate (if present)	Modulation	RLC data blocks per Radio Block	Raw data per RLC data block (bit)	Family	Data rate (kbit/s)
DAS-12	0.96 0.99*	0.38	0.38		3	658	B	98.4
DAS-11	0.80 0.83*	038	0.38	32QAM	3	546	A	81.6
DAS-10	0.64 0.66*	0.33	0.38		2	658	B	65.6
DAS-9	0.68 0.70*	0.34	0.38		2	546	A	54.4
DAS-8	0.56 0.58*	0.34	0.38	16QAM	2	450	B	44.8
DAS-7	0.54 0.57*	033	0.38		1	658	B	32.8
DAS-6	0.45 0.48*	0.33	0.38	8PSK	1	546	A	27.2
DAS-5	0.37 0.39*	0.33	0.38		1	450	B	22.4
DBS-12	0.98 1.00*	0.37	0.54		4	594	A	118.4
DBS-11	0.91 0.93*	0.37	0.38	32QAM	4	546	A	108.8
DBS-10	0.72 0.75*	0.34	0.38		3	594	A	88.8
DBS-9	0.71 0.73*	0.34	0.38		3	450	B	67.2
DBS-8	0.60 0.63*	0.31	0.38	16QAM	2	594	A	59.2
DBS-7	0.47 0.48*	0.31	0.38		2	450	B	44.8
DBS-6	0.63 0.69*	0.31	0.38		1	594	A	29.6
DBS-5	0.49 0.53*	0.31	0.38	QPSK	1	450	B	22.4
UAS-11	0.95 0.99*	0.36	0.38		3	514	A	76.8
UAS-10	0.84 0.87*	0.36	0.38		3	450	B	67.2
UAS-9	0.71 0.74*	0.36	0.38	16QAM	2	594	A	59.2
UAS-8	0.62 0.64*	0.36	0.38		2	514	A	51.2
UAS-7	0.55 0.57*	0.36	0.38		2	450	B	44.8

Table 2.16   (*Continued*)

Scheme	Code rate (payload)	Code rate (header)	PAN code rate (if present)	Modulation	RLC data blocks per Radio Block	Raw data per RLC data block (bit)	Family	Data rate (kbit/s)
UBS-12	0.96 0.99*	0.35	0.38		4	594	A	118.4
UBS-11	0.89 0.91*	0.35	0.38	32QAM	4	546	A	108.8
UBS-10	0.71 0.73*	0.35	0.36		3	594	A	88.8
UBS-9	0.70 0.72*	0.32	0.36		3	450	B	67.2
UBS-8	0.60 0.61*	0.33	0.38	16QAM	2	594	A	59.2
UBS-7	0.46 0.47*	0.33	0.38		2	450	B	44.8
UBS-6	0.62 0.67*	0.35	0.38		1	594	A	29.6
UBS-5	0.47 0.51*	0.35	0.38	QPSK	1	450	B	22.4

Note: *Code rate of the payload when a PAN is included.

The PAN format in SSN-Based FANR is similar to that of an Ack/Nack message and is shown in Figure 2.32. It contains an address field (TFI[14]) to identify the TBF being acknowledged.[15] It includes a bitmap of which each bit indicates the reception status of a data block. And a starting sequence number (SSN) indicates the starting point of the bitmap and thus the exact data blocks for which a reception status is provided in the bitmap. The reported bitmap (RB) is constructed so that the RLC receiver reports the most recent unreported negative acknowledgement. The size of the reported bitmap is determined by that of the ShortSSN field which contains the least significant bits of the starting sequence number. The number of least significant bit included is determined based on the RLC window size so that the ShortSSN uniquely identifies data blocks within that window. The size of the ShortSSN field varies between 7 and 11 bits. If the reported bitmap starts at the beginning of window, then the RLC receiver indicates this by setting the Beginning Of Window (BOW) bit accordingly.

2.5.3.3   Time-Based Fast Ack/Nack Reporting

While SSN-based FANR uses the RLC sequence numbering as a reference to identify the RLC data blocks for which an acknowledgement status is provided in the PAN, the Time-Based Fast Ack/Nack reporting uses a time reference, between the transmission of an RLC data block in

---

[14] TFI: Temporary Flow Identifier.

[15] The number of information bits fed to the channel coder is 20 bits. The TFI is encoded by scrambling the TFI with the five last parity bits added to the PAN information bits in the encoder.

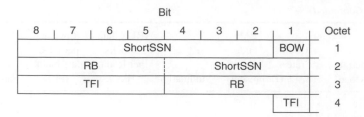

Figure 2.32 PAN format in SSN-based FANR. Reproduced by permission of 3GPP TS 44.060 v7.18.0 figure 10.3a.5.1. © 2009. 3GPP™ TSs and TRs are the property of ARIB, ATIS, CCSA, ETSI, TTA and TTC who jointly own the copyright in them. They are subject to further modifications and are therefore provided to you "as is" for information purposes only. Further use is strictly prohibited.

uplink and the transmission of a PAN in downlink, to identify an RLC data block. Thus while the SSN-based FANR is intended for a single mobile station, the time-based FANR is intended for a multiplicity of mobile stations.

In time-based FANR, the PAN provides an acknowledgement status for all RLC data blocks transmitted on one or more uplink time slots. The transmission time of the most recent RLC data block reported in the PAN is defined by a time reference. The time reference is represented as a time shift (TSH) between the first TDMA frame (N) of a given radio period in which the PAN is transmitted and the first TDMA frame of the radio block[16] in which the reported RLC data block has been transmitted (N − TSH or N − TSH − 1) modulo 2715648.[17] The possible values of TSH are listed in Table 2.17 (see Section 2.5.2 for details of RTTI).

The Time-Based PAN contains one code point for each RLC/MAC block transmitted in the uplink on the reported time slot. The code points for RLC/MAC blocks transmitted during the same transmission period are ordered in increasing order of the time slot numbers or PDCH-pairs (see Section 2.5.2.1) in which they were sent. The code points are bit strings of Huffman code and are described in Table 2.18. If more than one RLC data block was transmitted in a single radio block, then the data blocks are split into two groups as follows. For a radio block with two RLC data blocks, each group contains one RLC data block. For a radio block with three RLC data blocks, the first group contains the first RLC data block and the second group contains the second and third RLC data block. For a radio block with four RLC data blocks,

Table 2.17 Definition of the Time Shift as a function of TTI configuration

| TSH (2 bit field) | TTI configuration | |
	BTTI	RTTI
0 0	4 TDMA frames	2 TDMA frames
0 1	8 TDMA frames	4 TDMA frames
1 0	12 TDMA frames	6 TDMA frames
1 1	16 TDMA frames	8 TDMA frames

---

[16] A radio block consists of four bursts.
[17] Maximum frame number + 1, see [15].

Table 2.18 PAN code-points in Time-Based FANR. Reproduced by permission of 3GPP TS 44.060 v7.18.0 table 9.1.15.1.1 © 2009. 3GPP™ TSs and TRs are the property of ARIB, ATIS, CCSA, ETSI, TTA and TTC who jointly own the copyright in them. They are subject to further modifications and are therefore provided to you "as is" for information purposes only. Further use is strictly prohibited.

Bit string	Meaning (radio block with one RLC data block)	Meaning (radio block with two or more RLC data blocks)
0 1 0	Failed header decoding	Failed header decoding; or Header correctly received but failed decoding of all RLC data blocks
0 0	Header correctly received but failed decoding of RLC data block	Header correctly received, successful decoding of the first RLC data block group, failed decoding of the second RLC data block group
0 1 1	Reserved	Header correctly received, successful decoding of the second RLC data block group, failed decoding of the first RLC data block group
1	Header correctly received and successful decoding of RLC data block	Header correctly received and successful decoding of the payload of both RLC data block groups

the first group contains the first two RLC data blocks and the second group contains the second two RLC data blocks.

An example in Figure 2.33 shows how the Time-based PAN is created in a scenario where mobile stations using RTTI and BTTI configurations (see Section 2.5.2.3) are multiplexed on the same radio resources. The PAN received in a radio block using RTTI configuration (respectively BTTI configuration) conveys acknowledgement information only for blocks transmitted in RTTI configuration (respectively BTTI configuration) in the uplink. The PAN shown in the example includes a report for time slots 0 and 1 (one PDCH-pair). The PAN received in a radio block transmitted in RTTI configuration contains dummy information bits

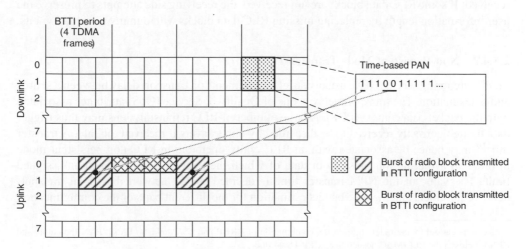

Figure 2.33 Example construction of a PAN in time-based FANR.

for uplink radio block periods in the BTTI configuration. The dummy information bits are the second and third leftmost "1" in the bitmap shown in the example.

2.5.3.4   Multiplexing with Legacy Terminals

Mobile stations using FANR may be multiplexed on the same radio resource with mobile stations not using FANR (i.e. legacy mobile stations, or mobile stations capable of FANR but ordered not to use FANR). In this case, a mobile station using FANR may detect the presence of a PAN field in an EGPRS RLC/MAC block for data transfer intended for a legacy mobile station (or a mobile station not using FANR) even if there is none.[18] Therefore, the network should avoid those polling periods, although the probability of a false positive decoding is very low.

2.5.4   RLC Non-Persistent Mode

2.5.4.1   Concept

The RLC non-persistent mode, originally introduced for Multimedia Broadcast/Multicast Services (MBMS) in Release 6, was adapted for point-to-point connections in Release 7. It is aimed at error-tolerant delay-sensitive services such as media streaming. The transfer of RLC data blocks in RLC non-persistent mode includes non-exhaustive (non-persistent) retransmissions (using selective type I ARQ or type II hybrid ARQ) controlled by the RLC window size, the non-persistent mode transfer time and acknowledgement information. RLC non-persistent mode can be seen as a hybrid between RLC acknowledged and unacknowledged modes where the RLC performance is significantly better than unacknowledged mode while maintaining a delay budget unlike RLC acknowledged mode.

The RLC transmitter operating in non-persistent mode is allowed to advance its RLC transmit window and transmit new RLC data even if the oldest block in the window has not been positively acknowledged by the receiver. This prevents the RLC window from stalling and ensures new requests for data transmission are served as soon as the radio resource is allocated. As a consequence, it is expected that some RLC data blocks will not be correctly received. If some RLC data blocks are not received, the receiving side attempts to preserve the user information length by replacing missing RLC data blocks with dummy information bits.

2.5.4.2   Non-Persistent Mode Transfer Time

Media streaming services have usually strict requirements on transport delay between a source and a destination. The transport delay is one of Quality of Service (QoS) attributes associated with the packet data context (PDP). A large number of RLC retransmissions may thus trigger data though correctly received to be discarded by upper layers if received too late. A transfer time[19] may hence be associated with an RLC entity operating in RLC non-persistent mode in order to avoid the transmission of data of which the delivery would exceed an assigned limit. The value of the NPM transfer time is decided by the network based on transport delay requirement and the time budget estimation for the different parts of the communication

---

[18] This is caused by the redefinition of the combination of RRBP and ES/P fields to PANI (PAN Indicator) and CES/P fields in the RLC/MAC header. See 3GPP TS 44.060.

[19] NPM transfer time.

channel. It is also possible for the network to assign different value of NPM transfer time to the RLC transmitter and the RLC receiver.

When the RLC transmitter begins to transmit a new upper layer PDU[20] and an NPM transfer time limitation is assigned, a timer set to this value is started, at the expiry of which no RLC data blocks containing segment(s) of the upper layer PDU are (re)transmitted any longer (except if an RLC data block also contains segment(s) of any other upper layer PDU for which the timer is still running). While the timer is running, if all RLC data blocks including a segment of an upper layer PDU have been positively acknowledged, the RLC transmitter stops the timer.

On the RLC receiver side, the boundaries of an upper layer PDU are unknown until all the RLC data blocks containing segments of the upper layer PDU are actually received. Therefore, the RLC receiver is able to start a timer with the value of the NPM transfer time only based on the status of RLC data blocks. The RLC receiver starts the timer for an RLC data block if the following conditions are met:

- the RLC data block has not been received yet; and
- an RLC/MAC block containing an RLC data block with equal or higher BSN has been received (this addresses the case when the RLC/MAC header is correctly decoded but the decoding of the RLC data block fails).

The timer at the receiver side is started only once for each RLC data block when the conditions above are met for the first time. The RLC receiver stops the timer when the RLC data block is correctly received.

The minimum and maximum values of the NPM transfer time are 30 ms and 5 s respectively.

2.5.5   Specification Aspects

RTTI, Fast Ack/Nack reporting and RLC non-persistent mode are specified as shown in Table 2.19.

2.5.6   Implementation Aspects

Implementation of latency reduction requires changes on both terminal and network sides. In order to support latency reduction, the network and the terminal must support new block formats and corresponding MCSs used for radio blocks carrying PAN, transmission and reception of blocks on PDCH-pair, new procedures such as event-based reporting. The terminal implementation also has to comply with new requirements on reaction times (e.g. a time needed for retransmission of a data block after it has been reported as erroneously received by the network). RLC non-persistent mode for non-MBMS services is a feature separate from latency reductions and as such its implementation is independent of latency reductions.

The terminal signals the support of latency reduction via a single bit in MS Radio Access Capability information element. This means that the terminal has to implement all the features included under the umbrella of latency reduction. However, there is no signaling of the support

---

[20] PDU: Packet Data Unit.

Table 2.19   3GPP specifications relevant to RTTI, FANR and RLC NPM

Specification	Aspects
24.008 [3]	Signaling of the mobile station's support of FANR and RTTI, as well as Non-persistent RLC mode
43.064 [4]	Functional description: – Fast Ack/Nack Reporting – RTTI configuration – Non-persistent RLC mode
44.018 [6]	Assignment of TBF using FANR, RTTI or RLC non-persistent mode, in particular for dual-transfer mode operation.

44.060 [7]	RLC/MAC protocol description of:
	– Fast Ack/Nack Reporting:
	  – Time-based
	  – Event-based
	  – RLC/MAC blocks including PAN field
	– Reduced TTI
	  – Configurations and assignments
	  – RTTI USF mode in dynamic and extended dynamic allocations
	– RLC non-persistent mode operation during EGPRS TBF

45.001 [8]	General description: – RTTI configuration – PDCH-pair
45.002 [9]	RTTI configuration Channel mapping in RTTI configuration 52-multiframe structure in RTTI configuration
45.003 [10]	Modulation and coding schemes for latency reduction
45.005 [13]	Radio performance requirements for RTTI and FANR
45.008 [14]	Link quality measurements for RTTI configurations
45.010 [15]	Reaction times for mobile stations using latency reduction features
51.010 [17]	MS conformance specification: definition of radio test methods and conformance requirements for RTTI, FANR and RLC non-persistent mode

from the network side which gives freedom to network vendors to implement a sub-set of the latency reductions, for example Fast Ack/Nack reporting only.

References

[1] 3GPP TS 45.912 v7.2.0, "Feasibility Study for Evolved GSM/EDGE Radio Access Network (GERAN)", February 2007.

[2] 3GPP TS 45.913 v1.3.0, "Optimized Transmit Pulse Shape For Downlink Enhanced General Packet Radio Service (EGPRS2-B)", February 2009.

[3] 3GPP TS 24.008 v7.14.0, "Mobile Radio Interface Layer 3 Specification; Core Network Protocols; Stage 3", March 2009.

[4] 3GPP TS 43.064 v7.13.0, "General Packet Radio Service (GPRS); Overall Description of the GPRS Radio Interface; Stage 2", May 2009.

[5] 3GPP TS 43.055 v7.6.0, "Dual Transfer Mode (DTM); Stage 2", August 2007.

[6]   3GPP TS 44.018 v7.18.0, "Mobile Radio Interface Layer 3 Specification; Radio Resource Control (RRC) Protocol", September 2009.

[7]   3GPP TS 44.060 v7.18.0, "General Packet Radio Service (GPRS); Mobile Station (MS) – Base Station System (BSS) Interface; Radio Link Control/Medium Access Control (RLC/MAC) Protocol", September 2009.

[8]   3GPP TS 45.001 v7.9.0, "Physical Layer on the Radio Path; General Description", September 2009.

[9]   3GPP TS 45.002 v7.7.0, "Multiplexing and Multiple Access on the Radio Path", May 2008.

[10] 3GPP TS 45.003 v7.10.0, "Channel Coding", September 2009.

[11] 3GPP TS 25.212 v7.11.0, "Multiplexing and Channel Coding (FDD)", September 2009.

[12] 3GPP TS 45.004 v7.3.0, "Modulation", August 2008.

[13] 3GPP TS 45.005 v7.19.0, "Radio Transmission and Reception", September 2009.

[14] 3GPP TS 45.008 v7.16.0, "Radio Subsystem Link Control", September 2009.

[15] 3GPP TS 45.010 v7.7.0, "Radio Subsystem Synchronization", November 2009.

[16] 3GPP TS 45.015 v7.0.0, "Release Independent Downlink Advanced Receiver Performance (DARP); Implementation Guidelines", August 2007.

[17] 3GPP TS 51.010-1 v8.3.0, "Mobile Station (MS) Conformance Specification; Part 1: Conformance Specification", September 2009.

[18] 3GPP TS 51.021 v7.7.0, "Base Station System (BSS) Equipment Specification; Radio Aspects", September 2009.

[19] Press release "Nokia Siemens Networks successfully conducts world's first, live EDGE Evolution Downlink Dual Carrier trial", Nokia Siemens Network, Beijing, China, September 14, 2009.

3

3GPP Release 8

Jürgen Hofmann, Vlora Rexhepi-van der Pol, Guillaume Sébire
and Sergio Parolari

3.1 Introduction

The rapid increase of data traffic over wireless networks and the increasing demand for real-time applications required cellular networks to evolve further. To cope with this trend in the longer term, enhancements of 3GPP access technologies to provide high data rates while enabling better spectrum efficiency over the radio interface were needed in the packet-switched domain. The feasibility studies on the evolution of cellular networks, which started in December 2004, focused on bringing these enhancements to UTRAN technology and 3GPP system architecture. The results of this work led to the definition in 3GPP Release 8 of "Evolved UTRA" generally known as LTE, and that of Evolved Packet Core (EPC) also referred as System Architecture Evolution (SAE). EPC is a flat core network architecture that enables the integration of LTE with legacy GERAN and UTRAN-HSPA networks as well as other non-3GPP wireless technologies. As defined in [1], the LTE and SAE/EPC jointly form the evolved packet system (EPS). LTE will boost mobile broadband and enable, for example real-time applications including Voice-over IP (VoIP) with high-performance in highly mobile environments and with bandwidth scalability up to 20 MHz.

In a landscape where other 3GPP RATs[1] are already deployed (i.e. GERAN and/or UTRAN), the deployment of LTE must follow an essential requirement: the ability to interwork with these technologies to ensure service continuity and service availability. Interworking mechanisms between GERAN/UTRAN and E-UTRAN have thus been specified in 3GPP Release 8.

Following another trend towards IP-based transport solutions in the wireless network, A interface over IP (AoIP) is another significant topic specified in Release 8. AoIP enables the A interface user plane data to be transported over IP. The support for the A interface control plane signaling over IP (SIGTRAN [2]) was already introduced in Release 7, while user plane data

---

[1] RAT: Radio Access Technology.

GSM/EDGE: Evolution and Performance   Edited by Mikko Säily, Guillaume Sébire and Eddie Riddington
© 2011 John Wiley & Sons, Ltd

was based only on TDM [3]. Implementing the A interface over IP allows efficient transport solutions, with higher flexibility and lower cost. AoIP enables simpler network design between the BSS and the MSCs, also allowing the CS core to utilize the same IP backhaul for different 3GPP radio access technologies. Furthermore, with AoIP, support for TrFO (Transcoder-free operation) becomes possible in GSM networks thus leading to bandwidth savings and improved voice quality.

RF performance specification for a GSM Multicarrier BTS (MCBTS) is a further topic handled in Release 8. In this context, two MCBTS classes have been introduced supporting multicarrier operation by means of a single wideband transceiver, that is without using passive TX combiners, thus improving power efficiency and allowing less complex antenna systems. However, the multicarrier architecture requires some relaxation compared to the stringent RF requirements specified for a state-of-the-art Normal BTS. Major impacts are caused by higher intra BSS TX intermodulation distortion (IMD) on the transmitter side and lower inband blocking performance due to the use of a wideband receiver architecture on the receiver side. The functional description and the standardization of GSM MCBTS are described at the end of this chapter.

3.2   Interworking with LTE

The design of LTE as a packet-switched only system brings a number of challenges in defining interworking mechanisms with other systems. The absence of the traditional circuit-switched domain [5] (Figure 3.1) means in particular that new solutions are needed to ensure service continuity of CS services such as voice in GERAN and UTRAN. Not only radio access technology (RAT) change needs to be executed, but also domain transfer between PS and CS domains is required.

Another design constraint is to limit the impact on the already deployed GERAN networks while taking into account the different sets of GERAN features supported in these networks.

Hence, the primary objective is to define adequate solutions that satisfy the LTE interworking requirements while having a minimum impact on the existing GERAN networks.

3.2.1   Interworking Requirements and Mechanisms

3.2.1.1   Interworking Requirements

The set of requirements for interworking between E-UTRAN and GERAN defined in [6] focuses on mobility and service continuity in idle and active modes for real-time and non-real-time applications. A multi-mode mobile station supporting E-UTRAN, GERAN and possibly UTRAN should be able to support measurements, cell change and possibly handover[2] between the two (three) systems, according to a fundamental design principle: "single radio". The principle of "single radio" consists of having at any point in time at most a single active radio (GERAN, UTRAN, E-UTRAN) which greatly minimizes the complexity of terminals as well as battery consumption. In addition, the impact on existing network elements would be

[2] A handover consists of allocating resources in the target cell prior to the mobile station being handed over to that cell, thus reducing the interruption which otherwise occurs at cell change. A handover keeps the mobile station in active mode, while a cell change forces a mobile station to idle mode in the target cell.

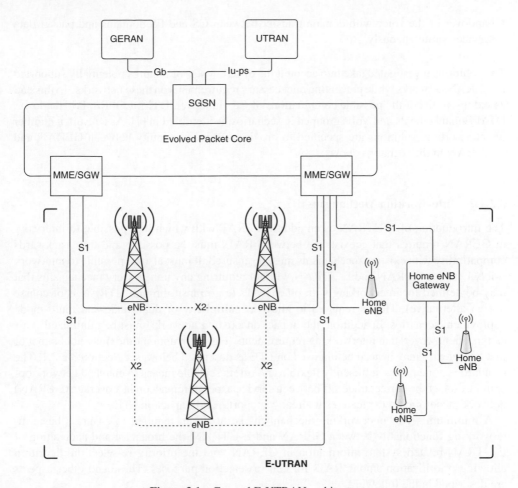

Figure 3.1   General E-UTRAN architecture.

minimized to reduce the deployment costs of E-UTRAN. Finally, performance criteria were defined as follows [6]:

- The interruption time of real-time services should be less than 300 ms.
- The interruption time for non-real-time services should not exceed 500 ms.

In case of inter-RAT handover between GERAN and E-UTRAN, three main scenarios are identified:

- handover of packet data services including voice service that is Voice over IP (VoIP);
- handover of voice services with domain transfer between CS and PS domains;[3]

---

[3] It should be noted that this scenario was considered only for the direction from PS (LTE) to CS (UMTS, GSM). Indeed, for the reverse direction it was assumed that a call ongoing in the CS domain (e.g. in GSM) would not be handed to LTE.

• handover of the voice with domain transfer between CS and PS domains and packet data services simultaneously.

The solutions investigated take into account the different sets of features potentially supported in GERAN networks while preventing unnecessary requirements on these networks. To this end an extensive list of the possible combinations of CS Handover, DTM, EGPRS, PS Handover, DTM Enhancements and VoIP, grouped in scenarios, is identified in [7]. As a result, a number of interworking solutions are specified to provide service continuity between GERAN and E-UTRAN in the scenarios above.

3.2.1.2 Interworking Mechanisms

The introduction of E-UTRAN as an additional RAT with which it is possible to interwork in GERAN requires that coexistence between RATs must be possible and thus backwards compatibility with existing mechanisms must be ensured. It must also be possible to interwork solely between GERAN and E-UTRAN without requiring any unnecessary mechanisms that may be defined for interworking with other RATs (e.g. transit through UTRAN is of course not an alternative). This is required to allow a multiplicity of dual-mode and multi-mode deployment scenarios. In addition, the impact on existing networks must be minimized. This can mean, however, that interworking requirements (e.g. interruption time) may not be met or that complexity may instead be imposed on LTE as described below, see Section 3.2.2.1. The standard is nonetheless sufficiently flexible and offers scalable means to enable a network operator to select the correct trade-off between service aspects, impact on its coverage (GERAN) network (according to the features it already supports) and impact on LTE.

A minimum set of interworking mechanisms is, however, required to ensure a basic interworking functionality between GERAN and E-UTRAN: the broadcast and acquisition of E-UTRAN-related system information in GERAN, and the priority re-selection algorithm allowing prioritization among RATs for cell re-selection purposes. These and other aspects are described in the following.

3.2.2 Service Continuity

3.2.2.1 Voice-Call Continuity with Domain Change between CS and PS

The concept of domain change was introduced in Release 7. It provides voice call continuity (VCC) between the IP Multimedia-Subsystem (IMS) and the CS domains [8]. The VCC application provides the necessary functions for originating and terminating calls while enabling domain transfer between the CS domain and the IMS (PS domain).

Release 7 VCC supports handovers between Wireless LAN (WLAN) VoIP and CS calls. This is known as "dual radio VCC" where two radios are simultaneously active, that is WLAN and for example GSM. IMS and CS registrations are concurrently active, and the mobile station with an ongoing call in one domain can start handover in the other domain simultaneously.

Inter-RAT handover of voice services between LTE and GERAN networks with no support for PS conversational services requires PS to CS domain transfer. As opposed to Release 7 VCC, the VCC solution between E-UTRAN and GERAN is based on single radio; Single

Radio VCC (SRVCC) specified as of Release 8 in [9]. SRVCC provides a solution for voice call continuity in the direction from E-UTRAN to GERAN only, seeing that this direction is identified as the most important and relevant scenario for early (typically spotty) LTE deployments [7]. It enables a PS to CS handover from E-UTRAN to GERAN by means of enhanced functionality in the MME and the MSC server. SRVCC is based on the assumption that the voice sessions are anchored in the IMS. The MSC server thus, upon receiving the handover request from the MME, initiates both the session transfer with the IMS and the CS handover with the target cell. Simultaneous handover of voice services and other packet data is handled such that the MME distinguishes the bearers to enable the handover of the data through the PS domain.

SRVCC is designed to provide voice service continuity with GERAN networks that support only CS handover (among the features listed in Section 3.2.1.1) and makes no changes to the GERAN networks. Instead SRVCC requires added functionality to the standard EPS functionality and importantly voice support in LTE. The complexity of SRVCC motivated the introduction of a CS fallback [10] solution developed to cope with early LTE deployments where voice is simply not supported. This solution enables the handling of traditional CS-domain services such as CS voice and SMS in E-UTRAN. For example, CS fallback enables a mobile station initiating/receiving a voice call in E-UTRAN to be redirected to GERAN where the call will be established and take place. Upon call termination, the mobile station may be redirected back to E-UTRAN by the network. The CS fallback in EPS is enabled by the introduction of the SGs interface, based on the Gs interface, defined in [11] between the MSC Server and the MME.

3.2.2.2 Packet Data Service Continuity

Packet data service continuity is enabled by inter-RAT PS handover between GERAN and E-UTRAN. While PS handover provides service continuity upon cell change, it is not always available. In this case cell re-selection occurs which involves service interruption. As opposed to a handover mechanism where resources in the target cell are assigned prior to the mobile station arriving in that cell and resources in the source cell are released only upon the mobile station successfully accessing the target cell, cell re-selection proceeds by first releasing the connection in the source cell, re-selecting to the target cell and upon successful access to that cell, establishing the connection to the cell.

In order to limit the service interruption at cell re-selection, the only way is to minimize the time needed for the mobile station to successfully access the new cell. To this end, the Network Assisted Cell Change (NACC) procedure [12, 13, 1] was introduced in Release 4 between GERAN cells. NACC enables the source cell to provide the mobile station with system information about the target cell necessary to access that cell, faster than it would take the mobile station to acquire that information from broadcast channels in the target cell. In Release 5, NACC was extended to cover the direction from UTRAN to GERAN only, while for the other direction it is faster to acquire the information in the UTRAN cell itself than transferring it while in GERAN. In Release 8, NACC has been extended for the same reason to support a cell change only in the direction from E-UTRAN to GERAN.

Although NACC enables better cell re-selection performance for packet data services through minimizing service interruption, it does not meet the timing requirements of LTE [14].

3.2.3   E-UTRAN Information in GERAN System Information

Mobility in a cellular system relies on the availability in each cell of neighboring cell information:

- to avoid mobile stations camping on a given cell to blindly search for neighboring frequencies and cells;
- to control the outbound mobility from a given cell and thus the inbound mobility to a given cell.

This neighbor cell information is typically broadcast as part of the system information of a cell. The conventional manner is for the network to broadcast a list of neighboring frequencies and possibly cells that a mobile station in this cell considers for cell re-selection and measurement reporting. This principle, generally known as a white list, has not been applied with the introduction of E-UTRAN.

Different from GERAN and UTRAN where the neighbor cell list is always a white list, in E-UTRAN the concept of black list was introduced. It consists of listing for given frequencies cells that are not to be considered for cell re-selection and measurement reporting. All other unlisted cells can thus be considered as potential candidates for cell change and measurements. The use of a black list for E-UTRAN cells was chosen due to the broadcast capacity savings it yields as less system information bits are transferred compared to the white list. A disadvantage of the black list, however, is that some cells not listed in the black list might not be allowed for some users, for example they may belong to a forbidden Tracking Area (TA). A mobile station re-selecting such a cell will, as in GERAN and UTRAN, be forbidden to re-select any cell on the same frequency for a certain time. In GERAN, in the case of E-UTRAN as well as UTRAN, this time is 20 minutes [15]. To alleviate this problem, adding the Tracking Area Identity (TAI) to the Physical Layer Cell Identity (PCI or PLCID) and the center frequency of an E-UTRAN cell was considered. It would have resulted in broadcast capacity utilization and therefore was not seen as desirable – it would also have led to a "gray" list. However, in order to identify such cells and avoid or minimize cell re-selection attempts to forbidden TAs and the resulting frequency ban, the grouping of E-UTRAN cells per TA is used in GERAN networks. The E-UTRAN neighbor cell information broadcast in GERAN system information hence consists of groups of cells (PCIs) belonging to the same TA and optionally include the related TAI. A device receiving this information would then be able to immediately re-select to a cell on a given frequency after it has re-selected a cell belonging to a forbidden TA on that same frequency. In connected mode, TA grouping saves also on the radio resources utilization as unnecessary measurement reporting of forbidden cells is prevented.

3.2.3.1   E-UTRAN-Related System Information Broadcast and Acquisition

The transmission of E-UTRAN neighbor cell information in GERAN follows similar principles to that of UTRAN neighbor cell information. E-UTRAN neighbor cell information is introduced in the System Information Type 2quater (SI2quater) message [15] broadcast on BCCH, and in other existing dedicated signaling messages (e.g. Measurement Information on SACCH).

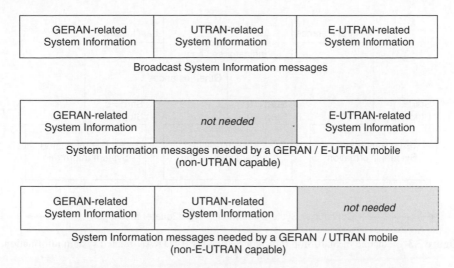

Figure 3.2 Fragmentation of System Information due to Multi-RAT MSs.

However, adding the E-UTRAN neighbor cell information to these messages, although based on the black list, increases the time to acquire the system information, in particular when a mobile station supports all three RATs (see Figure 3.2) but also in case when it only supports two RATs. In this case, it needs to read a significant amount of system information, some of which is irrelevant (e.g. UTRAN information when the mobile station is not UMTS-capable). This introduces delays in cell re-selection and measurement reporting. In some case, a reduction of several seconds is possible [16].

For a dual-mode GERAN/E-UTRAN mobile, the acquisition of E-UTRAN neighbor cell information (including measurement parameters) is optimized by enclosing this information in adjacent instances of the System Information Type 2quater message (or dedicated signaling message) with no other RAT information in between. UTRAN information is arranged in the same way. Each instance then contains:

- an indicator identifying the RAT for which it carries the information;
- a START indicator to identify the instance where the information starts;
- a STOP indicator to identify the instance where the information stops.

For example, if a message instance is the first one containing E-UTRAN information, the START field is set to 1, while if the E-UTRAN information were the last one, then the STOP field would be set to 1. If the message instance is neither the first or the last one containing E-UTRAN or UTRAN information, then the respective START and STOP fields will be set to 0. Figure 3.3 shows an example of the usage of the START and STOP indicators.

This optimization although requiring additional bits for the START/STOP bit indicators enables faster acquisition of system information allowing for the reduction in cell re-selection and measurement reporting time as well as reduced power consumption for mobile stations that do not support all three RATs and therefore can skip the reading of those instances belonging to the RAT they do not support.

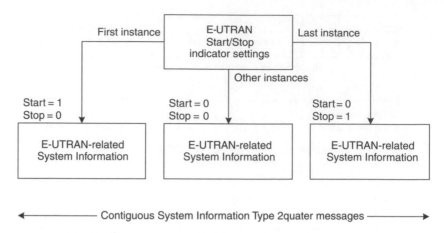

Figure 3.3   Example of Start/Stop Indicator Setting for E-UTRAN-related system information.

3.2.4   *Priority-Based Inter-RAT Cell Re-Selection in GERAN*

3.2.4.1   RAT Prioritization

The availability of E-UTRAN in addition to UTRAN forced the need for prioritization between RATs so that a multi-RAT mobile station, when cell re-selection is in its control, has the means to differentiate between the different RATs and re-select to the favorable RAT as indicated by the network, that is the priority RAT.

The algorithm based on thresholds/offsets [17] already utilized for GERAN/UTRAN inter-working could provide prioritization between cells of different RATs. For example, some of the parameters that control the cell re-selection in GERAN, if set appropriately, could influence the selection of a certain RAT so that cell re-selection to that RAT would more likely happen than re-selection to another RAT. Although feasible, this mechanism was not seen as advantageous due to the varying offsets.

Instead in Release 8, an algorithm based on absolute priorities is used. The priority-based inter-RAT cell re-selection algorithm operates with either common priorities or individual priorities (and associated thresholds).

Common priorities define the basic inter-RAT cell-reselection principles in a given cell. They are typically provided in the system information of a cell for general-purpose operation in that cell. Common priorities apply to mobile stations that do not have any individual priorities.

Individual priorities, on the other hand, last in time and across cells. They override any common priority. They are typically valid in a given entire PLMN, and are subscriber-specific, thus provided individually to mobile stations. Individual priorities may be used, for instance, to prioritize given RAT frequencies to premium data users or even make such frequencies accessible only to such users.

The network assigns individual priorities, for example, based on subscription-related information provided by the core network. This subscription-related information is known as the Subscriber Profile ID for RAT/Frequency priority (SPID [5, 13] or as RSFP [1]). SPID is an index referring to user information such as mobility profile, service usage, etc. The SPID is sent from the core network to the BSS (or eNB or RNC) and can be used by the BSS (or eNB or RNC) both in idle mode or connected mode, for example:

- to derive mobile station-specific cell re-selection priorities to control idle mode camping; or
- to control the redirection or handover to other RATs/frequencies in connected mode.

3.2.4.2 Cell Re-Selection Algorithm

The definition of cell re-selection criteria based on absolute priorities allows an operator, according to given radio conditions, to prioritize RAT or frequencies over others, as well as prevent access to a given RAT or frequencies. The priority-based inter-RAT cell re-selection algorithm only applies in the case of mobile station autonomous cell re-selection. It provides a means to control the mobility of the mobile station even if no network-controlled mechanism is used (i.e. a mechanism by which the network orders the mobile station to move to a given cell) by ensuring that the mobile station always camps on the highest priority RAT or frequency (provided that the given radio criteria are fulfilled).

Priorities are given on an E-UTRAN or UTRAN frequency basis. For GERAN, a single priority is used.[4] No two priorities belonging to two different RATs are equal. If for a given frequency no priority is given, this frequency is not considered for cell re-selection. The basic algorithm is described and illustrated hereafter, considering GERAN as the serving RAT. It should be noted that priorities and thresholds cannot be de-correlated. The algorithm is specified in detail in [17]:

- If the measurement quantity of a higher priority non-GERAN RAT frequency is higher than a given (high) threshold for that RAT during a time interval Treselection, the mobile station operates cell re-selection among higher priority cell candidates consisting of cells in decreasing order of priority and, when on the same frequency, in decreasing order of measurement quantity. This is illustrated in Figure 3.4.
- If the serving cell level is lower than a given (low) threshold and all measured GERAN cells during a time interval Treselection, the mobile station operates cell re-selection among lower priority cell candidates consisting of cells of which the measurement quantity is higher than a given threshold for the corresponding RAT, in decreasing order of priority and, when on the same frequency, in decreasing order of measurement quantity. This is illustrated in Figure 3.5.
- If the serving cell level is lower than a given (low) threshold for all measured GERAN cells during a time interval Treselection, and there is no cell re-selection candidate, of which the measurement quantity is higher than a given threshold for the corresponding RAT, the mobile station operates cell re-selection among non-GERAN cells of which the corresponding

---

[4] The mobile station may be assigned different priorities per GSM carrier while in UTRAN or E-UTRAN. If in this case the mobile station moves to a GSM cell on a given carrier, it applies while in GSM only the priority applicable to that carrier.

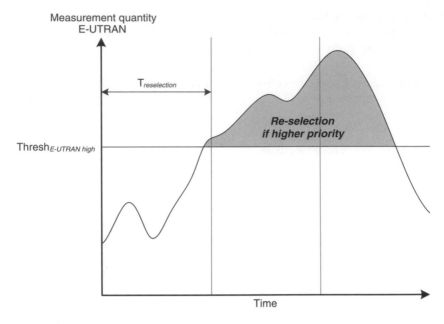

Figure 3.4   Re-selection to a higher priority RAT frequency (example of E-UTRAN).

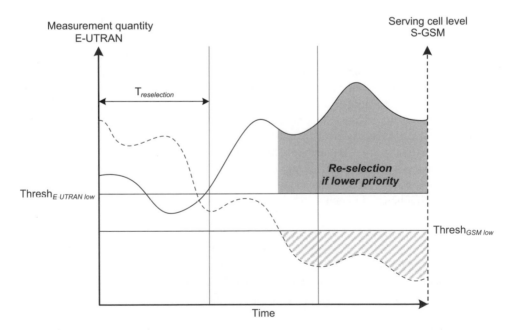

Figure 3.5   Re-selection to a lower priority RAT frequency (example of E-UTRAN).

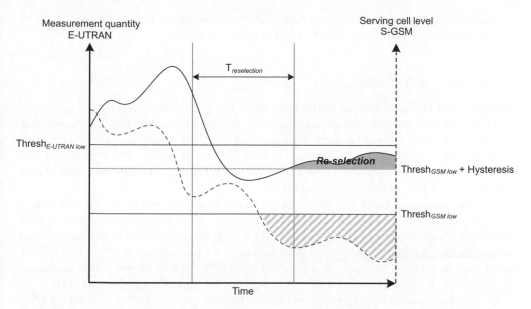

Figure 3.6   Re-selection to an inter-RAT cell when losing GSM coverage (example of E-UTRAN).

measurement quantity is higher than a minimum required serving cell threshold plus an hysteresis. This is illustrated in Figure 3.6.

3.2.5   *Inter-RAT Measurement Reporting and Control*

When interworking with LTE was introduced, some scenarios were identified where the monitoring of LTE cells was not seen as necessary or reporting was not needed, for example:

- LTE being a PS-only system, when a CS voice call is ongoing in GERAN, there is no possibility of handing over that call to E-UTRAN given there is no single-radio voice call continuity in the direction from GERAN/UTRAN CS domain to E-UTRAN. In such scenarios it is not necessary to measure (and report) LTE cells until the call is ended.
- When Closed Subscriber Group (CSG) cells (see Section 3.2.8) are deployed on a specific CSG frequency, it should be possible for CSG-capable mobile stations to measure cells on this frequency while ensuring no reports are sent and for non CSG-capable mobile stations not to measure cells on this frequency.

Means have thus been defined for the network to selectively control the measurement and reporting of inter-RAT cells by mobile stations. The mechanisms specified control finely and independently whether or not to take measurements (possibly on a frequency basis) of UTRAN and E-UTRAN neighboring cells, and if so whether only provided the serving cell signal level is below or above given levels. It is also possible to selectively control whether measurements are reported. These mechanisms allow tuning of the neighboring cell measurements and associated reporting on a case-by-case basis, and as a consequence to mitigate the impact on the MS battery consumption and on the radio interface signaling.

3.2.6   Inter-RAT PS Handover

The Inter-RAT PS handover between GERAN and E-UTRAN, defined in [1, 18], is applicable to all PS data services, real-time and non-real-time. It is the only mechanism in Release 8 that enables service continuity to fulfill the maximum service interruption times in both directions: GERAN to E-UTRAN and E-UTRAN to GERAN. It is based on the same principles as the inter-RAT PS handover defined between GERAN and UTRAN, thus with minimum impact on the specifications and on deployed networks.

This mechanism is under the control of the network and defined in such a way that it is the source system that decides on both the initiation and the execution of the handover. It follows the "source adapts to target" principle which means that the source system provides the information as expected by and defined in the target system, for example in a PS handover from E-UTRAN to GERAN information is provided by E-UTRAN to GERAN within a "Source BSS to Target BSS transparent container" (as used in a PS Handover in GERAN) [18].

In order for inter-RAT PS Handover to operate, for example to E-UTRAN, it requires that in the source RAT, for example GERAN, the terminal capabilities pertaining to the target RAT must be available prior to handover initiation. In addition, the following conditions must be met in the direction from GERAN to E-UTRAN:

- The mobile station is involved in an ongoing packet data session (and if in GERAN, not in a CS voice call).
- The BSS supports PFM (Packet Flow Management) procedures [13].

Simplified illustrations of the PS handover signaling from GERAN to E-UTRAN and E-UTRAN to GERAN as specified in [1] are depicted in Figure 3.7 and Figure 3.8.[5]

3.2.6.1   E-UTRAN Capability Transfer

The mechanism for enabling the transfer of the MS UTRAN capabilities from the mobile station to the BSS is re-used to support the transfer of the MS E-UTRAN capabilities as well. As described in [18] and specified in [13, 19], the E-UTRAN capabilities needed for PS handover to E-UTRAN are transferred from the MS to the BSS by means of NAS (Non-Access Stratum) signaling. An SGSN, as depicted in Figure 3.9, supporting inter-RAT PS handover to E-UTRAN will request the MS to transfer its E-UTRAN capabilities in Attach Complete and Routing Area Update Complete messages. The SGSN will then in turn include this container upon every PFC creation within the CREATE-BSS-PFC message sent to the BSS, that is when data transfer is initiated. A BSS which receives these capabilities will then forward

---

[5] The signaling between MME and SGSN is not illustrated. MME and SGSN are gathered under "EPC Network". For details on signaling, see [1].

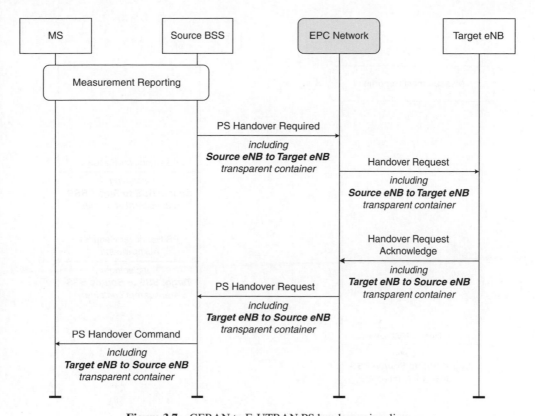

Figure 3.7   GERAN to E-UTRAN PS handover signaling.

them in the GERAN network, that is to a target BSS upon a subsequent (intra-GERAN) PS handover.

However, upon inter-RAT PS handover to GERAN, it cannot be guaranteed that an eNodeB will forward the MS UTRAN capabilities or that, likewise, an RNC will forward the MS E-UTRAN capabilities. This is because a RAT may not obtain the MS capabilities of a third RAT (i.e. if this RAT is not involved in the handover) and therefore cannot send those during handover. In order to cover this case in Release-8, PS handover is extended so that the missing capabilities of another RAT can be requested in GERAN upon successful PS handover. This is enabled by extending the PS Handover Complete procedure [13]. With this addition it is guaranteed that the E-UTRAN and UTRAN capabilities are available at the BSS in order to support a subsequent PS handover following the "source adapts to target" principle.

3.2.7   Mobile Station's Interworking Capabilities with E-UTRA

In order to simplify early LTE implementations in mobile stations a subset of the GERAN/E-UTRAN interworking mechanisms is mandated as a default mechanism. Other mechanisms may be supported as options.

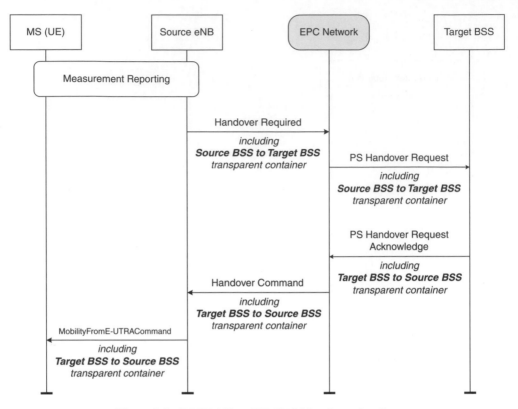

Figure 3.8   E-UTRAN to GERAN PS handover signaling.

Idle Mode
In idle mode, the default set of capabilities contains:

- E-UTRAN Neighboring cell measurements [12];
- autonomous cell re-selection to E-UTRAN using the priority-based reselection algorithm.

Dedicated Mode and Dual Transfer Mode
In dedicated mode and dual transfer mode, provided the mobile station supports CS services, the default capability is:

- redirection (e.g. to E-UTRAN) at channel release for example upon termination of a voice call previously established at CS fallback.

The optional capabilities are:

- E-UTRAN neighboring cell measurements; and optionally
- reporting of these measurements to the network [15].

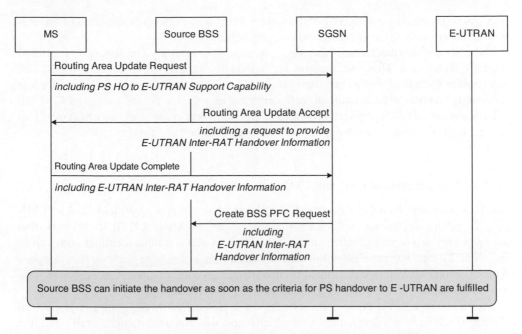

Figure 3.9 Transfer of E-UTRAN Capabilities to a GERAN BSS.

Packet Transfer Mode

In packet transfer mode only optional capabilities are defined, in an incremental manner,[6] as follows:

- E-UTRAN neighboring cell measurements and autonomous cell re-selection to E-UTRAN;
- Cell Change Notification (CCN) towards E-UTRAN, E-UTRAN Neighbor Cell measurement reporting and Network controlled cell re-selection to E-UTRAN;
- PS Handover to E-UTRAN.

3.2.8   CSG Mobility from GERAN

In EPS there is a possibility that some of the LTE users (respectively 3G users) have access to EPS services through eNodeBs (respectively NodeBs) located in private homes, commercial premises or enterprises. In LTE these eNBs are known as Home eNodeBs (HeNB) while in UTRAN they are referred to as Home NodeBs (HNB).

Users having access to cells under control of H(e)NBs belong to a Closed Subscriber Group (CSG), that is a closed group of subscribers. Such group may be, for example, that of premium data users. Hence these cells are also known as CSG cells. The primary goals with the introduction of CSG cells are the improvement of the coverage, for example, in areas that

---

[6] The support for these capabilities is incremental, that is the support of a capability bullet in the bulleted list implies the support of the preceding capabilities in that list.

may not be covered by macro cells of the same RAT as well as the differentiation of services between users with different billing and charging schemes.

The CSG mobility mechanisms specified in Release 8 provide only a basic support for HeNB or HNB deployment. These mechanisms are limited to CSG cell identification and CSG cell selection/re-selection for CSG members only. CSG mobility mechanisms are defined for idle mode only and limited to autonomous cell search and re-selection as well as manual CSG cell selection mechanisms. The priority reselection algorithm is also applicable to CSG cells. This implicitly allows some form of network control over CSG cell re-selection.

3.2.8.1   Identification of CSG and CSG Cells

A CSG is identified by a CSG identity (CSG-ID) which is a unique identifier within a PLMN [20]. It gathers a collection of CSG cells, within an E-UTRAN or a UTRAN network, that are open only to a certain group of subscribers. The CSG-ID is a static identifier with a fixed length of 27 bits. For the CSG ID of an H(e)NB, a human readable name is also assigned which is broadcast in free text format [23] and used as an aid for manual CSG cell selection. This is comparable to the SSID of a WLAN access point, for example.

Like a macro cell, the home cell controlled by a Home eNB is assigned a Physical Cell Identity (PCI) on the carrier frequency on which it operates. But unlike a macro cell's PCI, the PCI of a home cell is a variable identifier, although changes are expected to be infrequent. It may, however, change at each HeNB start-up or reconfiguration, although it is recommended that a HeNB uses its previously used PCI if no collision is detected.[7] It may also happen that the PCI changes, for example due to the detection of a PCI collision. In addition, the carrier frequency of the HeNB, selected from a preconfigured limited range of frequencies, may also be modified but only at power up [5]. In spite of changes that could occur to these basic yet crucial parameters, interworking with those cells must not be impaired. Moreover these changes should not require any effort or intervention from the CSG users or the managers of the HeNB – they should be handled by the network and mobile stations.

Moreover, it is also important to note that a home cell will be deployed on a carrier frequency either shared with (non-CSG) macro cells, or dedicated to CSG cells. In the former case, a range of PCIs dedicated to CSG cells is broadcast by CSG cells. This enables a terminal to identify "immediately" whether a cell on a carrier frequency is a CSG cell or not simply from the PCI (and frequency). When received, this range is stored for 24 hours [5, 22], and used for cell (re-)selection purposes. It is valid only within the network (PLMN) in which it is sent. This PCI range[8] is used in GERAN as well, but is indicated directly within the system information of GERAN cells. The PCI range enables mobile stations that are non-CSG members to ignore CSG cells, for example when in idle mode, thus preventing cell re-selection failures to any such cell in that frequency [15]. In connected mode, it permits, in an easy manner, these mobile stations not to measure CSG cells. This, in turn, prevents wasting measurement reporting capacity used instead for non-CSG E-UTRAN or UTRAN cells.

---

[7] It is indeed expected that an HeNB has the capability to detect PCIs in its vicinity so a non-colliding PCI (on a given frequency) can be allocated.

[8] For UTRAN CSG cells, a similar range of PSC is also used.

When deployed on a dedicated carrier frequency, all corresponding PCIs are reserved for CSG cells, thus yielding the same benefits as when a PCI-range is used on a shared frequency.

3.2.8.2   CSG Cell Selection and Re-Selection in Idle Mode

CSG cells are managed by the owner of the H(e)NB and the operator. Under the supervision of the operator, the H(e)NB owner may add or remove CSG members from its CSG cells. The network operator controls the H(e)NB deployment and (re-)configuration by verification of its identity, configuring its settings, checking the regulatory requirements and determining its location. The addition of H(e)NB does not impose/require any network changes in the operator's network, other than the identification aspects described in Section 3.2.8.1.

Each CSG user has a list of allowed CSG cells[9] stored in its USIM [20]. This "Allowed CSG list" prevents camping on and cell re-selection attempts to forbidden CSG cells. It is either provided to the mobile station from the network by means of NAS signaling or obtained by manual selection.

In the case of manual selection, the CSG user scans for the available CSG identities within the registered PLMN and selects only one of its identified allowed CSG cells from its stored list [23, 24]. If the request to search the available CSG IDs comes from the network through NAS, the mobile station reports these cells back to the network.

Only a mobile station that contains at least one CSG ID in its "Allowed CSG list" proceeds in addition to normal Inter-RAT cell re-selection, with E-UTRAN or UTRAN CSG cells search. A cell is considered to be a CSG cell indicated as such through a CSG Indicator (set to TRUE) [17]. Once the strongest detected suitable cell found on a frequency is a CSG cell, the mobile station will re-select to this cell. The search function and measurement rules are implementation dependent.

Further details and additions regarding inbound mobility to CSG cells are described in Chapter 4.

3.3   A Interface over IP

The A interface over IP (AoIP) refers to the solution allowing the transport of CS user plane data over an IP protocol stack [25]. Note that the standardization of the A interface control plane over IP was already introduced in Release 7, with the specification of the SIGTRAN protocol stack for BSSAP signaling.

3.3.1   Architecture

3.3.1.1   Legacy Architecture: A Interface User Plane over TDM

Before the introduction of the A User Plane interface over IP feature in Release 8, the architecture of the CS domain part of a GERAN network was that described in Figure 3.10.

In this case, the A User Plane interface only uses TDM transport solutions. The codec used on the A interface for voice services is G.711 and then transcoding takes place in the BSS, thanks to the Transcoders (TC) located in some TRAU (Transcoder and Rate Adaptation

---

[9] Cells it is allowed to access.

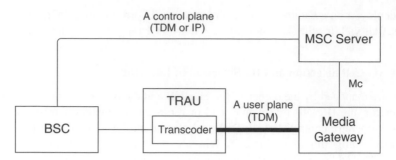

Figure 3.10   Legacy architecture: A interface User Plane over TDM.

Unit). Alternatively, TFO (Tandem Free Operation) is also possible, allowing the tunneling of compressed speech frames through the PCM links connecting the BSS with the Core Network. This greatly helps to improve the voice quality but cannot save bandwidth on the A interface.

The A Control Plane interface also can either use the same TDM transport solution used for the User Plane, or it can alternately make use of an IP-based solution, thanks to the adoption of the SIGTRAN protocol stack, in Release 7.

3.3.1.2   A over IP with Transcoders in BSS

One of the first options introduced in Release 8 was the basic possibility of using IP-based transport solutions for the A User Plane interface, without really changing the overall architecture and the work split between the BSS and the Core Network, as shown in Figure 3.11.

This basically means that the transcoding functionality remains in the BSS, the codec used on the A interface is still G.711 but then the solution foresees that the 64 kbit/s flows carrying

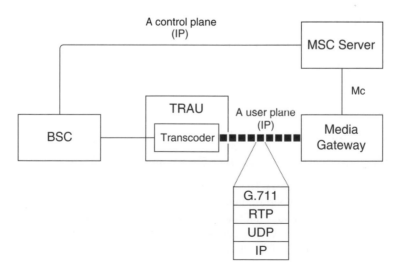

Figure 3.11   Architecture for AoIP with Transcoders in BSS.

the CS User Plane between the BSS and the MGW are chunked into RTP/UDP/IP packets, each carrying a 20 ms voice sample. TFO is possible also using this approach, but also in this case this would not lead to any bandwidth savings. In general, considering that the amount of payload bits does not change but the header of the IP protocol stack needs to be added, the overall bandwidth requirement with this solution is not expected to decrease with respect to the A over TDM case, unless some other optimizations are considered. One standardized option to improve the bandwidth efficiency in this case is to adopt IP multiplexing techniques.

The main benefit of this solution is that it allows the re-use of already deployed equipments (e.g. existing TRAUs could continue to be used, without requiring additional transcoding resources in the MGW) at the same time adding the full flexibility provided by IP, in terms of connectivity, network configuration, etc. One important example is the one of "MSCs in pool", where the connectivity of the same BSS to several MGWs is greatly simplified by the use of an underlying IP infrastructure.

3.3.1.3   A over IP without Transcoders in BSS

Besides the introduction of the plain IP transport solution described in Section 3.3.1.2, the A over IP Release 8 feature also allows the transition to the same network architecture defined for 3G networks, for the CS domain.

The solution in fact allows a network configuration where the functional split between the BSS and the Core Network is changed with respect to legacy architecture and where transcoders are completely removed from the BSS and only located in the MGW, controlled by the MSC Server, as shown in Figure 3.12.

This approach makes it possible to replace the use of PCM (G.711) on the A interface with the more efficient transmission of compressed speech, through the packetization into 20 ms voice samples included in RTP/UDP/IP packets. One first benefit of this approach is that the bandwidth utilization on the A interface is heavily reduced. Compared to the TFO scenario of

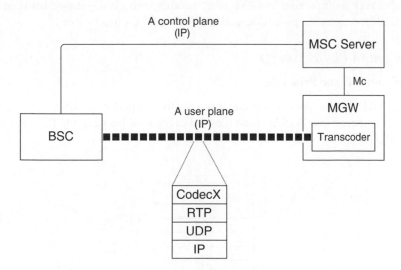

Figure 3.12   Architecture for AoIP without Transcoders in BSS.

previous solutions, in this case there is no need to embed the compressed speech in the higher bandwidth PCM stream: the speech is directly formatted into IP packets of the right size.

This also means that TrFO (Transcoder Free Operation) is made possible by this approach: the codec to be used on both the radio and the A interfaces is negotiated between the BSS and the MSC Server, and then – through the Core Network – with the peer system as well with the aim of establishing a TrFO operation, using the same codec through the whole mobile-to-mobile path and not requiring any transcoding in any part of the network. Whenever this is possible (TrFO is established), the end-to-end voice quality greatly improves, thanks to the lack of unnecessary transcoding operations and the reduction of the end-to-end delay.

With this solution, transcoders are only needed to cover scenarios where TrFO is not used, that is when the codecs used on the two radio interfaces of the two legs of a mobile-to-mobile call are incompatible with each other, so that transcoders have to be inserted.

The fact that transcoders are not always needed with this architecture highlights one more benefit of this approach: although additional transcoder resources might be needed in the MGWs (for instance, to take care of the needed transcoding functionality for all mobile-to-PSTN calls), the total amount of transcoder resources in the overall network (considering the BSS as well) is certainly lower than in legacy networks today.

3.3.2   User Plane Aspects

For both architectural A over IP solutions, the user plane information is transported via RTP/UDP/IP packets (see Figure 3.13), each carrying a 20 ms voice sample. No specific solution is standardized for the Link (L2) and Physical (L1) layers [27].

Especially when compressed speech samples are transferred over this protocol stack, this can be quite inefficient in terms of bandwidth utilization, given that the overall header size can be higher than the payload size. To overcome this problem an optional solution has been defined, allowing the multiplexing of several RTP payload units of different user plane connections within the same UDP/IP packet. As a further option, the RTP header can also be compressed.

The use of RTP multiplexing and RTP header compression is negotiated between the BSS and the MGW for each user plane connection by using RTCP mechanisms.

3.3.3   Control Plane Aspects

3.3.3.1   Control Plane Principles

First of all, SIGTRAN support is a prerequisite for the support of an IP-based A User Plane interface. This means that BSSAP messages are exchanged between the BSS and the MSC using the SIGTRAN protocol stack, see Figure 3.14.

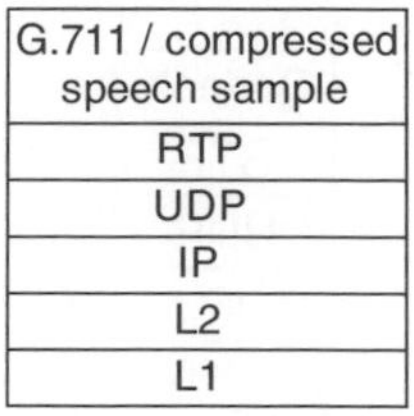

| G.711 / compressed speech sample |
| RTP |
| UDP |
| IP |
| L2 |
| L1 |

Figure 3.13   Protocol stack for the A Interface User Plane over IP.

| BSSAP |
| SCCP |
| M3UA |
| SCTP |
| IP |
| L2 |
| L1 |

Figure 3.14   Protocol stack for the A Interface Control Plane over IP (SIGTRAN) [26].

Then some new principles for the A Interface Control Plane signaling have been defined [26]. The first modification is the introduction of a Transport Layer Address information element in all the relevant BSSMAP messages. The Transport Layer Address indicates the IP Address and the UDP Port Number used by a given endpoint to support the IP connection. An IP connection on the A interface involves two Transport Layer Addresses: one at the BSS side and one at the MGW side. The exchange of the Transport Layer Address information on the A interface is realized by including the corresponding information elements in the relevant messages between the BSS and the MSC Server, namely in the messages for the assignment procedure at call set-up and for the different handover types.

Another modification is the introduction of a Call Identifier information element, as a MSC-unique value by which an IP connection can be identified in both the MSC Server and the BSS. For calls using an IP-based A User Plane interface, the Call Identifier replaces the CIC (Circuit Identification Code) currently used in legacy TDM-based A User Plane interfaces. Call Identifiers are assigned by the MSC server and communicated to the BSS during assignment and inter-BSS handover procedures.

One more important principle introduced with A over IP, specifically required by the solution of getting rid of transcoders in the BSS, is that the codecs to be used on the radio interface need to be carefully negotiated between the BSS and the Core Network. In architectures without transcoders in the BSS, G.711-encoded RTP packets cannot be sent on the A interface. In this case RTP packets have to carry voice samples encoded with exactly the same compressed speech codec used on the radio interface, and the codec choice therefore needs to be agreed between the BSS and the MSC Server. Furthermore, in order to achieve TrFO for mobile-to-mobile calls, codec negotiation should happen through the Core Network up to the distant end as well, so that the same common codec is eventually selected on both radio legs, both A interfaces and in the internal network interfaces as well. New information elements specifying the speech codecs have therefore been added in the relevant BSSMAP messages. In particular, at call set-up/handover, the BSS indicates to the MSC Server – in a Codec List (BSS Supported) IE – all the codec it supports, without any specific order. On the other hand, also the MSC Server indicates to the BSS – in a Codec List (MSC Preferred) IE – all the codecs it supports, in a priority order, to provide an indication of its preference (e.g. for codecs that may enable TrFO or TFO).

3.3.3.2   Procedures

A number of BSSMAP procedures and messages have been introduced or modified to support A over IP. In particular, as anticipated in Section 3.3.3.1, a number of new information elements have been added at least to the messages exchanged during the assignment procedures at call set-up and the messages used during handovers.

For each individual call, during the set-up procedure, an independent codec negotiation procedure is performed. The BSS indicates the list of codecs it supports, in the specific cell, at that specific time, in the first message to the MSC Server (the one conveying the service request from the mobile station). The MSC Server then determines its own ordered list of preferred codecs and sends it back to the BSS in the assignment request message, taking into account at least BSS, MS and MGW capabilities. However, when the goal is to establish TrFO for a mobile-to-mobile call (which requires the architecture without transcoders in the BSS), an end-to-end codec negotiation needs to be performed within the Core Network up to the terminating side, and the outcome of this procedure needs to be considered when creating the list of preferred codecs to be included in assignment request messages on the A interface. The other new information element included by the MSC Server in the assignment request messages is the Transport Layer Address at the MGW side, so that the BSS knows the IP/UDP endpoint to be considered at the Core Network side for the ongoing call. If the call can be successfully established, the BSS then answers with the assignment complete message, including the indication of the codec which was finally chosen on the radio interface and the Transport Layer Address at the BSS side. At this point in time, the IP connection between the BSS and the MGW can be established, since both IP/UDP endpoints are known to each peer, as well as the codec and the payload format to be used for RTP packets.

A new codec negotiation and a new exchange of Transport Layer Addresses need to be performed during inter-BSS handovers as well. This is done using the handover request and handover request acknowledge messages exchanged between the MSC Server and the target BSS during an inter-BSS handover procedure.

On the other hand, for intra-BSS handovers, no new procedures are foreseen on the A interface, as long as the codec used in the target cell is the same or compatible with the one used in the source cell. In this case the payload format to be used for RTP packets stays the same, as well as the Transport Layer Addresses used by the IP connection on the A interface.

The exception is the case of intra-BSS handovers, including intra-cell ones, requiring a codec in the target cell which is incompatible with the one in the source. This can happen, for instance, in overload situations when a transition from EFR to HR codecs is required. In this case, a new codec needs to be negotiated and new Transport Layer Addresses need to be exchanged. A new optimized three-way signaling procedure has been defined to cover this scenario, called "BSS Internal Handover with MSC Support", to be used in place of the existing inter-BSS handover procedure. The BSS starts with a new Internal Handover Required message, indicating the target cell, the updated list of supported codecs and the new Transport Layer Address at the BSS side for the new radio channel termination. The MSC Server would then respond with a new Internal Handover Command message, indicating the chosen codec and the new Transport Layer Address at the MGW side. The BSS can finally confirm the success of the handover procedure by sending the Handover Complete message.

3.4 Multi-Carrier BTS (MCBTS)

Throughout the operation of GSM/EDGE networks reducing capital and operational expenditures has been a main focus of mobile operators to increase their operating profit. Since the population of BTSs with multiple single carrier transceivers is large compared to other network elements, for example a single Base Station Controller (BSC) may serve about 200 to 500 BTSs, cost improvements in the BTS largely impact an operator's CAPEX and OPEX figures. The introduction of base stations based on a multicarrier architecture – Multi-carrier BTS – is expected to yield cost improvements as further described in Section 3.4.2. It also provides the adequate hardware platform for operating multiple 3GPP radio technologies simultaneously.

This section provides an overview of MCBTS, an evolution of Base Transceiver Station (BTS) hardware platforms; and of its standardization. The functional description is then introduced, followed by detailed standardization and regulatory aspects. It should be noted that the standardization of GSM MCBTS in 3GPP Release 8 is closely related to the standardization of multicarrier multi-standard base stations in 3GPP Release 9 supporting multiple 3GPP radio access technologies by a single wideband transceiver. This is further addressed in Chapter 14.

3.4.1   Overview

The concept of applying a multicarrier architecture in GSM was investigated several years ago. Driven by progress in component technology on one hand, and the introduction of wideband mobile radio systems such as WCDMA, on the other, wideband power amplifiers (PA) have become widely available. Approved for deployment in the US by the FCC in the year 2000 for GSM 850 MHz and PCS 1900 MHz frequency bands, both for single GSM deployment mode and for mixed deployment mode with TIA/IS-54 systems using 30 kHz narrowband carriers, relaxations of unwanted emissions based on the usage of wideband power amplifiers were introduced in GSM Release 99 specifications.

GSM Multicarrier BTS was introduced in 3GPP Release 8 specifications in 2009. The discussion started in 2006 for deployment in the GSM 900 MHz and DCS 1800 MHz frequency bands [28] and was extended to cover the GSM 400 MHz, GSM 700 MHz, GSM 850 MHz and PCS 1900 MHz frequency bands as well.

3.4.2   Functional Description

In the past, the architecture of GSM/EDGE base stations, hereafter referred to as "GSM Normal BTS", generally followed the concept of one transceiver (TRX) per carrier comprising a transmitter and a receiver unit. BTSs with multiple carriers were realized by combining the carriers behind the power amplifiers with passive components, for example using 3 dB hybrids or tunable RF filters, and by passively splitting the received antenna signal towards multiple receivers. Figure 3.15 shows the architecture of a GSM Normal BTS operating four carriers. This architecture is also referred to as single carrier BTS (SCBTS).

Technological progress in transmitter and receiver design permitted implementation of BTSs for 2G and 3G radio access technologies (GSM, EDGE, UMTS, cdma2000®), that are capable of processing multiple carriers within a single wideband transmitter/receiver unit. In the case of GSM/EDGE, such BTSs are called GSM Multicarrier Base Stations (GSM MCBTS). The differences between the Multicarrier BTS architecture and the Normal BTS architecture are obvious from a comparison of Figure 3.15 and Figure 3.16.

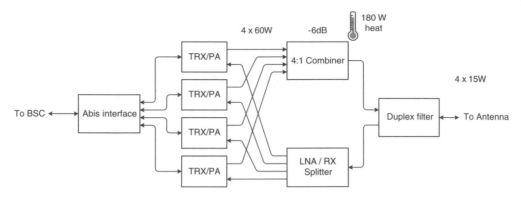

Figure 3.15 Traditional architecture of a Normal BTS with single carrier transceivers.

In particular, the MCBTS architecture consisting of a wideband transmitter and receiver has several advantages:

1. *Higher integration level of RF components*. The analog RF components, that is mixers, amplifiers, oscillators, filters, etc., are needed only once per antenna instead of once per carrier. The per-carrier circuitries are shifted to the digital part where they can be integrated into a few chips of silicon.
2. *Improved power efficiency*. The huge loss of energy in the passive TX combiner (75% in case of traditional SCBTS architecture, see Figure 3.15) is avoided because the carriers are combined in the digital domain, that is by a mathematical ADD operation on digitized carrier signals, and the multicarrier PA output goes directly to the antenna via the duplex filter. This results in better power efficiency of the BTS when multiple carriers are to be transmitted over one antenna, which is the more efficient, the more carriers are sent per antenna and the more combiner stages would be needed in the traditional architecture.
3. *Reduced complexity of antenna systems*. Operators benefit from higher integration level of GSM MCBTS architecture reducing site requirements, in particular reduced complexity of the antenna systems. This finally translates into reduced costs of MCBTS installation and maintenance per each radio site.
4. *Flexibility to split the total available output power between carriers*. The composite output power of the GSM MCBTS is split between all active carriers, thus the maximum output power per carrier depends on the number of active carriers. A GSM MCBTS, for example,

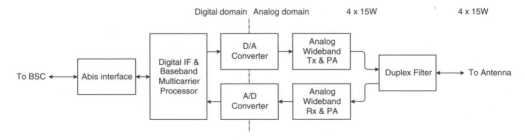

Figure 3.16 Multicarrier BTS architecture.

may have a composite power of 80 watts and thus support different carrier configurations, for instance, 1 carrier with up to 80 watts output power, 2 carriers with up to 40 watts per carrier, 4 carriers with up to 20 watts per carrier and 8 carriers with up to 10 watts per carrier. In addition, the carrier power can be assigned on a more flexible basis, for example higher carrier power can be assigned to the BCCH carrier or to the carriers belonging to the frequency layer operating data services requiring a high CIR, while the carriers belonging to the frequency layer for voice services may require less output power per carrier due to lower required CIR.

The concept of multicarrier BTS can be realized in different ways: either the multicarrier transceiver is included in the ground-mounted BTS cabinet or it is tower-mounted as part of a Remote Radio Head (RRH) solution. Implementation and operation of MCBTSs nevertheless require a few relaxations and changes of the emission characteristics for GSM MCBTS that are described in Sections 3.4.3 and 3.4.4.

3.4.3   Impact of Usage of MCBTS on Unwanted Emissions

The usage of wideband transmitters for multicarrier transmission in GSM affects the level of unwanted emissions outside the wanted channel, in that these emissions are higher than for traditional single carrier BTS. This section describes the sources of these increased emissions and their contribution to the unwanted emissions.

3.4.3.1   Main Sources of Unwanted Emissions in the MCBTS Transmitter

The main sources of the unwanted emissions in the MCBTS transmitter are:

1. *D/A converter*. Due to the incoming digital wideband multi-carrier signal, the D/A converter needs to operate a wider signal bandwidth and a higher input dynamic range compared against a narrowband signal input to a TRX of a SCBTS. Hence a larger portion of amplitude and phase noise is generated leading to broader noise spectrum, which contributes to a higher level of wideband noise emissions over the maximum supported RF bandwidth of the MCBTS and several MHz beyond. In addition, spurious emissions occur at certain frequencies, that is at harmonics of the D/A converter clock frequencies and their intermodulation products.
2. *RF-modulator*. The dominant source of unwanted emissions in the analog RF modulator is the phase noise generated in the I/Q modulator, in the TX variable gain amplifier and also in the RF synthesizers. Due to the incoming wideband signal this phase noise is translated into a broader noise spectrum which contributes to higher level of wideband noise emissions. In addition, spurious emissions are generated due to oscillator leakage and intermodulation products of the mixer stages.
3. *Multi-carrier power amplifier*. A major technological challenge for using wideband power amplifiers for transmission of multiple narrowband GSM signals with a bandwidth of 200 kHz is the nonlinear characteristic of such amplifiers. This characteristic affects the level of unwanted emissions, introduced by those nonlinearities typically characterized by AM/AM and AM/PM compression curves. In contrast to a GMSK single carrier having constant amplitude, in the multicarrier case, due to the AM component, these nonlinearities

yield to intermodulation products transmitted in addition to the multiple carriers over the air interface. The generation of these intermodulation products is considered in more detail in the following section.

3.4.3.2   Transmitter Intermodulations[10]

Transmitter intermodulation products are a major source of unwanted emissions. They are generated due to nonlinearities in the transmitter, when several signals are processed in a common active element [29]. For MCBTS, the multicarrier power amplifier is the dominating source of these intermodulation products. Among different orders of the generated intermodulation products, third, fifth and seventh order intermodulation products are most relevant since they may fall into or are close to the transmit band and hence are an additional source of interference to mobile receivers.

The transfer function of an active element input with M signals S_i can be described by a Taylor series of powers as follows:

$$S_{out} = \sum_{n=1}^{\infty} c_n \cdot (S_1 + S_2 + \ldots + S_M)^n \tag{3.1}$$

The value of the power index n, equals at the same time the order of intermodulation, as this describes the number of signals involved in the process. Normally the factor c_n is decreasing rapidly with the power index that is with the intermodulation order and only odd intermodulation orders, that is 3rd, 5th, 7th, etc. may fall inside the GSM transmit or receive bands. Thus the most important intermodulation products to consider are the 3rd, 5th and 7th order.

The third order intermodulation products from M carriers located between f_1 and f_M may be found at

$$|2 \cdot f_a - f_b|$$

and also at

$$|f_a + f_b + f_c|$$

for any combination of a, b and c in the range 1 to M and where f_a, f_b and f_c are the center frequencies of carriers a, b and c. These products can be found for offsets up to $|f_1 - f_M|$ outside the group of carriers. Another characteristic of these products is the broadening of spectrum. When assuming GMSK signals and 200 kHz channels, the 3rd order intermodulation (IM3) products have a bandwidth of about 600 kHz and thus will occur at the adjacent channels as well, with a 3 dB lower power than on the channels at the center frequency of the intermodulation.

---

[10] This section is largely based on 3GPP TS 25.816 v8.0.0, Annex A.1. © 2009. 3GPP TM TSs and TRs are the property of ARIB, ATIS, CCSA, ETSI, TTA who jointly own the copyright on them. They are subject to further modifications and are therefore provided to you "as is" for information purposes only. Further use is strictly prohibited.

Similarly, the 5th order intermodulation products may be found at

$$|3 \cdot f_a - 2 \cdot f_b|, |3 \cdot f_a - (f_b + f_c)|, |2 \cdot f_a + f_b - 2 \cdot f_c|, |2 \cdot f_a + f_b - (f_c + f_d)|,$$
$$|f_a + f_b + f_c - 2 \cdot f_d| \text{ and also on } |f_a + f_b + f_c - (f_d + f_e)|$$

for any combination of a, b, c, d and e in the range 1 to M and where f_a, f_b, f_c, f_d and f_e are the center frequencies of carriers a, b, c, d and e. These products may be found for offsets up to $2 \cdot |f_1 - f_M|$ outside the group of carriers. For this case the spectrum broadening is higher, since the 5th order intermodulation (IM5) products have a bandwidth of about 1000 kHz, however, the power is 2 to 2.5 dB lower in the adjacent 200 kHz channels than on the channels at the center frequency of the intermodulation.

Finally, 7th order intermodulation (IM7) products may be found for offsets up to $3 \cdot |f_1 - f_M|$ outside the group of carriers. In general, they have a much lower power than 3rd and 5th order intermodulation products.

If the carriers are equally spaced in frequency, several 3rd order and 5th order and higher order intermodulation products will coincide on the same frequencies and add up, thus creating the maximum amplitude of the intermodulation products. The highest amplitude can be found closest to the carriers, that is next to the outermost carrier of the MCBTS allocation, and the amplitude decreases by increasing frequency offset.

An example of a power spectrum of a GSM MCBTS, normalized to the carrier power level, including wideband noise emission and intermodulation products with 3 equally spaced carriers using 1 MHz spacing and equal power is shown in Figure 3.17.

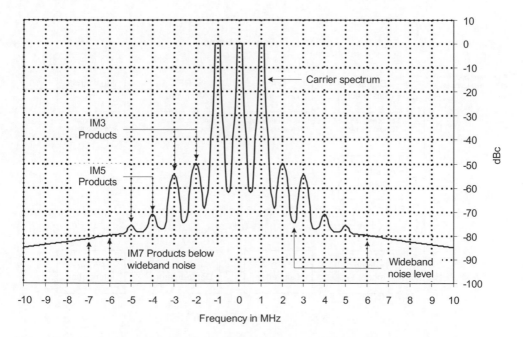

Figure 3.17   Exemplary power spectrum of MCBTS operating 3 GSM carriers (no predistortion active).

In Figure 3.17 the 3rd order intermodulation products identifying the dominant intermodulation products are located between −3 and 3 MHz, that is at 1 and 2 MHz frequency offsets on both sides of the outermost carriers. The 5th order intermodulation products are located between −5 and 5 MHz, that is between 1 and 4 MHz frequency offsets on both sides of the outermost carriers. Higher order intermodulation products, such as the 7th order IM located between −7 and 7 MHz, that is between 1 and 6 MHz frequency offsets on both sides of the outermost carriers, are much lower compared to 3rd and 5th order IM and hence may even not exceed the wideband noise level.

If the number of active carriers M is reduced, while keeping the same output power for the active ones, the power of 3rd and 5th order intermodulation products is rapidly reduced as the coefficients c_3 and c_5 remain the same and fewer products add up.

3.4.3.3   Mitigation of Unwanted Emissions in the MCBTS Transmitter

To mitigate distortions in the transmitter one essential technique is the usage of predistortion in the transmitter branch. Recent fast digital signal processing devices execute the predistortion in the digital domain. An exemplary block diagram of the digital predistortion in a multicarrier base station is depicted in Figure 3.18.

The predistortion is executed by means of a control loop including a feedback branch and a nonlinear Digital Adaptive Pre-Distortion (DAPD) of the digital signal. The distorted multicarrier signal output from the multicarrier PA (MCPA) is demodulated into baseband and converted into the digital domain. The sampled distorted transmit signal is then equalized by a training model by applying the estimated inverse transfer function of the nonlinear distortion

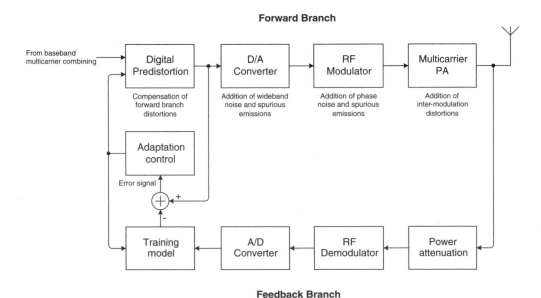

Figure 3.18   Exemplary block diagram of the digital adaptive predistortion in the MCBTS transmitter.

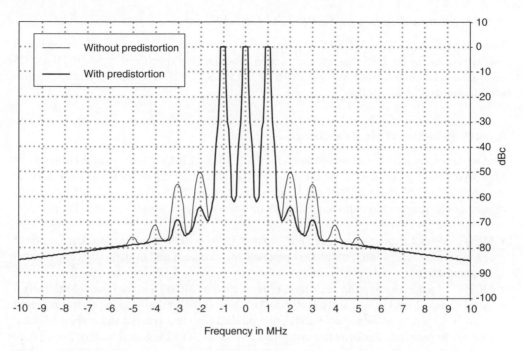

Figure 3.19 Exemplary power spectrum of GSM MCBTS with three carriers using predistortion means.

in the forward branch. The error between ideal, that is the undistorted transmit signal and the equalized distorted transmit signal is then calculated. The error signal is input into a control block which dynamically adjusts the digital predistortion block in the transmitter forward branch by adding a time variant digital correction signal, so that the error signal is minimized. An exemplary power spectrum of a GSM MCBTS with three carriers using predistortion means is shown in Figure 3.19.

It can be seen that the level of unwanted emissions, in particular of the dominant 3rd order and 5th order intermodulation products can be remarkably decreased.

3.4.3.4 Spurious Emissions

Spurious emissions are generated at distinct frequencies due to oscillator leakage of the RF synthesizers, intermodulation products generated in the analog TX stages and unwanted mixing products of the D/A converter as depicted above. Due to the increased operating bandwidth of the transmitter compared to a narrowband single carrier transmitter, more spurious emissions fall into the operating bandwidth and hence the MCBTS generates a higher level of spurious emissions in the operated bandwidth and at frequencies close to it. This has been considered in the standardization by the definition of relaxed requirements for inband and out-of-band spurious emissions as depicted in the next section.

3.4.4   Standardization in 3GPP TSG GERAN

The operation of a GSM MCBTS requires a few relaxations and changes of the emission characteristics for MCBTS that were introduced into the core specification 3GPP TS 45.005 and to the BTS conformance test specification 3GPP TS 51.021 in Release 8. These changes are summarized in this section.

3.4.4.1   Definition of MCBTS Classes

Due to the expressed interest of operators in 3GPP TSG GERAN to allow different levels of improved power efficiency of MCBTS, thereby impacting the linearization performance of the MCBTS, 3GPP TSG GERAN has standardized two MCBTS classes with different levels of relaxation compared to radio requirements for state-of-the-art single carrier BTS, namely:

- MCBTS class 1 allowing unwanted emissions of up to −70 dB relative to the carrier power at frequencies where intermodulation products are expected and at their adjacent channel frequencies.
- MCBTS class 2 allowing the same level of relaxation as MCBTS class 1 and additionally a further relaxed level of unwanted emissions of up to −60 dB relative to the carrier power at frequencies where 3rd order intermodulation products can be expected and at their adjacent channel frequencies. Performance compliant to this MCBTS class is shown in Figure 3.19 (with predistortion).

Due to the relaxed intermodulation attenuation requirements for MCBTS class 2 compared to MCBTS class 1, MCBTS class 2 is expected to yield a higher power efficiency, that is a reduced power consumption of the wideband transmitter, than MCBTS class 1.

In addition, the manufacturer needs to declare the maximum supported RF bandwidth for the MCBTS, the supported carrier configurations with specified maximum power level per carrier, assuming equal power distribution among the carriers, and the supported modulation types.

The different intermodulation attenuation requirements for both MCBTS classes specified in sub-clause 4.7 of 3GPP TS 45.005 led to different requirements for the spectrum due to modulation and wideband noise applied to the frequency range close the MCBTS carrier allocation inside the transmit frequency band (see Section 3.4.4.3) as well as for out-of-band spurious emissions outside the transmit frequency band (see Section 3.4.4.4). The MCBTS performance specification is based on equal or even frequency spacing for all active MCBTS carriers as described later in this section. In most test cases the carriers have equal power, but the performance requirements are also verified by applying an unequal power profile to all declared configurations of active MCBTS carriers. For example, for the case of four active MCBTS carriers, two carriers have a transmit power 2 dB higher and two carriers a transmit power 4 dB lower than for the equal power profile.

3.4.4.2   Study of System Impacts Through the Introduction of GSM MCBTS

System impacts due to the introduction of GSM MCBTS have been investigated in 3GPP TSG GERAN for different coexistence scenarios by simulating a GSM aggressor network with MCBTS base stations transmitting on all carriers with maximum transmit power and a non-coordinated GSM victim network employing single carrier BTS and RF power control in the adjacent frequency allocation being interfered with by the aggressor network both in downlink and in uplink. In particular, system impacts caused by higher intra BSS TX intermodulation distortion on the transmitter side of the aggressor network and lower inband blocking performance due to use of a wideband receiver architecture on the receiver side of the GSM MCBTS network have been analysed. The conclusion of these studies was that the relaxations due to the introduction of both MCBTS classes did not yield higher call drop rates in case of voice calls or yielded a noticeable degradation of EGPRS/EGPRS2 throughput performance [30].

3.4.4.3   Spectrum Due to Modulation and Wideband Noise

Being part of the unwanted emissions close to the allocation of the wanted carrier, the spectrum due to modulation and wideband noise, specified in sub-clause 4.2.1 of 3GPP TS 45.005, is measured for state-of-the-art SCBTS on both sides of the allocation of the wanted carrier with up to 2 MHz outside the operated BSS transmit band being typically a subset of the relevant transmit band, for example a SCBTS may operate the frequency range 940–955 MHz as part of the P-GSM 900 band. For MCBTS, the spectrum due to modulation and wideband noise is measured for several configurations, that is for all supported number of carriers, on both outer sides of the group of carriers belonging to a contiguous MCBTS frequency allocation up to 10 MHz outside the RF bandwidth excluding the measurement locations where intermodulation products are expected. In fact, the relaxation of this radio requirement for MCBTS is only minor, since the method used when creating a transmission mask for MCBTS with M carriers is derived from the equivalent requirement resulting from the aggregation of M single-carrier transceivers connected to one antenna port:

- for frequency offsets up to 1.8 MHz from the outermost MCBTS carrier the transmission mask is identified by the aggregated wideband noise of all MCBTS carriers based on the existing mask for spectrum due to modulation and wideband noise for single carrier and taking into account the offset between each carrier allocation and the measurement point for wide band noise;
- for frequency offsets beyond 1.8 MHz from the outermost MCBTS carrier, the transmission mask is identified by determining the mask for the dominant outermost MCBTS carrier and adding a multicarrier offset factor $10 \times \log_{10}(M)$ where M is the number of active carriers.

Note, most of the relaxation of this radio requirement for MCBTS compared against SCBTS stems from an increased number of exceptions, where a higher level of unwanted emissions is granted than for SCBTS, that is the level is based on the relative intermodulation attenuation of the respective MCBTS class (see Section 3.4.4.1) rather than on the absolute power level of −36 dBm as defined for SCBTS.

For verification purposes, the requirement for spectrum due to modulation and wideband noise and the intermodulation requirement are tested together. The specification prescribes using different carrier spacings for the test, that is a minimum carrier spacing based on 600 kHz and a maximum carrier spacing, resulting from the spread of the carriers over the maximum RF bandwidth of the MCBTS, which typically lies between 10 and 20 MHz. In the latter case, the spread of carriers is even, that is two carriers are allocated at the lowest

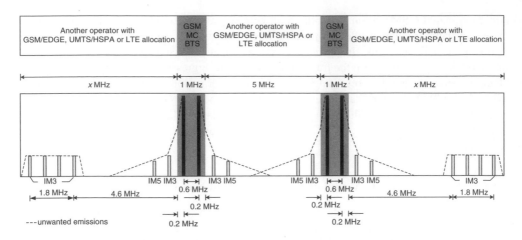

Figure 3.20   Test of MCBTS for non-contiguous frequency allocation.

and highest carrier frequencies of the maximum RF bandwidth and the remaining carriers are located evenly in between, so that the spacing between adjacent carriers may only vary in a range $+/-200$ kHz.

The requirements for spectrum due to modulation and wideband noise need to be met also for the case of non-contiguous frequency blocks. This applies to deployments where the operator has a split frequency allocation, the carriers of which are operated by MCBTS. This scenario is relevant in several European countries, in India, in the US and in other countries. The core specification 3GPP TS 45.005 prescribes that the MCBTS manufacturer declares the support of split frequency allocation. This capability needs to be verified through test of a configuration with four active MCBTS carriers, where two pairs of carriers, each using the minimum carrier spacing of 600 kHz, are separated so that an interim UMTS, HSPA or LTE carrier with 5 MHz bandwidth can be allocated between the pairs of carriers. Figure 3.20 depicts this test scenario. Note that in this figure, unwanted emissions consist of spectrum due to modulation and wideband noise as well as the dominating 3rd and 5th orders intermodulation products. The measured range of the unwanted emissions is x MHz.

3.4.4.4   Spurious Emissions

Spurious emissions specified in sub-clause 4.3 in 3GPP TS 45.005 can be distinguished into inband and out-of-band spurious emissions. Inband spurious emissions are measured within the transmit frequency band (e.g. 935 MHz to 960 MHz for P-GSM MCBTS and 925 MHz to 960 MHz for E-GSM MCBTS). For MCBTS, no separate requirement for inband spurious emissions, measured from 2 MHz beyond the outermost carrier until the edge of the transmit frequency band on both sides, exists as for SCBTS; instead the requirements defined for spectrum due to modulation and wideband noise apply in this frequency range for MCBTS. Out-of-band spurious emissions are measured outside the transmit frequency band with an offset greater than 2 MHz and need to comply with absolute power levels. A relaxation of up to 10 dB is granted for MCBTS class 1 and of up to 20 dB for MCBTS class 2 compared to SCBTS for offsets lower than 5 MHz from the transmit frequency band edge, while for larger

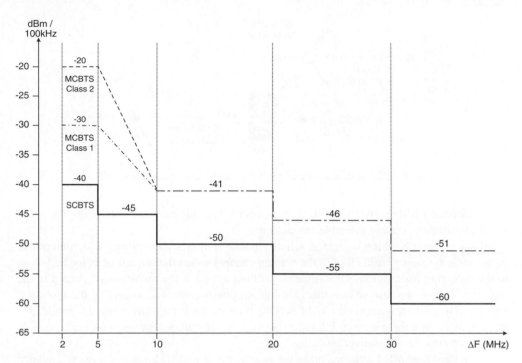

Figure 3.21   Allowed out-of-band spurious emissions for single carrier BTS and both MCBTS classes 1 and 2 (GSM 900 MHz band, normalized to 100 kHz measurement bandwidth with average detector method).

offsets beyond 10 MHz the size of the relaxation is at most 9 dB compared to SCBTS due to different used measurement methods (average detector instead of peak-hold). For offsets between 5 MHz and 10 MHz a linear interpolation of the out-of-band spurious emissions requirements is specified, which takes into account the additional attenuation of the duplexer filter in the MCBTS. This is depicted in Figure 3.21.

3GPP TSG GERAN agreed that the out-of-band spurious emissions requirements are based on the most stringent requirement between the absolute power mask, as shown in Figure 3.21 for GSM 900, and the unwanted emission requirements defined by the cumulated mask composed by the spectrum due to modulation and wideband noise and the intermodulation products. Thus, for a MCBTS operating far from the band edge the unwanted emission requirements as depicted in Sections 3.4.4.1 and 3.4.4.3 apply, while the absolute power mask is valid for a MCBTS operating close or at the band edge.

3.4.4.5   Inband Receiver Blocking

A MCBTS including a MCPA may also be equipped with a multicarrier wideband receiver as shown in Figure 3.16, or else it may be operated with multiple single carrier receivers. In the first case the selectivity of the analog receiver front-end is decreased compared to a narrowband single carrier receiver which yields reduced inband receiver blocking performance. Another limitation exists in the restricted dynamic range of current A/D converters, which in the case

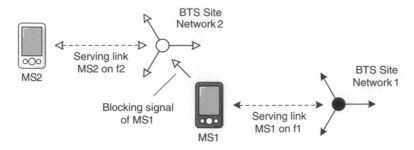

Figure 3.22   Receiver blocking in the case of uncoordinated networks.

of a wideband GSM receiver need to operate with a dynamic range of about 100 dB, being about 20 dB above current available technology.

Inband receiver blocking is created, when a mobile station is transmitting with high power in the same frequency band close to the wanted channel and at the same time is located close to the receiving base station. This situation will not appear if the mobile is registered to the same network as the base station thanks to uplink power control. However, if the mobile is registered to a different network and the serving base station is far away from the mobile, this effect may occur if there is no coordination between the operators of the networks. Figure 3.22 depicts the situation of receiver blocking.

A mobile station MS 1 attached on frequency f_1 to a distant BTS site of Network 1 owned by operator A is located close to a BTS site of Network 2 owned by operator B, serving a mobile station MS 2 on frequency f_2 and no network coordination between operator A and B is in place. Thus the MS 1 transmits a high power signal on uplink on frequency f_1 which may be close or even adjacent to the frequency f_2, thus yielding potential blocking of the reception of the signal, coming from MS2, at the BTS site of the Network 2. The impact of such blocking signals is the more severe the closer it is transmitting to the wanted channel in the frequency domain. Such high power levels will occur mainly in dense urban areas but with a rather small probability. However, due to complete blocking of the BTS receiver, they can lead to an amount of dropped calls that is not acceptable within GSM networks. For MCBTS with a wideband receiver, this problem is more evident than for SCBTS or for MCBTS with single carrier receivers due to the reduced receiver selectivity and the restricted input dynamic range as depicted above. Among the different alternatives, 3GPP TSG GERAN agreed to specify for a MCBTS with a wideband receiver two additional relaxation levels for inband receiver blocking, allowing a larger desensitization than for SCBTS, that is a larger degradation of the receiver sensitivity. This means that in case of strong RX blockers, the receiver AGC is biased towards reception of strong signals, thus allowing for the usage of available A/D converter technology with restricted dynamic range.

SCBTS receiver blocking performance is stated in 3GPP TS 45.005 subclause 5.1, for example for GSM 400 MHz and GSM 900 MHz frequency bands and for a blocking signal offset by 3 MHz or more from the wanted channel, the receiver reference sensitivity performance must be met with a blocker signal up to -13 dBm and with a wanted signal level at -101 dBm (3 dB above the BTS receiver reference sensitivity level of -104 dBm) as stated in 3GPP TS 45.005. For a blocker signal closer to the wanted signal, that is between 800 kHz and 3 MHz, the receiver reference performance needs to be met with a blocker signal up to -16 dBm at the input signal level of -101 dBm.

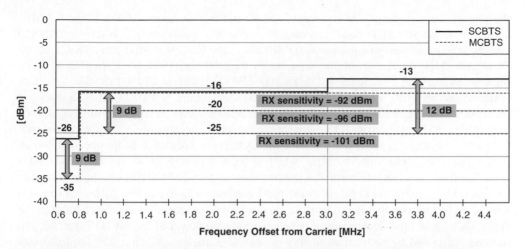

Figure 3.23   Inband RX blocking level mask for SCBTS and MCBTS in 900 MHz band depending on the frequency offset of the blocking signal from the wanted channel.

For SCBTS, MCBTS class 1 and MCBTS class 2, the receiver blocking performance is shown in Figure 3.23. Three blocking limits are defined for GSM 400 MHz and GSM 900 MHz frequency band both for MCBTS class 1 and class 2:

1. *First blocking limit at 3 dB desensitization.* The receiver reference performance must be met for the same conditions as for SCBTS, that is for a wanted signal at -101 dBm with a blocker signal up to -25 dBm, yielding a relaxation of 9 dB in case the blocker is 0.8–3 MHz apart from the wanted channel and a relaxation of 12 dB if it is more than 3 MHz apart.
2. *Second blocking limit at 8 dB desensitization.* The reference receiver performance must be met with a blocker level of up to -20 dBm and a signal level of -96 dBm, that is allowing a further desensitization of 5 dB related to first blocking limit.
3. *Third blocking limit at 12 dB desensitization.* The reference receiver performance must be met with a blocker level of up to -16 dBm and a signal level of -92 dBm, that is allowing a further desensitization of 9 dB related to first blocking limit.

Note, for the low frequency bands GSM 700 MHz and GSM 850 MHz as well as for the high frequency bands such as DCS 1800 MHz and PCS 1900 MHz, the same inband RX blocking levels are specified as for SCBTS. For instance, for DCS 1800 MHz and PCS 1900 MHz, a blocking limit of -25 dBm for blocker offsets at 800 kHz and above and a limit of -35 dBm below 800 kHz is required. Hence there is no relaxation introduced for both MCBTS classes for these frequency bands.

3.4.5   Status in Regulatory Bodies

While 3GPP TSG GERAN has investigated in particular inband coexistence of GSM MCBTS with other public GSM networks, as depicted in Section 3.4.4.2, coexistence studies of GSM MCBTS with services in the same band such as UMTS 900 MHz or UMTS 1800 MHz and in the adjacent band such as GSM Railway (GSM-R) and Private Access Mobile Radio (PAMR) below the GSM 900 MHz band, aeronautic systems like radio navigation service (ARNS) and distance measurement equipments (DME) above the GSM 900 MHz band, DECT above DCS 1800 MHz band and military systems in the 900 and 1800 MHz bands have been undertaken in 3GPP TSG RAN 4 (UMTS 900, UMTS 1800), in CEPT ECC SE7 (military systems) and in CEPT ECC PT1 (the remainder), concluding that GSM MCBTS does not yield additional degradation for coexistence with other systems in adjacent bands compared to existing GSM single carrier BTS. Some restrictions have been identified for the coexistence with DME systems, in particular in case DME systems are deployed in European countries in the frequency range 962–977 MHz, which is located directly above the GSM 900 MHz downlink frequency band. In particular, the coexistence study in ECC PT1 concluded that only in case DME systems are deployed using the lower edge channel at 962 MHz, interference mitigation techniques, that is usage of RF power control on downlink by the operator using GSM MCBTS or other techniques to reduce the radiated power of the MCBTS over the air, may be prescribed by national regulators to achieve sufficient isolation. For the European telecommunication market, the European Telecommunication Standards Institute (ETSI) and technical committee TC MSG is responsible for issuing the new Harmonized Standard for GSM base stations including MCBTS. Recently ETSI TC MSG has approved the Release 8 version of the Harmonized Standard for GSM base stations [31] including both GSM features EGPRS2 and MCBTS. For the latter one ETSI TC MSG has approved one set of requirements corresponding to MCBTS class 2 in 3GPP TSG GERAN. The authors' expectation is that the new Harmonized Standard for GSM base stations will presumably be published during the second half of 2010.

References

[1]   3GPP TS 23.401 v8.8.0, "GPRS Enhancements for E-UTRAN Access", December 2009.
[2]   L. Ong, I. Rytina, M. Garcia, H. Schwarzbauer, L. Coene, H. Lin, I. Juhasz, M. Holdrege, and C. Sharp "Framework Architecture for Signaling Transport", in *IETF RFC2719*, October 1999.
[3]   3GPP TS 48.001 v8.0.0, "Base Station System – Mobile-Services Switching Centre (BSS-MSC) Interface; General Aspects", September 2009.
[4]   3GPP TS 45.050 v8.1.0, "Background for Radio Frequency (RF) Requirements", August 2008.
[5]   3GPP TS 36.300 v8.10.0, "Evolved Universal Terrestrial Radio Access (E-UTRA) and Evolved Universal Terrestrial Radio Access Network (E-UTRAN); Overall description; Stage 2", September 2009.
[6]   3GPP TS 25.913 v8.0.0, "Requirements for Evolved UTRA (E-UTRA) and Evolved UTRAN (E-UTRAN)", January 2009.
[7]   LGW-070038, "3GPP Workshop on LTE GSM Handovers", Source: 3GPP Workshop on LTE GSM Handovers, Sophia-Antipolis, France, January 2007.
[8]   3GPP TS23.206 v7.5.0, "Voice Call Continuity (VCC) between Circuit Switched (CS) and IP Multimedia Subsystem (IMS); Stage 2", December 2007.
[9]   3GPP TS 23.216 v8.6.0, "Single Radio Voice Call Continuity (SRVCC); Stage 2", December 2009.
[10] 3GPP TS 23.272 v8.6.0, "Circuit Switched (CS) Fallback in Evolved Packet System (EPS); Stage 2", December 2009.
[11] 3GPP TS 23.060 v8.7.0, "General Packet Radio Service (GPRS); Service Description; Stage 2", December 2009.
[12] 3GPP 3GPP TS 44.060 v8.6.0, "General Packet Radio Service (GPRS); Mobile Station (MS) – Base Station System (BSS) Interface; Radio Link Control/Medium Access Control (RLC/MAC) Protocol", September 2009.
[13] 3GPP TS 48.018 v8.4.0, "General Packet Radio Service (GPRS); Base Station System (BSS) – Serving GPRS Support Node (SGSN); BSS GPRS Protocol (BSSGP)", September 2009.

[14] GP-071308, "Defining E-UTRAN GERAN External Network Assisted Cell Change (NACC)", Nokia Corporation, Nokia Siemens Networks, TSG GERAN#35, Dublin, Ireland, August 2007.

[15] 3GPP TS 44.018 v8.8.0, "Mobile Radio Interface Layer 3 Specification Radio Resource Control (RRC) Protocol", September 2009.

[16] G2-080134, "Alternatives for E-UTRAN Neighbor Cell Information", Telefon AB LM Ericsson, GERAN2#37bis, Sophia-Antipolis, France, March 2008.

[17] 3GPP TS 45.008 v8.4.0, "Radio Subsystem Link Control", October 2009.

[18] 3GPP TS 43.129 v8.1.0, "Packed-Switched Handover for GERAN A/Gb Mode; Stage 2", March 2009.

[19] 3GPP TS 24.008 v9.0.0, "Mobile Radio Interface Layer 3 Specification; Core Network Protocols; Stage 3", September 2009.

[20] 3GPP TS 23.003 v8.6.0, "Numbering, Addressing and Identification", September 2009.

[21] 3GPP TS 22.011 v8.9.0, "Service Accessibility", October 2009.

[22] 3GPP TS 25.367 v8.2.0, "Mobility Procedures for Home Node B (HNB); Overall Description; Stage 2", September 2009.

[23] 3GPP TS 36.304 v8.7.0, "Evolved Universal Terrestrial Radio Access (E-UTRA); User Equipment (UE) Procedures in Idle Mode", September 2009.

[24] 3GPP TS 25.304 v8.7.0, "User Equipment (UE) Procedures in Idle Mode and Procedures for Cell Re-Selection in Connected Mode", September 2009.

[25] 3GPP TR 43.903 v8.3.0, "A Interface over IP Study (AINTIP)", November 2008.

[26] 3GPP TS 48.008 v8.7.0, "Mobile Switching Centre – Base Station System (MSC-BSS) Interface; Layer 3 Specification", May 2009.

[27] 3GPP TS 48.103 v8.2.0, "Base Station System – Media GateWay (BSS-MGW) Interface; User Plane Transport Mechanism", September 2009.

[28] GP-071542, "Work Item Description: Introduction of a New Multicarrier BTS Class", August 2007.

[29] 3GPP TR 25.816 v8.0.0, "UMTS 900 MHz Work Item Technical Report (Release 8)", September 2009.

[30] 3GPP TS 45.050 v8.1.0: "Background for Radio Frequency (RF) Requirements", August 2008.

[31] MSG-09-059r2, "Draft ETSI EN 301 502 V9.1.0 (2009-12) Harmonized EN for Base Station Equipment Covering the Essential Requirements of Article 3.2 of the R&TTE Directive", ETSI TC MSG#21 meeting, Sophia-Antipolis, France, December 2009.

4

3GPP Release 9 and Beyond

Jürgen Hofmann, Eddie Riddington, Vlora Rexhepi-van der Pol,
Sergio Parolari, Guillaume Sébire and Mikko Säily

4.1 Introduction

Given the huge market penetration of GSM/EDGE worldwide and the interest of numerous
operators to keep running the technology for at least the coming decade; given its increasing
footprint in countries such as India, China and other emerging markets where demographics
and/or network topology pose new challenges; given the coming deployment of LTE, the
expansion of 3G/HSPA and the growing interest of mobile operators to enter the home environment; given all these and other reasons, new evolutionary steps are being defined for GSM
within 3GPP Release 9. These involve both the radio access and the core parts of the network
and can be classified into four major categories:

1. *Evolution of voice and data services.* The evolution of voice services, described in Section
   4.2, is addressed by the introduction of "VAMOS", a feature aiming to double the hardware
   and spectral efficiencies for speech. The evolution of (packet) data services, treated in Section 4.3, targets further EGPRS and EGPRS2 throughput increase by means of a spectrally
   wider transmit pulse shape.
2. *Enhanced mobility with Home NodeBs and Home eNodeBs.* While mobility between
   GERAN cells (typically macro cells) and UMTS home cells (Home NodeB) or LTE home
   cells (Home eNodeB) is restricted to autonomous reselection by the mobile station in Release 8, it is extended to cover network-controlled mechanisms in Release 9, that is during
   an ongoing data session and/or an ongoing voice call. The mechanisms introduced are
   depicted in more detail in Section 4.4.
3. *Security enhancements.* Since the introduction of GSM, security mechanisms have been in
   place to protect network operators and users against malicious attacks from third parties.
   Section 4.5 summarizes the evolution of security in GSM from the early days to 3GPP
   Release 9.

GSM/EDGE: Evolution and Performance   Edited by Mikko Säily, Guillaume Sébire and Eddie Riddington
© 2011 John Wiley & Sons, Ltd

4. *Network architecture evolution.* Network architecture evolution is based on the standardization of local switching for local calls as part of 3GPP Release 9. The feature "Local Call Local Switch" switches calls locally in the Base Station Subsystem (BSS) with minimal impact on the core network (CN) and thus leads to considerable OPEX savings in terms of reduced bandwidth, especially in the BSS internal interfaces, but possibly also on the A interface and on the core network interfaces. It is dealt with in Section 4.6.

These main 3GPP Release 9 features are described in the following sections in more detail and for each of them an outlook is provided. It is noted that the introduction of multicarrier multi-standard base stations (MSR) identifies a further Rel-9 feature in 3GPP GERAN. It is described in more detail in Section 14.4.

4.2   Voice Evolution

4.2.1   Introduction

A continuous evolution of voice services has taken place ever since GSM Phase 1 (see Chapter 1), in terms of both speech codecs and radio interface efficiency. It is summarized here.

Voice Evolution before 3GPP Release 9

Speech in GSM started with the support of the Full Rate speech codec (FR) [2] yielding a source bit rate of 13 kbit/s. FR is based on the RPE/LTP-LPC[1,2] speech encoding algorithm. It was selected by CEPT in 1987 for GSM Phase 1 and later used in network deployments in the early 1990s. Through the introduction of GSM Half Rate codec (HR) [3] in GSM Phase 2 in 1995 with a source bit rate of 5.6 kbit/s based on the VSELP[3] speech encoding algorithm, voice capacity could be doubled compared to FR. However, the speech quality suffered, as witnessed by a MOS[4] up to 3.3, compared to FR speech which achieved a MOS up to 3.5 [7]. Thus operators primarily use HR for mitigation of traffic peaks, for example during busy hours.

It was later in GSM Phase 2 that a new speech codec was introduced: Enhanced Full Rate (EFR) [4]. EFR has a source bit rate of 12.2 kbit/s and is based on CELP[5] speech encoding algorithm. It brings a considerable improvement in speech quality compared to FR with a MOS up to 4.0 [7].

Another evolutionary step was taken with the introduction of Adaptive Multi-Rate speech codec (AMR) [5] in GSM Phase 2+ in Release 98. AMR is based on the A-CELP[6] speech encoding algorithm. It supports source bit rates between 4.75 and 12.2 kbit/s enabling link and codec adaptation to suit not only the radio environment but also capacity requirements. Indeed, AMR channel coding is defined both for GMSK full-rate channels (TCH/AFS) for all codec modes and for GMSK half-rate channels (TCH/AHS) for 4.75 kbit/s to 7.95 kbit/s

---

[1] LTP: Long Term Predictor/Prediction.

[2] LPC: Linear Predictive Coding.

[3] VSELP: Vector-sum Excited Linear Prediction.

[4] Mean Opinion Score or MOS is a metric which for speech measures the quality of a given codec, as evaluated by subjective tests. MOS ranges from 1 to 5 and indicates a quality expanding from Bad (1) to Excellent (5) (and going through Poor (2), Fair (3) and Good (4)) [1].

[5] CELP: Code Excited Linear Prediction.

[6] A-CELP: Algebraic CELP.

codec modes. Like FR and EFR, AMR operates narrowband speech (i.e. at POTS[7] wireline voice quality). Voice is sampled at 8 kHz and filtered to just 3.1 kHz (0.3–3.4 kHz), hence its name "narrowband AMR" (AMR-NB). AMR-NB speech quality not only outperforms EFR with a MOS slightly above 4.1 but it is also much better in degrading radio conditions [7].

Later, in Release 5, AMR-NB support was defined on 8-PSK half-rate channels, for all codec modes (O-TCH/AHS). In the same release, the use of the Wideband AMR speech codec (AMR-WB) [6] was introduced in GERAN specifications. Voice is sampled at 16 kHz, and filtered to 7 kHz (50 Hz–7 kHz) thus designed to support audio signals such as music. It has nine codec modes,[8] five of which can be used in GSM (6.60, 8.85, 12.65, 15.85 and 23.85 kbit/s). AMR-WB is supported through the use of 8-PSK full-rate channels (O-TCH/WFS) for all codec modes. Codec modes up to and including 12.65 kbit/s are supported through the use of GMSK full-rate channels (TCH/WFS) and 8-PSK half-rate channels (O-TCH/WHS). Speech quality with AMR-WB reaches a MOS up to 4.2 [8].

Voice Evolution in 3GPP Release 9
Given the high demand for voice services in current GSM/EDGE networks and the general interest of mobile network operators to reduce operational costs, a new feature to boost the spectral efficiency of voice in GSM is, at the time of writing, being specified in 3GPP Release 9. It should indeed be noted that Release 7, as seen in Chapter 2, focused on data. By fitting more speech users on given radio resources, not only does this feature free radio resources in order to serve more data users with higher data rates (e.g. using Downlink Dual Carrier EGPRS and/or EGPRS2), it can also be used to help release part of a GSM network's spectrum while not affecting its capacity, for example for re-farming purposes.

4.2.2 Objectives

The objectives of this later evolution as agreed by 3GPP [9] are classified into performance objectives and compatibility objectives as summarized below:

- *Performance objectives.* Capacity improvements at the BTS: increase of Hardware Efficiency by doubling voice capacity per BTS transceiver for various (GMSK) full rate and half rate speech traffic channels (and associated signaling channels) for FR, HR, EFR, AMR-NB and AMR-WB, that is TCH/FS, TCH/HS, TCH/EFS, TCH/AFS, TCH/AHS and TCH/WFS. Capacity improvements at the air interface: increase of Spectral Efficiency by enhancing voice capacity by means of multiplexing at least two users simultaneously on the same radio resource both in downlink and in uplink for the channel types listed previously.
- *Compatibility objectives.* In order to allow deployment in potentially all GSM/EDGE markets and to minimize the impact of the new feature on existing networks and mobile stations while avoiding additional costs for voice-only handsets, five compatibility objectives are defined:
  - *Maintenance of voice quality*: voice quality as perceived by the user should not decrease and a voice quality better than GSM HR should be ensured, taking into account the

---

[7] POTS: Plain Old Telephone Service.
[8] 6.60, 8.85, 12.65, 14.25, 15.85, 18.25, 19.85, 23.05 and 23.85 kbit/s.

influence of the inter-channel interference (ICI) on voice quality, originating from the simultaneous activity of users in the same radio resource.

- *Support of legacy mobile stations*: no implementation impact should be required for legacy mobile station types. Besides, these mobiles should be included in the multiplexing. Given the nature of the feature, the priority is given to multiplexing of legacy DARP Phase 1 mobiles. However, mobiles not supporting DARP Phase 1 are also considered.

- *Implementation impact on new mobile stations*: hardware changes for new mobile stations should be avoided. Additional complexity in terms of processing power and memory should be kept to a minimum.

- *Implementation impact on BSS*: hardware upgrades to the BSS should be avoided. BTS transceivers need to support new and legacy channel types serving all kinds of legacy mobile stations simultaneously as per the above objectives. Impact on the dimensioning of resources on the BTS-BSC interface (Abis) should be minimized.

- *Impact on network planning:* impact on network planning and frequency reuse should be minimized. Impact on legacy MS interfered on downlink through the deployment of the new technique should be avoided in the case of usage of a wider transmit pulse shape in the downlink. Also scenarios related to the usage of the feature at the band edge should be considered, that is at the edge of an operator's band allocation and in country border regions without frequency coordination.

4.2.3 Functional Description

As agreed in 3GPP, a new feature meeting the objectives listed above is being specified at the time of writing. "VAMOS" or Voice over Adaptive Multi-user channels on One Slot is based on the concept of Orthogonal Sub-channels (OSC) and on the use, in the downlink, of the QPSK modulation of which the symbol constellation can be adjusted (Adaptive Symbol Constellation). VAMOS aims at multiplexing on the same radio resources twice as many speech users per channel mode, thus allowing on a single GSM time slot either:

- up to four half-rate voice traffic channels; or
- up to two half-rate voice traffic channels and one full-rate voice traffic channel; or
- up to two full-rate voice traffic channels.

The functional concept of this technique is described for downlink and uplink in the following subsections.

4.2.3.1 Downlink

Orthogonal Sub-Channels (OSC)
The OSC concept aims at multiplexing two users on the same radio resource, that is on the same frequency and same time slot, by assigning each user onto a given sub-channel. These sub-channels are then multiplexed by means of orthogonal sequences onto a common carrier signal and transmitted. On the receiver side, each sub-channel can be detected knowing which of the orthogonal sequences was used.

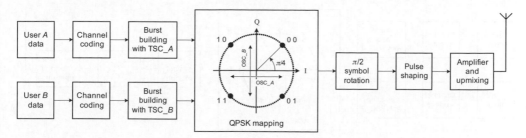

Figure 4.1   Baseband transmitter block diagram for orthogonal sub-channels technique.

In particular, OSC is based on the usage of orthogonal midambles, part of the GSM normal burst, and thus operates with user specific training sequences each indexed by a Training Sequence Code (TSC). The composite baseband signal is generated by multiplexing two user signals with orthogonal TSC onto one baseband signal employing a quaternary modulation scheme such as ordinary QPSK. On the receiver side, the signal detection applying the known training sequence indexed by the user-specific Training Sequence Code, is executed. Figure 4.1 depicts the baseband transmitter in downlink using the OSC technique.

The data from User A and User B are mapped onto QPSK constellations where for each constellation symbol, the first bit is assigned to User A, in the first sub-channel OSC_A, and the second bit is assigned to User B, in the second sub-channel OSC_B. Since the signal constellations are based on ordinary QPSK and are always symmetric, the orthogonal sub-channels are transmitted with equal power, that is the I and Q branches have the same power. The addition of 8-PSK compatible signal constellations allowing higher signal power in a sub-channel is described in more detail in Chapter 8.

VAMOS Adaptive Symbol Constellation
The modulation scheme applied for VAMOS is called Adaptive QPSK (AQPSK) and is based on an α-QPSK (alpha-QPSK) modulation, where "α" adjusts the symbol constellation and consequently the power of each sub-channel in a flexible manner. OSC can thus be seen as using α-QPSK with a fixed α equal to $\pi/4$.

The modulating symbols c_k for AQPSK, using the normal symbol rate of 270.83 ksym/s, are defined as a function of the incoming data bits a_k of the first sub-channel (OSC_A), b_k of the second sub-channel (OSC_B) and the argument α as shown in Equation 4.1:

$$c_k = (1 - 2 \cdot a_k) \cdot \cos\alpha + j(1 - 2 \cdot b_k) \cdot \sin\alpha \qquad (4.1)$$

with $a_k = \{0;1\}$, $b_k = \{0;1\}$, the symbol index $k = 0, 1, 2 \ldots$ and $0 \leq \alpha \leq \frac{\pi}{2}$

Figure 4.2 depicts the baseband transmitter in downlink for the adaptive symbol constellation technique applied for VAMOS.

The adaptability of the modulation scheme is achieved by the time-varying argument α. The value of α can range between 0 and $\pi/2$, these limits representing the BPSK[9] signal constellation points, that is either suppressing the second sub-channel (OSC_B) on the Q

---

[9] BPSK: Binary PSK.

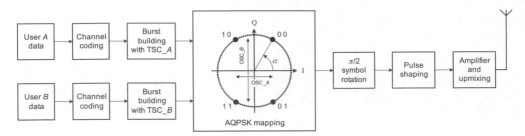

Figure 4.2 Baseband transmitter block diagram for VAMOS adaptive symbol constellation technique – AQPSK.

branch ($\alpha = 0$) or the first sub-channel (OSC_A) on the I branch ($\alpha = \pi/2$). The α value may be changed on a burst-by-burst basis according to the requirements for power assignment to both sub-channels. The power assignment is based on a continuous α range, but can be set discretely by the network. The embedding of the power assignment in the downlink power control procedure is described hereafter in this section. Like OSC, a rotation between symbols of $\pi/2$ is then applied.

Training Sequences Used in VAMOS
VAMOS applies to GSM normal bursts, keeping the burst structure unchanged. Only a second Training Sequence is defined for the paired sub-channel. To allow multiplexing of legacy mobile stations in the field the first sub-channel carries by definition a training sequence code out of the existing (legacy) eight TSCs. This legacy TSC set is referred to as "TSC set 1".

The TSC applied to the paired sub-channel belongs to "TSC set 2" which is designed both to have optimum autocorrelation properties and to achieve the lowest cross-correlation with the Training Sequences of TSC set 1. The training sequence codes of this second set can only be applied to the second sub-channel, that is when new VAMOS-aware mobiles (see the section below) are served. The two TSC sets are depicted in Table 4.1.

The multiplexing is done in such a way that a TSC pair is selected by taking the two TSC with the same index in both sets. This TSC pairing applies if either a legacy mobile is paired with a VAMOS-aware mobile or if two VAMOS-aware mobiles are paired.

VAMOS also includes a legacy mode, multiplexing two legacy mobiles. In this case, both training sequences belong to TSC set 1.

In all cases, the TSC assigned is signaled to the mobile station in the channel assignment message at call set-up or after handover.

VAMOS Operation in Discontinuous Transmission
If DTX is activated in the downlink and one of the sub-channels enters DTX mode (due to e.g. a silent period), only the active sub-channel is transmitted. This allows the use of GMSK modulation with linearized GMSK pulse shape as for legacy channels. This has the advantage that the power of the GMSK transmission compared to AQPSK can be reduced during this period by for example 3 dB, since the signal energy for the remaining active user doubles compared to ordinary QPSK transmission when both users are active. Once the sub-channel in DTX mode needs to transmit a silence indicator description (SID FIRST, SID-UPDATE, ONSET, NO-DATA) or reenters the next speech activity period, the AQPSK modulation scheme is selected.

Table 4.1 First TSC set (TSC Set 1) applied to first sub-channel and second TSC set (TSC Set 2) applied to second sub-channel

Training Sequence Code (TSC)	Training sequence bits for TSC Set 1
0	(0,0,1,0,0,1,0,1,1,1,0,0,0,0,1,0,0,0,1,0,0,1,0,1,1,1)
1	(0,0,1,0,1,1,0,1,1,1,0,1,1,1,1,0,0,0,1,0,1,1,0,1,1,1)
2	(0,1,0,0,0,0,1,1,1,0,1,1,1,0,1,0,0,1,0,0,0,0,1,1,1,0)
3	(0,1,0,0,0,1,1,1,1,0,1,1,0,1,0,0,0,1,0,0,0,1,1,1,1,0)
4	(0,0,0,1,1,0,1,0,1,1,1,0,0,1,0,0,0,0,0,1,1,0,1,0,1,1)
5	(0,1,0,0,1,1,1,1,0,1,0,1,1,0,0,0,0,0,1,0,0,1,1,1,0,1,0)
6	(1,0,1,0,0,1,1,1,1,1,1,0,1,1,0,0,0,1,0,1,0,0,1,1,1,1,1)
7	(1,1,1,0,1,1,1,1,1,0,0,0,1,0,0,1,0,1,1,1,0,1,1,1,1,0,0)

	Training sequence bits for TSC Set 2
0	(0,1,1,0,0,0,1,0,0,0,1,0,0,1,0,0,1,1,1,1,1,0,1,0,1,1,1)
1	(0,1,0,1,1,1,1,0,1,0,0,1,1,0,1,1,1,0,1,1,1,0,0,0,0,1)
2	(0,1,0,0,0,0,0,1,0,1,1,0,0,0,1,1,1,0,1,1,1,0,1,1,0,0)
3	(0,0,1,0,1,1,0,1,1,1,0,1,1,1,0,0,1,1,1,1,0,1,0,0,0,0)
4	(0,1,1,1,0,1,0,0,1,1,1,1,1,0,1,0,0,1,1,1,0,1,1,1,1,1,0)
5	(0,1,0,0,0,0,0,1,0,0,1,1,0,1,0,1,0,0,1,1,1,1,0,0,1,1)
6	(0,0,0,1,0,0,0,0,1,1,0,1,0,0,0,0,1,1,0,1,1,1,0,1,0,1)
7	(0,1,0,0,0,1,0,1,1,1,0,0,1,1,1,1,1,0,0,1,0,1,0,0,0,1)

VAMOS Pulse Shaping

Pulse shaping is performed using the linearized Gaussian Minimum Shift Keying (LGMSK) pulse shape, introduced for 8-PSK signals in EDGE. In addition, further (optimized) pulse shaping forms that have a wider bandwidth than the LGMSK pulse and which aim at yielding additional spectral efficiency and hardware efficiency gains are, at the time of writing, under discussion in 3GPP:

- RRC pulse shape with 3 dB-bandwidth of 240 kHz and rolloff of 0.3;
- synthetic pulse shape with 3 dB-bandwidth of about 210 kHz.

These pulse shapes are depicted in Figure 4.3.

Both optimized pulse shapes are considered only if both sub-channels are active, while in DTX mode of one sub-channel the usage of the LGMSK pulse shape is proposed on the other sub-channel.

Mobile Station Types for VAMOS

VAMOS-aware Mobile Stations

Two types of VAMOS-aware mobile stations are defined:

1. *VAMOS Level 1*: VAMOS-aware mobile stations that are DARP Phase 1 capable (see Chapter 1) and implement TSC Set 2 (in addition to TSC Set 1). These mobiles need to comply with performance requirements for VAMOS Level 1 mobiles.

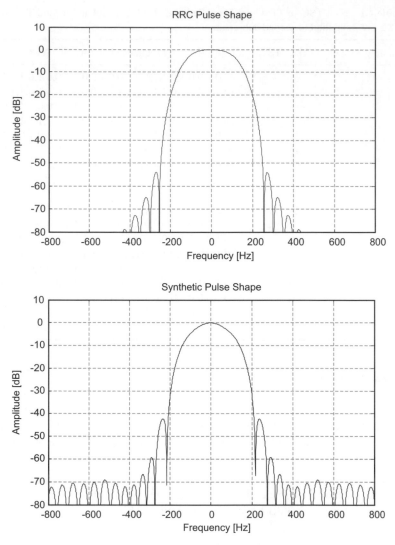

Figure 4.3 Transmit pulse shapes under discussion for VAMOS in downlink. Reproduced by permission of 3GPP TR 45.914 v8.2.0, Figures 7.2a and 7.2b. © 2009. 3GPP™ TSs and TRs are the property of ARIB, ATIS, CCSA, ETSI, TTA and TTC who jointly own the copyright in them. They are subject to further modifications and are therefore provided to you "as is" for information purposes only. Further use is strictly prohibited.

2. *VAMOS Level 2*: VAMOS-aware mobile stations that are DARP Phase 1 capable, implement TSC Set 2 (in addition to TSC Set 1) and use an advanced receiver architecture to comply with tighter performance requirements than for VAMOS Level 1 mobiles.

These new mobiles are designed to be multiplexed on VAMOS sub-channels. In addition, multiplexing of these mobile stations with legacy mobile stations is foreseen.

Legacy Mobile Stations
As explained earlier, legacy mobile stations are always assigned to the first sub-channel bearing legacy TSCs. Two types of legacy mobile stations are considered, namely:

1. Legacy mobiles stations that support DARP Phase 1.
2. Legacy non-DARP Phase 1 mobiles that comply with existing performance specifications in 3GPP TS 45.005 and support at least FR and HR legacy speech channels.

VAMOS thus allows multiplexing between two VAMOS-aware mobiles and between a VAMOS-aware mobile and a legacy mobile. It also supports multiplexing between two legacy mobiles provided they both support DARP Phase 1. Implementation aspects for mobile stations are described in Section 4.2.5.2.

Power Control
Power control in downlink for VAMOS is done in two successive stages:

1. Determination of the required transmit power levels for both mobile stations (MS_A and MS_B) according to the radio link measurement reports (signal level – RXLEV and signal quality – RXQUAL) received from these mobiles. The BSS determines the power level P_MS_A required for MS_A in the first sub-channel and P_MS_B for MS_B in the second sub-channel.
2. Determination of the corresponding AQPSK signal constellation and output power for the AQPSK signal. A control unit in the BTS computes a combination of output power P and α that gives the required combination of P_MS_A and P_MS_B in downlink based on the following relationship:

$$P = P\_MS_A + P\_MS_B = P \times \cos^2 \alpha + P \times \sin^2 \alpha \qquad (4.2)$$

Both steps are part of the radio link control in the BSS. The procedure is depicted in an exemplary implementation in Figure 4.4.

Associated Control Channels
An important aim with VAMOS is to preserve the integrity of the associated control channels (FACCH and SACCH), which in downlink direction identify the basis for reliable neighboring cell descriptions, power control and timing advance commands for uplink. Hence a balanced performance between traffic channels and associated control channels is targeted. To achieve this, a metric has been defined such that the relative performance between traffic channels and associated control channels should not degrade in comparison to legacy traffic channels in non-VAMOS mode.

While a robust FACCH performance is achievable by transmitting a FACCH frame using GMSK modulation, one of the drawbacks in VAMOS mode is that speech frames will need to be stolen from both sub-channels simultaneously in order to transmit a FACCH frame to one user. This will increase the FACCH signaling load and may lead to artifacts audible in the speech. To minimize this impact, *FACCH soft stealing* may be employed, whereby the power

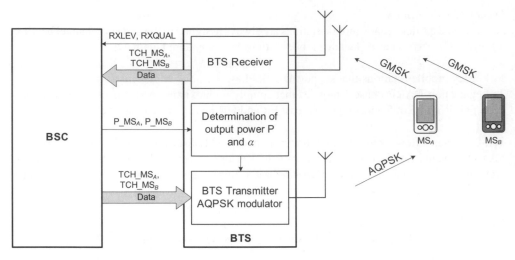

Figure 4.4 Exemplary implementation of power control in downlink for VAMOS [7].

between two users sharing the same radio resource is apportioned in favor of the user to which the FACCH is associated, but so that the other user's speech frame is not entirely stolen. This is depicted in Figure 4.5, where the first constellation shows equal power for two users sharing the same radio resource, the second shows a power imbalance in favor of user B and the third constellation shows a power imbalance in favor of user A.

To overcome temporary severe channel degradations, Downlink Repeated FACCH and Repeated SACCH, described in Chapter 7, need to be supported by VAMOS-aware mobile stations.

4.2.3.2 Uplink

The uplink for VAMOS is based on the OSC concept.

Burst Structure and Training Sequence Codes
In contrast to downlink, the physical layer is not changed for uplink. This means that two independent uplink transmissions are simultaneously received at the BTS. The legacy GMSK

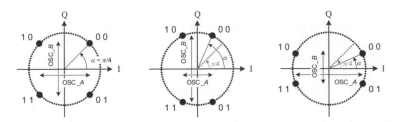

Figure 4.5 FACCH soft stealing for two users sharing the same radio resource: TCH for both users A and B, FACCH for user B, FACCH for user A.

modulation and normal burst are used. Thus the transmitter side is the same as for legacy GMSK-based voice traffic channels. The same TSC as applied in downlink is, however, used in uplink.

Associated Control Channels

Since uplink power control is executed on a dedicated basis for both sub-channels, there is no difference to the legacy non-VAMOS channel mode.

With regards to the SACCH transmission, Repeated SACCH defined in Release 6 must be supported by VAMOS-aware mobiles to overcome severe channel degradations.

Reception of VAMOS Sub-channels at the BTS

The BTS receiver needs to process signals from two mobiles on simultaneous sub-channels with individual propagation paths, which may include time offsets of one or two symbols. Hence the uplink for VAMOS can be seen as a 2×2 Multi User MIMO system, where different propagation paths from two users provide the basis to fully utilize the degree of freedom of two receive antennas in typical base station (BTS) implementations. The BTS receiver types are further described in a section below.

4.2.4   Specification Aspects

VAMOS is being specified in 3GPP Release 9 at the time of writing. Table 4.2 provides an overview of the related 3GPP GERAN specifications.

4.2.5   Implementation Aspects

The implementation impacts are listed hereafter for networks and mobile stations.

4.2.5.1   Impact on the Network

Four major network impacts due to VAMOS need to be considered as described in this section.

BTS Receiver

With regards to the BTS receiver, different receiver algorithms may be used, that is Space Time Interference Rejection Combining (STIRC), Successive Interference Cancellation (SIC) or Joint Detection (JD) to receive both orthogonal sub-channels distinguished by their individual training sequences. Another option is to use two independent GMSK receivers for each sub-channel. The concept of a SIC receiver implementation is depicted in Figure 4.6.

For efficient operation of the uplink, the base station sub-system (BSS) should apply uplink power control possibly interworked with Dynamic Frequency Channel Allocation (DFCA) scheme to keep the difference of received uplink signal power levels of paired sub-channels within a certain power window, for example within a 15 dB window. Performance for VAMOS uplink reception is specified in TS 45.005 for discrete Sub-channel Power Imbalance Ratios (SCPIR), see Section 4.2.5.2. The exact SCPIR figures are subject to agreement at the time of writing.

Table 4.2   Related 3GPP Specifications for VAMOS

Specification	Title and description of impact
45.001	Physical layer on the radio path – General description Functional description of VAMOS − Channel organizations for TCH and ACCH − Network and Mobile Station Support − Modulation, burst format, associated control channels in downlink/uplink − Channel mode adaptation
24.008	Mobile radio interface Layer 3 specification – Core network protocols Stage 3 − Mobile Station capability indication for support of VAMOS within MS classmark 3 information element
44.018	Mobile radio interface layer 3 specification – Radio Resource Control (RRC) protocol − signaling support for TSC set 2 in channel descriptions
45.002	Multiplexing and multiple access on the radio path − channel configurations for VAMOS − new training sequences − mapping of associated control channels
45.004	Modulation − alpha constellations and alpha range − symbol rotation − TX pulse shape for VAMOS
45.005	Radio transmission and reception − test scenarios for VAMOS − modulation accuracy and power versus time mask for AQPSK − radio performance requirements for MS and BTS
45.008	Radio subsystem link control − power backoff for AQPSK modulation on broadcast carrier − sub-channel specific power control − impacts on measurement quantities for AQPSK

Radio Resource Management

Radio Resource Management (RRM) is responsible for optimum pairing of users in channels supporting the VAMOS mode. This is executed according to the available information on path loss, location and mobility of the users. In general, it is preferable to pair users with a similar path loss, since uplink reception is improved in this case. Hence RRM needs to balance the received uplink signal levels of both sub-channels within a power window of 15 dB range.

Handover to channels in non-VAMOS mode also needs to be provided if the channel quality remarkably degrades for one of the sub-channels. This applies to both channel modes: full-rate and half-rate voice traffic channels.

RRC protocol signaling also indicates at channel assignment the used TSC set on this sub-channel and by this whether the first or second VAMOS sub-channel is assigned. In

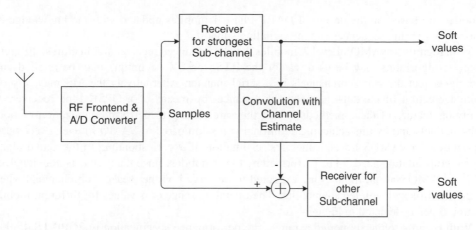

Figure 4.6   Successive Interference Cancellation (SIC) Receiver.

addition, AMR channel rate and codec mode adaptation are performed by RRM, if AMR-type channels are served.

Abis Interface

In the extreme case that all Abis time slots in all TRXs in the given BTS site are working at the same time in VAMOS mode, the Abis bandwidth needed for CS traffic must be doubled in comparison to the legacy non-VAMOS mode. Note that Abis bandwidth reserved for signaling must also be adjusted accordingly in such a case. However, unless a 100% penetration of VAMOS-aware mobiles is available in the network, it is unlikely to achieve 100% penetration of VAMOS channels. Thus, a suitable mix of VAMOS channels and legacy full-rate and half-rate channels has to be assumed for Abis dimensioning, taking into account that the mobiles may require channel mode adaptation from VAMOS half-rate channel mode or legacy half-rate channel mode to legacy full-rate channel mode in the case of degrading radio conditions. Thus, the actual impact of VAMOS on Abis resources, that need to be reserved for the CS domain, is proportional to the penetration of channels in VAMOS mode. A viable solution in this context is the migration of the transport network towards packet/IP-based Abis.

Frequency Planning

VAMOS channels are intended to be employed both on BCCH carrier and on TCH carriers. As performance investigations during the study phase have shown, VAMOS can exhibit the best performance for hardware-limited scenarios and loose frequency reuses, while for tighter frequency reuses such as 1/1 and 1/3, the performance gains are rather moderate in general. This is described in more detail in Chapter 8. In general, impacts on frequency planning due to VAMOS introduction are rather limited, but nonetheless dependent on the mix of mobile station types in the network and the applied frequency reuse on traffic layer.

4.2.5.2   Impacts on the Mobile Station

As described in Section 4.2.2, VAMOS is expected to serve legacy mobile stations, as it is either non-DARP or DARP Phase 1 compliant. New VAMOS-aware mobile stations need to support the new training sequences of TSC Set 2 in downlink and uplink. VAMOS level 1 mobiles are based on the design of DARP Phase 1 mobiles updated with the knowledge of these new training sequences and the signaling support.

With regards to VAMOS Level 2, mobiles, having advanced receiver architectures, different receiver algorithms may be used: as for the BTS, the SIC technique may be applied and thereby detect the two sub-channels in a serial manner. Alternatively the MS may rely on Joint Detection of the sub-channels, for instance, by means of a QPSK type receiver. A common feature of these methods is that they are aiming to perform equalization of both sub-channels and by this achieving additional gains compared to VAMOS Level 1 receivers. In contrast to VAMOS Level 1, mobiles' estimation of α corresponding to the actual alpha signal constellation needs to be performed by these mobiles, since the used α is not signaled.

The VAMOS-aware mobiles are expected to be served on the weaker sub-channels when being multiplexed with legacy mobiles. The operating range of α values for different mobile station types is depicted in Figure 4.7.

With regards to the supported α range, the performance specification in 3GPP TS 45.005 for both downlink and uplink is based on the Sub-channel Power Imbalance Ratio (SCPIR). SCPIR is defined by Equation 4.3 assuming that MS_B receives the quadrature component and MS_A the in-phase component of the AQPSK signal as depicted in Figure 4.7:

$$SCPIR[dB] = 10 \times \log(\frac{P\_MS_B}{P\_MS_A}) = 10 \times \log \frac{|Q|^2}{|I|^2} = 20 \times \log(\tan \alpha) \qquad (4.3)$$

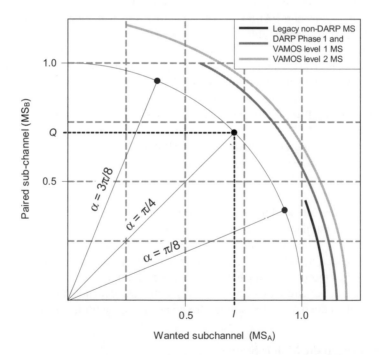

Figure 4.7 Discrete α values with operating ranges for legacy non-SAIC, VAMOS level 1 and VAMOS level 2 mobiles [10] (in-phase component "I" dedicated to MS_A, quadrature component "Q" dedicated to MS_B).

Note that VAMOS level 2 mobiles are specified to support a larger SCPIR range than VAMOS level 1 mobiles. The performance is specified to be tested for different SCPIR ratios with SCPIR = 0 dB being one of them. The exact SCPIR range is under discussion at the time of writing.

If further transmit pulse shapes other than LGMSK are defined in downlink, these and the corresponding capability signaling need to be supported by these mobile stations.

4.2.6   Future Work

VAMOS is being specified in Release 9. A number of enhancements are, at the time of writing, under discussion and no decision has been reached as to their standardization:

- Definition of optimized transmit pulse shape(s) in downlink in addition to legacy GMSK pulse shape.
- Introduction, in addition to Repeated SACCH, of an alternative SACCH mapping for the second sub-channel.

VAMOS has so far focused on single antenna mobiles. Dual antenna mobiles, that is compliant with DARP Phase 2 requirements standardized in 3GPP Release 7, have not been included in the investigations. It is expected that further performance improvements can be achieved especially in networks employing tight frequency reuses when introducing such mobiles. In addition, multiplexing of sub-channels as introduced by VAMOS could be studied for channel types other than voice traffic channels, for example for packet traffic channels or even for signaling channels, should a demand arise for spectral efficiency increase for these channels.

4.3   Data Evolution

4.3.1   Introduction

In December 2007, a major advance in the data rates of EDGE was achieved with the introduction of EGPRS2 in Release 7 of the 3GPP specifications. EGPRS2 doubles the peak data rate of EGPRS from 59.2 kbit/s to 118.4 kbit/s per time slot with the support of new modulations (QPSK, 16-QAM and 32-QAM) and a higher modulating symbol rate at 325 ksymb/s.

The new modulations and the higher modulating symbol rate were introduced in conjunction with a wide bandwidth transmit pulse that was optimized specifically for the higher modulating symbol rate for use in the uplink. This "wide pulse" is expected to increase throughput performance considerably thanks to its reduced inter-symbol interference. For more details about the wide pulse in uplink, see Chapter 2 on 3GPP Release 7.

A wider bandwidth pulse, while yielding gains in throughput performance, provides lower protection to adjacent channel users and this aspect needs particular attention on the downlink. This is because while in the uplink, any potential interference introduced by the wide pulse can easily be suppressed by the interference cancellation algorithms that are widely deployed within dual antenna base stations, in the downlink, the suppression capability in legacy mobile stations is restricted by the use of a single antenna and limited baseband complexity. For this reason, no wide pulse shape has been specified yet for the downlink (at the time of writing). Instead, a feasibility study into the performance of a wide pulse on the downlink and on its impact on legacy mobiles is ongoing within 3GPP, the results of which are being collated in 3GPP TR 45.913 [11].

In this section, two of the wide bandwidth pulse shapes under evaluation in 3GPP for use on the downlink are described, together with an evaluation of their performance in terms of their adjacent channel protection. For a more comprehensive evaluation in terms of link level and system level performance, see Chapter 6 on data evolution performance.

4.3.2   Wide Bandwidth Pulse Shape for Downlink

The optimization of a new pulse shape on the downlink can be viewed as a trade-off between the throughput performance and the level of interference introduced by the new pulse shape. This latter aspect is dependent not only on the adjacent channel protection provided by the pulse shape but also on a number of other factors:

- the level of protection provided to a co-channel user (co-located in frequency) as well as to an adjacent channel user (adjacent in frequency);
- the proportion of time slots supporting the EGPRS2 service and the pulse shape's time slot occupancy;
- the degree to which the frequencies on which the EGPRS2 service is deployed are interference limited (for example, frequencies allocated to a TCH layer tend to be more interference limited than those allocated to a BCCH layer due to a tighter frequency reuse).

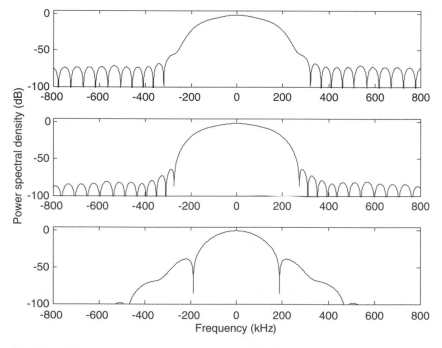

Figure 4.8   Two wide bandwidth pulse shapes under evaluation in 3GPP for use in the downlink (upper two figures) and the linearized GMSK pulse shape (lower figure).

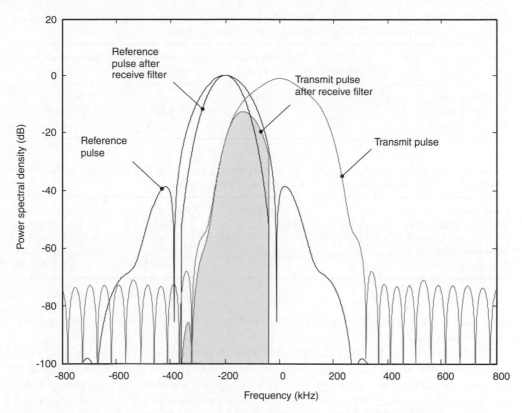

Figure 4.9 The protection ratio of a transmit pulse when on an adjacent frequency (200 kHz offset).

A number of pulse shapes are being evaluated in 3GPP for use on the downlink, two of which are depicted in Figure 4.8 (the linearized GMSK pulse shape – LGMSK – is also shown for reference).

The protection that a transmit pulse provides to a user who is on a channel that is co-located in frequency or to a user who is on a channel that is adjacent in frequency, can be defined as the ratio of power of that pulse shape to the power of the legacy pulse shape (LGMSK), after both have passed through the legacy mobile's front-end filter. This is depicted in Figure 4.9 for the case when the transmit pulse is adjacent (with 200 kHz offset). If we assume the front-end filter of that mobile has a frequency spectrum which is similar to the linearized GMSK truncated to −160–+160 kHz, then the protection ratios for the proposed transmit pulse can be calculated, as shown in Table 4.3. By comparing the protection ratios with the LGMSK pulse shape, we can see that the wide bandwidth pulse shapes offer a lower protection to users in the adjacent channel (at 200 kHz offset) and a higher protection to users in the co-channel.

A transmit pulse should offer sufficient protection in the adjacent frequencies. For example, in the case of GMSK modulation, the adjacent channel protection at offsets 200 kHz, 400 kHz and 600 kHz (the 1st, 2nd and 3rd adjacent channels) is specified in 3GPP TS 45.005 [12] at −18 dB, −50 dB and −58 dB respectively. The greater protection at the higher offsets

Table 4.3 The protection ratios of two of the wide bandwidth pulse shapes that are being evaluated in 3GPP and the linearized GMSK pulse shape (excluding Tx impairments)

Pulse shape protection ratios (dB)	Co-channel	Adjacent channel (200 kHz)	Adjacent channel (400 kHz)	Adjacent channel (600 kHz)
LGMSK pulse	0	−18.4	−63.7	−106
1st wide pulse	−0.7	−13	−68	−73.9
2nd wide pulse	−1.0	−12.1	−68.8	−83.6

Table 4.4 The protection ratios of two of the wide bandwidth pulse shapes that are being evaluated in 3GPP and the linearized GMSK pulse shape (including Tx impairments)

Pulse shape protection ratios (dB)	Co-channel	Adjacent channel (200 kHz)	Adjacent channel (400 kHz)	Adjacent channel (600 kHz)
1st wide pulse	−0.7	−13	−52.3	−66.5
2ndwide pulse	−1.0	−12.1	−50.1	−65.5

corresponds to the expected proximity of the interferer in frequency as a result of frequency reuse planning.

These 400 kHz and 600 kHz requirements were used as boundary conditions in the optimization of the 1st and 2nd pulse shapes in Figure 4.1. To take into account transmitter impairments such as spectral re-growth coming from the non-linear characteristics of the power amplifier, a model of the gain and phase characteristics of a typical base station power amplifier was used during the pulse shape optimization process. The resulting protection ratios for the proposed transmit pulse when including transmitter impairments are shown in Table 4.4. It can be seen that the transmit pulse meets the requirements at the 400 kHz and 600 kHz offsets.

At the 200 kHz offset, the 18 dB requirement has been relaxed to enable the investigation into the trade-off between adjacent channel protection at the 200 kHz offset and throughput performance. In Chapter 6, the system impact of this trade-off, both in terms of the throughput performance for the EGPRS2 users and impact to other users is evaluated.

4.4   H(e)NB Enhancements

In Release 9 the CSG mobility mechanisms specified in GERAN in Release 8 (see Chapter 3) are extended to provide support for inbound mobility to Home NodeB and Home eNodeB (H(e)NB) in connected mode thus enabling service continuity and ensuring improved user experience towards home cells.

In addition, a Release 9 Home (e)NB may now provide access not only to the mobile subscribers belonging to the Closed Subscriber Group (CSG) to which the H(e)NB belongs but also to other subscribers. This is known as hybrid access [13, 14]. A set of service requirements for Home (e)NB enhancements for Rel-9 is given in [14]. It should be noted

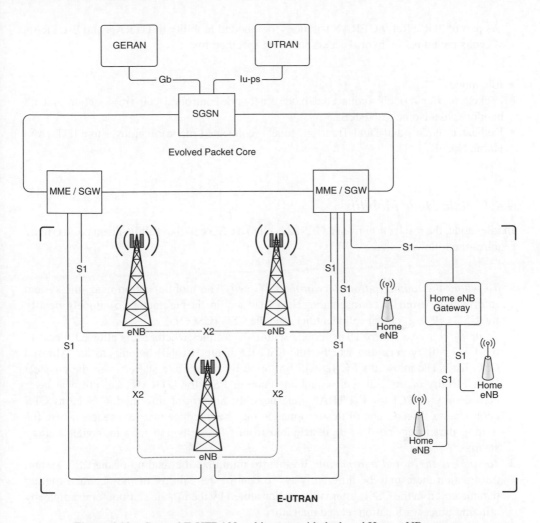

Figure 4.10 General E-UTRAN architecture with deployed Home eNB gateway.

that H(e)NB enhancements are being specified at the time for writing. For this reason only concepts and guiding principles are described hereafter.

The architecture for Home eNodeB support in E-UTRAN is depicted in Figure 4.10. The H(e)NB Gateway (GW) ensures support for a large number of HeNB through a single S1 interface to the EPC, the same interface as specified between an eNB and the EPC. The H(e)NB GW serves thus as a collection point for the control plane signaling, while the user plane signaling may either stop at the H(e)NB GW or be sent directly to the Serving Gateway (SGW) [14]. The HeNB provides the same functionality as the eNB with the addition of H(e)NB GW discovery when available. A HeNB can only be connected to one HeNB GW or MME at a time. The MME provides access control for a CSG member (i.e. verifies whether the user is member of the CSG or not) as well as routing of handover messages [14].

As part of 3GPP Rel-9 GERAN solutions for inbound mobility to UTRAN and E-UTRAN CSG cells including the hybrid access are being specified for:

- Idle mode.
- Packet transfer mode[10] (connected mode): network-controlled cell (re)selection and PS handover to a Home (e)NodeB.
- Dedicated mode[11] and Dual Transfer Mode[12] (connected mode): handover to a (UTRAN) Home NodeB.

4.4.1   Idle Mode Mobility

In idle mode, the reselection from a GERAN cell to a CSG cell that fulfills given radio criteria requires the following steps:

1. *To obtain the routing parameters of the CSG cell*: The mobile station reads the system information[13] from the corresponding H(e)NB to acquire the parameters to uniquely identify the CSG cell (e.g. global cell identifier) and the CSG (CSG ID).
2. *To perform access control*: The mobile station determines whether it is allowed access to the CSG cell by verifying whether the CSG ID of the H(e)NB belongs to its "Allowed CSG list". The allowed CSG ID will be stored by the mobile station with the physical layer identity of the cell (i.e. scrambling code or PSC for UTRAN, and physical layer cell identity or PCI for E-UTRAN) and possibly fingerprint information (such as GPS coordinates). The storage of this information may help reduce the occurrences where the routing parameters need to be re-acquired from CSG cells and may thus yield battery savings.
3. *To validate the stored information*: In order to minimize the reading of the CSG system information there may be different ways to control the validity of the obtained routing parameters from the CSG system information stored by the mobile station, for example, by utilizing timers or location related indicators.
4. *To detect the physical layer identities confusion*: The deployment of H(e)NB within the coverage of a GERAN macro cell, makes it likely that the same physical layer identity (PSC/PCI) be used on the same frequency by two or more CSG cells under coverage of the GERAN cell – this is referred to as PSC/PCI confusion, that is the physical layer identifier coupled with the frequency do not allow to uniquely identify the cell itself. The mobile station detecting the PCI/PSC confusion could store this information and possibly utilize it later to inform the network and/or trigger the acquisition of routing parameters from a CSG cell where confusion exists.

---

[10] During an ongoing packet data transfer.

[11] During an ongoing CS connection for example a voice call.

[12] During an ongoing packet data transfer and CS connection.

[13] Contained in MIB (Master Information Block) and SIB (System Information Block) messages.

4.4.2   Connected Mode Mobility

The deployment of H(e)NBs, although under the supervision of network operators, is, unlike macro cells, uncoordinated or loosely coordinated. Their geographical location may also change. This increases the risk for PSC/PCI confusion as described in Section 4.4.1.

The challenge in GERAN networks for inbound mobility to CSG cells in connected mode is to cope with the CSG cell identification, PCI/PSC confusion and measurement reporting of CSG cells.

The only way to uniquely identify a CSG cell is by means of its global cell identifier (GCI). However, the GCI is only broadcast by the H(e)NB in the system information of the CSG cell. The acquisition of routing parameters and access control as described in Section 4.4.1 may thus be necessary steps in connected mode.

MIB/SIB reading in connected mode results in service interruption and increased battery consumption. While it is not reasonable to require the mobile station to perform MIB/SIB reading when the information is readily available, it is also essential to ensure this information is accurate when sent to the network. Means are thus being investigated to minimize the occurrences where MIB/SIB reading is necessary (e.g. CSG cell being the best suitable cell on the frequency) while ensuring the accuracy of the information reported to the network for use, for example in handover preparation. Inaccurate information could otherwise trigger handover failure [15]. The solution being defined in GERAN is based on these two principles.

In addition to minimizing MIB/SIB reading occurrences, measurement reporting of CSG cells is further restricted to CSG cells that belong to the mobile station's "Allowed CSG list" and that meet given radio criteria. Not only does this minimize the impact on signaling (due to signaling capacity limitations), it also ensures that only information potentially useful to the user is exchanged between the mobile station and the network. It is indeed useless for a mobile station to provide reports for cells to which no mobility is possible, be it for radio or membership reason.

Inbound mobility to CSG cells in connected mode is being specified at the time of writing.

4.5   Security Improvements

GSM security aims at protecting operators and subscribers against malicious attacks from third parties. It provides both authentication and confidentiality. Authentication ensures the verification and confirmation of a subscriber's identity by the network. Confidentiality ensures the privacy of a given piece of information so it is never made available or disclosed to any unauthorized individuals, entity or process.[14] The subscriber's identity, the transfer of user data (e.g. CS voice, CS/PS data) and some signaling information elements such as the identities of the subscriber (IMSI[15]) and of the mobile terminal (IMEI[16]) are all subject to confidentiality [16].

---

[14] Reproduced by permission of 3GPP TS 42.009 v4.1.0 section 3.1.1. © 2006. 3GPP™ TSs and TRs are the property of ARIB, ATIS, CCSA, ETSI, TTA and TTC who jointly own the copyright in them. They are subject to further modifications and are therefore provided to you "as is" for information purposes only. Further use is strictly prohibited.

[15] IMSI: International Mobile Subscriber Identity.

[16] IMEI: International Mobile Equipment Identity.

Three algorithms are used for GSM security: A3, A5 and A8. The introduction of GPRS included also the GPRS-A5 algorithm (or GEA[17]). The A3 algorithm is used for authentication and operates between the authentication center and the mobile station. The A5 algorithm is used for the confidentiality of CS connections over the radio interface and operates on the radio link between the radio access network (BTS) and the mobile station. The GPRS-A5 algorithm is used for confidentiality of PS connections over the radio interface and operates between the core network (SGSN) and the mobile station. Both A5 and GEA provide ciphering using a specific encryption key generated by the A8 algorithm [17].

Although A5 is referred to as an algorithm, there are in fact different variants ("A5/x") which use different algorithms and achieve different security levels. Likewise different variants ("GEAx") of GPRS-A5 exist. A5/x and GEAx are listed hereafter.

- A5/0 and GEA0 provide no encryption.
- A5/1 is the original mandatory algorithm designed in 1987 for the pan-European GSM system (see Chapter 1).
- GEA1 is not related to A5/1. It uses an algorithm which has been kept secret.
- A5/2 was a deliberate weakened export[18] version of A5/1. As it was so weak and as even its implementation made the terminal vulnerable, the GSM Association strongly recommended that it no longer should be implemented in terminals built as of 1 July 2006. Thereafter, following GSMA recommendation, 3GPP modified its specifications from R99 onwards to prohibit the implementation of A5/2 in mobile stations.
- GEA2 is not related to A5/2. It uses an algorithm which has been kept secret.
- A5/3 and GEA3 [18] were introduced in R99 alongside the specifications of EDGE and UMTS. They provide an increased level of security compared to A5/1 and GEA1. A5/3 and GEA3 use the same KASUMI [19] f8 algorithm as in UMTS but with a ciphering key of 64 bits instead of 128 bits.
- A5/4 and GEA4 [20] were introduced in Rel-6. A5/4 and GEA4 use the same algorithm as A5/3 and GEA3 but thanks to a ciphering key of 128 bits offer greater security. However, only the specification of the algorithm using a 128-bit key was completed in Rel-6. The remaining specifications, such as signaling changes, were finalized in Rel-9.

Ever since cryptographic methods have been used, studies have been made to identify breaches that could permit the corresponding codes to be broken. Historical examples are plenty and the security of GSM and of other systems is no exception. A5/2 was quickly dismissed as a result of this process. A number of publications have also been published on potential attacks to A5/1. Theoretical threats do exist, and sometimes also practical threats. Nonetheless the authors are not aware, at the time of writing, of any malicious attack having actually occurred in any live GSM network worldwide. It should also be noted that there is no known attack on

---

[17] GEA: GPRS Encryption Algorithm.

[18] Cryptography and in particular encryption methods have historically been a matter of national security for the countries in which they were developed. Systems using these means were thus bound to apply drastic export restrictions. Exporting such systems to other countries meant they could not support the original encryption method, but instead a weaker version, if any at all. An export version of an encryption algorithm is thus defined typically as a weaker version of an algorithm originally subject to export restrictions.

GEA. Though even in case when no theoretical threat exists, the remote possibility that there could be one has itself justified the specification and implementation of more and more secure mechanisms The gradual introduction both of stronger mechanisms with means to ensure their secrecy, and of products, especially mobile terminals, implementing those, guarantees the timely integrity of the system and of its users.

The implementation of A5/0, A5/1 and A5/3 in GSM terminals, and additionally that of GEA0, GEA1, GEA2 and GEA3 when GPRS capable, have been mandatory for a few years. With the recent completion of A5/4 and GEA4 specifications in Rel-9, it is expected their implementation in mobile terminals will soon become mandatory as well.

4.6 Local Call Local Switch

A Work Item for Local Call Local Switch was agreed at GERAN#41, with the goal of standardizing procedures, messages and information elements on the A interface to enable local switching in the BSS of the user plane of local calls. Local call refers to a mobile-to-mobile CS voice call involving two MSs connected to the same BSS, while the term local switching refers to the possibility of switching a local call in the BSC, or create a direct communication between the involved BTSs, see Figure 4.11. In any case the effect is that some resources on the BSS internal interfaces (Abis and Ater) can be saved. In actual implementations the specific solution is based on BSS network topology and remains implementation specific. The specification of LCLS is ongoing at the time of writing with expected completion in Release 10.

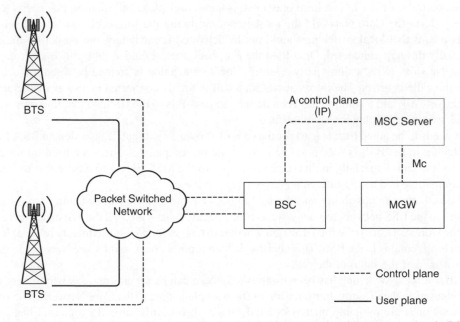

Figure 4.11 Control and User Plane paths for local switching, scenario without transcoders in BSS.

4.6.1   User Plane Aspects

The only user plane aspects that are standardized are the ones affecting the A interface [21]. To minimize changes to existing deployments using a legacy TDM-based A interface, and to AoIP networks being deployed, the impact on the A interface user plane handling is kept as small as possible. In particular, for A over TDM, no changes to the A interface are currently defined. Even if a call is (temporarily) locally switched, the two corresponding circuits always remain active, meaning that bandwidth savings on the A interface for locally switched calls are not possible, while of course savings are realized on the Abis/Ater interfaces. While a call is locally switched, the foreseen solution is that the TRAU will send some silence codeword on the A interface. Also for AoIP, the IP connections between the BSS and the MGW always remain active, that is the corresponding IP/UDP endpoints are not released. What happens is that while a call is locally switched, user plane transmission on the AoIP interface is suspended and the BSS/MGW does not expect to receive data through its IP/UDP endpoints.

4.6.2   Control Plane Aspects

To support local switching a number of control plane aspects have been addressed [22]. First of all, in order to trigger local switching, the BSS needs to be able to correlate the two legs of a local call, that is it needs to know who is talking to whom. A solution has been defined to pass this piece of information from the Core Network to the BSS during the assignment procedure at call set-up and during handovers. The BSS can then perform the correlation and inform the Core Network when the correlation has been found.

Second, even if a call is a local one, there is some user plane information that needs to be received via the Core Network, the most important being the ring-back tone at call set-up. This means that local switching should not be activated immediately, but only after the call is finally through-connected. Or at least the ring-back tone should still be conveyed from the Core Network to the calling party, even if a local connection is created between the calling and the called parties. The information that a call is finally connected is also available at the Core Network and a solution has been defined to pass this to the BSS, so that it knows when local switching can finally be established.

It needs to be noted that, at least for the case of A over IP without Transcoders in BSS, local switching in the BSS is only possible if the same or compatible codecs are used on the two legs of the call. Especially in this scenario it is up to the MSC Server to ensure that the same or compatible codecs are offered on the two legs of the call.

Then, since the possibility and the usefulness of establishing local switching are directly linked to the BSS network topology, it has been specified that – after all the assistance received by the Core Network – the final decision whether to trigger (and/or subsequently release) local switching remains in the BSS. This decision is then reported back to the Core Network to keep it informed of the status of the call.

After a local switching has been established, there can be several reasons why it needs to be released to then resume transmission of the user plane data to the Core Network. The most obvious ones are outgoing inter-BSS handovers and, for AoIP, also BSS internal handover procedures due to incompatible codecs. One more reason is the activation of a Supplementary Service that requires it to forward user plane data to the Core Network. The knowledge that some Supplementary Services are activated is also available at the Core Network, and is then passed to the BSS to command the release of local switching. More generally, all intra-BSS handovers where maintaining local switching is not practical, possible or useful, trigger the release of local switching. In all these cases the BSS informs the Core Network that local switching has been released and that user plane transmission on the A interface has been re-established.

References

[1] ITU-T Recommendation P.800, "Methods for Subjective Determination of Transmission Quality", August 1996.
[2] GSM 06.10 v3.2.0, "Full Rate Speech Transcoding", January 1995.
[3] GSM 06.20 v4.3.1, "Half Rate Speech Transcoding", May 1998.
[4] GSM 06.60 v4.1.1, "Enhanced Full Rate Speech Transcoding", August 2000.
[5] GSM 06.90 v7.2.0, "Adaptive Multi-Rate (AMR) Speech Transcoding", December 1999.
[6] 3GPP TS 26.190 v5.1.0, "AMR Wideband Speech Codec; Transcoding Functions", December 2001.
[7] 3GPP TR 26.075 v1.2.0, "Performance Characterization of the GSM Adaptive Multi-Rate (AMR) Speech Codec", October 1999.
[8] 3GPP TR 26.976 v8.0.0, "Performance Characterization of the Adaptive Multi-Rate Wideband (AMR-WB) Speech Codec", December 2008.
[9] 3GPP TR 45.914 v8.1.0, "Circuit Switched Voice Capacity Evolution for GSM/EDGE Radio Access Network (GERAN)", May 2009.
[10] –GP-090731, "Discrete Alphas for AQPSK", Telefon AB LM Ericsson, 3GPP GERAN#42, May 2009.
[11] 3GPP TR 45.913 v1.3.0, "Optimized Transmit Pulse Shape for Downlink Enhanced General Packet Radio Service (EGPRS2-B)", November 2009.
[12] 3GPP TS 45.005 v9.0.0, "Radio Transmission and Reception", September 2009.
[13] 3GPP TS 22.220 v9.2.0: "Service Requirements for Home Node B (HNB) and Home eNode B (HeNB)".
[14] 3GPP TS 36.300 v9.1.0: "Evolved Universal Terrestrial Radio Access (E-UTRA) and Evolved Universal Terrestrial Radio Access Network (E-UTRAN); Overall Description; Stage 2", September 2009.
[15] G2-090292 "Working Assumptions on 'Inbound Mobility to CSG Cells in Connected Mode' ", TSG GERAN WG2#43bis, Sophia Antipolis, France, October 2009.
[16] 3GPP TS 42.009 v4.1.0, "Security Aspects", June 2006.
[17] 3GPP TS 43.020 v9.0.0, "Security Related Network Functions", September 2009.
[18] 3GPP TS 55.216 v8.0.01, "Specification of the A5/3 Encryption Algorithms for GSM and ECSD, and the GEA3 Encryption Algorithm for GPRS; Document 1: A5/3 and GEA3 Specifications", December 2008.
[19] 3GPP TS 35.202 v8.0.0, "Specification of the 3GPP Confidentiality and Integrity Algorithms; Document 2: KASUMI Specification", December 2008.
[20] 3GPP TS 55.226 v1.0.0, "Specification of the A5/4 Encryption Algorithms for GSM and ECSD, and the GEA4 Encryption Algorithm for GPRS; Document 1: A5/4 and GEA4 Specifications", March 2004 (subject to French export control restrictions).
[21] 3GPP TS 48.103, "Base Station System – Media GateWay (BSS-MGW) interface; User plane transport mechanism", Release 9.
[22] 3GPP TS 48.008, "Mobile Switching Centre – Base Station System (MSC-BSS) interface; Layer 3 specification", Release 9.

Part II

GSM/EDGE Performance

5

Fundamentals of GSM Performance Evaluation

Mikko Säily, Rauli Järvelä, Eduardo Zacarías B. and Jari Hulkkonen

5.1 Introduction

The aim of this chapter is to give an overview of performance evaluation as a key input to the design, deployment, and utilization of telecommunication systems. The most common Key Performance Indicators (KPIs) are introduced and performance evaluation methods used throughout the book are explained.

Continued radio technology evolution makes it possible to take the GSM system to a whole new performance level. Enhancements specified by 3GPP standardization forum are bringing in benefits, which will improve the user experience and the overall system performance.

Operators and network infrastructure vendors are challenged to find new ways to improve the network services. New creative methods are needed to increase the coverage of a radio link and capacity at system level given the limitations of the operator spectrum and network resources. 3GPP Release 4 introduced highly successful EDGE technology and since then several new features have been added to the GSM standard. These new features are now reaching network equipment and user terminals.

Starting from 3GPP Release 7 new features for GSM Evolution have been discussed in the standardization. Downlink Dual Carrier, Mobile Receive Diversity, Higher Symbol Rate, Higher Order Modulation, lower latency, and many more are specified to add new features and improve the network service performance. GSM voice capacity has been significantly improved with Orthogonal Sub-Channels (OSC). These new technologies are designed to be backwards compatible and work seamlessly with the existing features. New performance requirements are needed to ensure compatibility of new features. Feature performance is specified on the link level, but also the system performance needs to be verified and further optimized to meet the performance targets.

GSM/EDGE: Evolution and Performance Edited by Mikko Säily, Guillaume Sébire and Eddie Riddington
© 2011 John Wiley & Sons, Ltd

5.2   On the GSM Radio System Performance Engineering

The primary goal of the cellular network performance engineering is to design a communications network, which meets the coverage, quality and capacity requirements set by the network operator to satisfy the customers ordering the network services. Today multiple cellular network operators are competing for the same market share in the same geographical area. The challenge for the radio performance engineers is to meet the customer expectations and requirements, and at the same time keep the operational costs low enough to be profitable, and being able to upgrade and maintain the investment to enable future service.

The competition for the market shares has a direct impact on the network performance requirements. When competition gets more intense, the cost of the service for the customer gets lower. User equipment and terminals are getting cheaper while the capability increases and more and more attractive functionalities can be implemented such as high average data rates for effortless internet browsing and gaming. The cheaper the calls, the more rapid increase of subscribers will be experienced. The cellular telephone market has been growing all over the world for more than fifteen years and there were four billion users in 3GPP specified systems by 2009.

The radio frequencies are fundamental to cellular systems, and also a limited resource, which cannot be increased. Therefore, resources become more and more limited while the cellular boom goes on. The available resources must serve the increasing number of service requests while the number of subscribers and cellular operators increases. If an operator cannot acquire more radio frequencies, then the spectrum efficiency needs to be improved. Spectrum efficiency is one of the key characteristics when selecting a system and planning the network roll-out and capacity performance.

The popularity of cellular services grows year by year and the amount of traffic rapidly increases. In particular, the recent popularity of data services has increased the absolute number of user data bits transmitted through the systems. Data services are equally popular in both rural and dense urban areas.

Operators are seeking new means and solutions to meet the increasing capacity requirements and demand for higher average data rates. Rural areas can be populated with more base station sites to narrow down the distance signal needs to travel. This is a simple but costly way to increase the average data rate. Upgrading the existing system to a more robust and higher performing link connections is in many cases a more viable option. There are also several options on how to build up the network capacity. In some cases, the operator can acquire a license to utilize additional frequencies, for example, on a higher frequency band. However, if frequencies are not available, the only solution is to use the existing frequencies more efficiently, for example improve the network capacity of voice and data services within the existing spectrum.

Spectral efficiency is one of the key the optimization criterion, where GSM has several radio engineering tools to work with. One traditional way to increase capacity is to decrease the cell size, which gives more freedom in increasing the frequency reuse. These so-called micro cells have smaller coverage but can potentially serve more traffic since more resources are available in the same geographical area. Micro cells are also good for office buildings and in general for indoors, since the attenuation from building walls does not affect the link budget and also the interference levels can be lower towards the rest of the network. One drawback with large number of micro cells is the higher cost of the network infrastructure and overall maintenance required to cover a certain area with more base stations. However, there are alternative ways to improve the capacity.

The GSM system has been continuously improved in the 3GPP standardization forum and contains several high performance features, which make GSM one of the most competitive systems in terms of the cost of the network and the number of subscribers using the network. GSM Evolution can address these challenges and issues. The huge number of GSM terminals, a large installed base of GSM base stations and infrastructure have such a massive volume that it makes sense to improve and maintain the network capability to make the most of the spectrum. A good GSM network performance also paves the way for other systems as users are more comfortable with using the latest services not only in hot spots but also in rural areas and while roaming on other operators' networks.

GSM Evolution provides service continuity to new systems such as HSPA and E-UTRAN long-term evolution. GSM Evolution in 3GPP Release 7 and Release 9 has standardized several new features which are designed to allow a more efficient use of the allocated frequency band and improve the network spectral efficiency and hardware efficiency. As always, there are also some proprietary features available in network equipment and cellular infrastructure that do not require standardization.

5.3 Simulation Tools

Cellular networks with radio transmitters and receivers are dynamic systems, which cannot be evaluated using exact analytical models and equations. One of the reasons for this is the random behavior of signal propagation. The fundamental reason behind this behavior is the complex interactions (reflection, absorption, diffraction, dispersion, etc.) of electromagnetic waves in different environments.

In link and network level performance evaluation, the chosen methodology and simulation tools have a significant role. A link level simulator is an efficient tool for studying the various transmission and reception related modules with new modeling ideas. System level simulators, on the other hand, are outstanding for studying overall performance of new link level features. The following chapters discuss the link and system level simulators, which have been used throughout this book.

5.3.1 Semi-Analytical Methods

Analyzing the performance via computer simulation works well for a large variety of communication systems, but can be time-consuming for very low error rates. Semi-analytic techniques allow the computation of error rates, which are small (in the order of 10^{-6} or less). The semi-analytic technique can provide analysis and results more quickly than methods relying on Monte Carlo simulations. The benefit of a semi-analytic technique comes from the combination of simulation and analytic methods to analyze the error rate. A computing environment can be used to implement the semi-analytic techniques to perform some of the initial analysis related to modulations used in GSM and GSM Evolution.

The semi-analytic technique works well for certain types of communication systems, but care should be taken when defining the methodology. Applicability of the semi-analytic technique depends on the type of processes to be modeled. The technique is applicable if a system has these characteristics:

- effects of multipath fading;
- quantization;
- the receiver is perfectly synchronized with the carrier;
- timing jitter is negligible.

Because phase noise and timing jitter are slow processes, they reduce the applicability of the semi-analytic technique to a communication system. Amplifier nonlinearities must precede the effects of noise in the actual channel being modeled. The noiseless simulation has no errors in the received signal constellation. Distortions from sources other than noise should be mild enough to keep each signal point in its correct decision region. If this is not the case, the predicted BER is too low. The semi-analytic function also assumes that the noise in the actual channel being modeled is following a Gaussian distribution.

The semi-analytic simulation (Figure 5.1) averages the error probabilities over the entire received signal to determine the overall error probability. If the error probability calculated in this way is a symbol error probability, the error probability can be converted to a bit error rate.

5.3.2   Link Level Simulator

One way to compute the bit error rate or symbol error rate for a communication system is to simulate the transmission of data messages and compare all the messages before and after transmission.

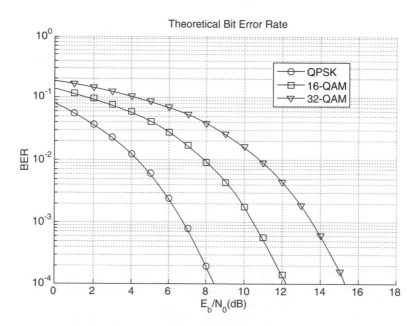

Figure 5.1   Example of analytical BER computation.

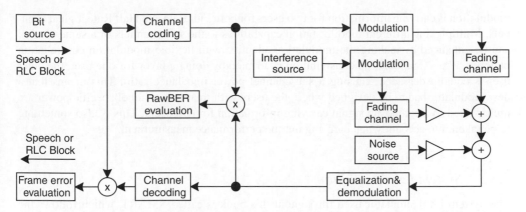

Figure 5.2   Link simulation flow diagram.

Link level performance is the baseline design criteria for a communications system. Link level simulations are performed between the transmitter and the receiver to evaluate the performance of the physical layer. There is always at least one directional connection where the modeled system contains a data source, a channel coding block, a modulator, a radio propagation channel, a demodulator, a channel decoding block and performance evaluation blocks (Figure 5.2). Details to run a simulation in different scenarios are dependent on the requested service.

The process of running link simulations consists of building parameter files, starting simulations, waiting for the simulation to end and examining the results. Typically a single simulation run can include any number of different simulation cases with different requested services and channel coding parameters. Those cases can be simulated in parallel, sharing most of the common processing between different cases.

5.3.2.1   Link Level Modeling

Starting from 3GPP Release 7 there are new GSM Evolution features, which require changes to the physical layer modeling. Higher symbol rate, an optimized pulse shaping filter combined with higher order modulation using 32-QAM and 16-QAM double the data rate compared to EGPRS. Introduced features need to be evaluated against the GPRS and EGPRS reference performances. New modulations and higher symbol rate usually mean that also the receiver algorithms need to be revised and optimized in terms of performance and complexity.

The link simulator for GSM Evolution is implemented using Matlab computing environment with MEX extensions to modules implemented using C-language. C-implementations are also used in the bit true modules, which are used to model a realistic signal processing platform with the impact of limited signal dynamic range, for example.

EGPRS2 introduced a new higher symbol rate, QPSK, 16-QAM and 32-QAM higher order modulations and new transmitter pulse shaping filters for both downlink and uplink. For the performance requirements all the baseband signal processing algorithms need to be upgraded to accommodate the new modulations and potentially the higher complexity of receiving end.

Voice evolution introduces multiplexing of two mobiles to a same radio resource. This is achieved by using Orthogonal Sub-Channels where one symbol from a higher order QPSK modulation is carrying information for two users for both Full Rate and Half Rate logical channels. Multiplexing different mobile station capabilities to the sub-channels of a same symbol can be enhanced by making the modulation adaptive with flexible modulation constellation patterns. When a modulation constellation is adapted by doing a favor for one user, the other user is equally degraded, causing a sub-channel power imbalance ratio. This means that a new modulator had to be designed where the users with their sub-channel specific power are multiplexed to a composite signal carrying information for two users. This is also something to be taken into account when carrying out the performance measurement.

5.3.3 System Level Simulator

The system level simulator used throughout this book is called SMART, which follows the principles and requirements for dynamic simulator described in [1], Appendix E. In SMART, cell phones are dynamically moving around the chosen network environment, which allows realistic modeling of radio resource management (RRM) algorithms (handovers, power control, channel/codec mode adaptation, etc.). Realistic algorithms with a realistic network environment make a dynamic simulator to give realistic performance results.

5.3.3.1 Introduction to System Simulations

SMART is a dynamic system level simulator which is continuously being updated with new features coming onto 3GPP standardization releases. Later on, features that are standardized are optimized to support studies for product development. Realistic algorithms corresponding to ones from actual BSS (Base Station Sub-system) software make SMART also capable of supporting customer network trials by realistic network simulations. First, this makes it possible to evaluate performance with the features that are not available in the real network. System parameters needs to be optimized for new algorithms and dynamic system level simulator with a realistic network set-up can be used for this task. Initial parameters can be achieved through performance simulations with theoretical simulation environments, but realistic initial values are only possible with simulations using real network data. Real network data in a certain format can be imported into the simulator and optimal parameters can be obtained by executing a simulation campaign. Hence, a dynamic network simulator is an excellent tool for studying new features at network level and can provide valuable input and cost savings to live network trials.

5.3.3.2 Basic Functionality

Simulation length is typically defined as simulation steps. The natural definition of a step in GSM is a TDMA frame consisting of 8 time slots. During one simulation's step, all the necessary actions are executed for each mobile phone in the network. At first, mobiles are moved and then received power for bursts of each link in the simulation is calculated. The interference situation in the network is then frozen for that time step, and carrier to interference and noise power ratio C/(I+N) values are calculated for each active transmission link. Transmission links are stored in an interference matrix. There are separate matrixes for both downlink and uplink. A matrix is found effective and essential for the interference accumulation loop over the active connections during a simulation step. Basic functionalities of a system level simulation platforms are described in [1], Appendix E.

5.3.3.3 Cellular Models

SMART models logical connections and physical links between base station and cell phones. The network topology has several options. Standard idealistic cellular models (hexagon macro cells, "Manhattan" grid micro cells, indoor office) are supported as well as digital terrain maps. A theoretical propagation model is used for standardization cases, such as the Okumura-Hata model. On the other hand, propagation data can be imported from customer networks. Correlated multipath fast fading and slow fading, for example shadowing, models are supported and these provide the expected performance of a radio link with feasible computational complexity. Vehicular and pedestrian models are used with theoretical topologies. Path and road mobility models are supported, but not commonly used. In the traffic model, calls arrive according to Poisson distribution and their traffic profile service time follows exponential distribution for circuit switched calls. Various packet data models are supported for the data service (web browsing, FTP, email, video streaming).

The following cellular models are supported:

- Multipath fading
- Shadowing
- Distance-based attenuation or imported propagation data
- Adjacent channel interference
- Antenna patterns (omni-directional, 3 sectors, 6 sectors)
- Burst level interference modeling (Actual Value Interface)
- Realistic mobility models
- Different network configurations (frequency reuse, number of frequencies, number of TRXs, layer modeling (BCCH/TCH)).

Radio resource modeling contains complete models of the following features:

- DTX
- Power control
- Handover algorithms
- Frequency hopping (cyclic/random hopping sequence).

The following GSM-specific models are supported:

- Channel mode adaptation for logical channels
- Codec mode adaptation
- Full Rate codecs EFR, AMR, 8PSK AMR, WB-AMR
- Half Rate codecs AMR, 8PSK AMR
- VAMOS 1/OSC: Double FR, Double HR
- VAMOS 2/OSC: Double FR, Double HR
- EGPRS2: complete model (normal symbol rate and higher symbol rate)
- Optimized transmit pulse shapes for EGPRS2 ("WIDER")

- Dual Transfer Mode
- Enhanced QoS.

The latest features are implemented for 3GPP standardization purposes. WIDER and VA-MOS/OSC models are implemented according to details proposed in 3GPP standardization forum.

The following EGPRS2 features are modeled explicitly:

- Modulation and coding schemes
- RLC block split
- Interleaving
- USF handling
- WIDER pulse shapes.

The following VAMOS/OSC features are modeled in detail:

- Sub-Channel Power Imbalance Ratio (SCPIR)
- Sub-Channel Power control via modulation constellation (constellation power control)
- Channel mode adaptation
- Optimized transmit pulse shape.

It is important to model cellular network features explicitly. One critical part of the dynamic system level simulator is the link-to-system interface (L2S). The system level simulator can be adapted to use new physical layer features via L2S modeling. One such model is the Actual Value Interface (AVI) table, where the expected link performance can be converted to a mapping table, which maps the system level C/I to BER and BER further to FER or BLER depending on the service. More information on the Link-to-System mapping can be found in Section 5.3.4.

5.3.3.4 Simulation Parameters

Basic parameters are usually the same for the different simulation cases. Propagation models, fast fading, slow fading and mobility models are not normally changed during the simulation campaign. The general parameters are kept the same in all the simulations in this book if not otherwise stated. The basic set-up for radio network engineering parameters is listed in Table 5.1.

There is a set of parameters that are defined for VAMOS/OSC and WIDER simulations in 3GPP standardization, see [2]. Scenarios are called MUROS according to the initial 3GPP working item, when the OSC concept was introduced. See Tables 5.2 and 5.3.

5.3.4 Link-to-System Interface

System level simulators operate on discrete steps, reflecting successive time periods when a number of mobile users move through the network. Thus, at any given step, it needs to evaluate the performance of each active link. However, executing the link-level simulator

Table 5.1 Basic simulation parameters

Parameter	Value	Comment
Cellular layout	Homogeneous	
BTS antenna height	30 m	
Antenna pattern	65 deg beam width	
Propagation model	UMTS 30.03	
MS speed	3 km/h	
Max BTS output power	20 W	~43 dBm
Max MS output power	2 W	~33 dBm

for each active link results in a prohibitive amount of calculation. For example, determining the link level performance for one link on a given step can take of the order of minutes, and the network simulator typically operates on thousands of steps and thousands of links. Furthermore, different network simulations are executed for different number of users. This motivates the need for a fast calculation of the expected performance of a given link and this interface is called the link-to-system interface (L2S).

5.3.4.1   Statistical Methods

Ultimately, the network simulator must estimate the throughput that a link has on a given step. This depends on factors such as the attenuation experienced by the data signal, the symbol modulation in use, the interference levels at the mobile position, the channel coding scheme employed, and to some extent on the mobile speed. A modular approach is used to decouple the effects of the fading channel and receiver, from the effects of the channel coding scheme. The basic idea is to compute a throughput (equivalently BLER or FER) estimate from burst-wise quality measures computed for the interleaving period [3, 4, 5]. Throughout this book, the

Table 5.2 MUROS parameter set-up

Parameter	MUROS-1	MUROS-2
Frequency band	900 MHz	1800 MHz
Cell radius	500 m	500 m
Bandwidth	4.4 MHz	11.6 MHz
Guard band	0.2 MHz	0.2 MHz
Number of channels excluding guard band	21	57
Number of TRX	4	6
BCCH frequency re-use	4/12	4/12
TCH frequency re-use	1/1	3/9
Frequency Hopping	Synthesized	Baseband
Length of MA (Number of FH frequencies)	9	5
Fast fading type	TU	TU
BCCH or TCH under interest	Both	Both
Network sync mode	Sync	Sync

Table 5.3   WIDER parameter set-up

Parameter	WIDER-1	WIDER-2	WIDER-3
Frequency band	900 MHz	900 MHz	900 MHz
Cell radius	500 m	500 m	500 m
Bandwidth	4.4 MHz	11.6 MHz	8.0 MHz
Guard band	0.2 MHz	0.2 MHz	0.2 MHz
Number of channels (excl guard band)	21	57	39
Number of TRX	3	6	4
BCCH frequency reuse	4/12	4/12	4/12
TCH frequency reuse	1/1	3/9	3/9
Frequency hopping	synthesized	baseband	baseband
Length of MA	9	5	4 (includes BCCH carrier)
BCCH or TCH under interest	BCCH and TCH	BCCH and TCH	BCCH and TCH
Network sync mode	sync[1]	sync[1]	sync[1]
Voice	3	7	3
Data	4	0	4
Network sync mode	sync[1]	sync[1]	sync[1]

Note: [1]Time slots are assumed to be aligned; TDMA frames are assumed to be aligned on intra-site level and randomly aligned on inter-site level.

two-stage approach is referred to the link to the system level simulator interface and takes the following general form:

1. Compute burst-wise signal-to-interference-plus-noise ratios (SINRs) for a group of bursts belonging to one interleaving period. Use a SINR to BER mapping to compute one BER per SINR.
2. Compute the average and standard deviation of the burst-wise BERs. Use a BLER mapping to read the expected BLER for the computed average and standard deviation.

The reliability of the simulated network performance largely depends on the accuracy of the link-level performance modeling. If the SINR to BER mapping is biased to give higher or lower error rates than the actual expected values, then the estimated user and cell throughputs are likely to be pessimistic or optimistic, respectively. Therefore, verification techniques that assess the accuracy of the whole mapping chain are typically employed. This ensures that the predicted throughputs match the link-level results, at least on average.

The following sections describe each of the mapping stages, and the verification strategies are summarized in Section 5.3.4.6.

5.3.4.2   First Stage Mappings: From Signal Powers to Bit Error Rates

A fundamental difference between these mappings for system simulation and the link level results is that the expected BER is associated with an "instantaneous" SINR (i.e., the average SINR over the particular burst under consideration), as opposed to the expected BER over a distribution of SINRs. For this reason, this mapping has been called the Actual Value Interface (AVI) [5].

For a single antenna receiver, the first stage is a SINR to BER mapping. This relationship depends on the modulation employed, receiver characteristics such as phase noise and power amplifier linearity, interference power statistics and spectral properties of the interfering signals. Indeed, for the same SINR, the expected BER is different for the cases of co-channel, adjacent channel and mixtures of both types. Additionally, the modulation employed by the interferers can have a large impact on the SINR to BER mapping. For example, the SAIC receiver has interference rejection capabilities for GMSK-modulated co-channel interferers, but this gain is largely lost for PSK or QAM-modulated interferers. Furthermore, user-multiplexing techniques, such as VAMOS, exhibit additional variability, depending on the orthogonality properties of the training sequences employed by the paired users.

A trade-off between modeling complexity and accuracy has been achieved with two alternative methods:

1. Derive the SINR to BER mapping from a link-level simulation where the statistics of the interference levels of different interferer types (e.g., co-channel/adjacent channel, narrow/wide band transmit pulse, GMSK/04/16/32QAM, single user/VAMOS multiplexed user pair) have been derived from the system level simulations. This is done iteratively by simulating with an initial SINR to BER mapping, then deriving an interference profile, upon which a new SINR to BER mapping is produced and fed to the system level simulator. After one or two iterations, the profiles and mappings do not change significantly. This is the approach used for EGPRS2 and SAIC/OSC in Chapters 6 and 8 respectively.
2. Derive SINR to BER mappings from link-level simulations representing scenarios where the mappings arc substantially different, and implementing a mapping switching logic in the network-level simulator. This logic can be based, for example, on the characteristics of the dominant interferer during the burst. While this approach does not require system-level simulations, it assumes a deep knowledge of the receiver performance characteristics. On the other hand, the switching criteria need to be defined and tuned, and ultimately subject to the verification procedure. This is the approach employed for the VAMOS study.

5.3.4.3   Interference Rejection Receivers

When more than one receive antenna is available, the receiver can use the associated diversity to process the signal before the detection step. The basic idea is to exploit the short-term correlation properties of the interfering signals, in order to maximize the signal to interference ratio after processing. This is done by applying a whitening filter, followed by a maximum-ratio combiner (MRC) filter. Both MRC and IRC concepts are well known and used extensively in the uplink of current GSM networks.

In terms of the first stage mapping, the number of variables has increased, compared to the single antenna case. It is known that the performance of the IRC receiver depends on the relationship between the different interfering powers suffered by the receiver. More specifically, the ratios of the dominant interfering power to the aggregated power of the rest of the interferers (dominant-to-rest interference ratio, DIR) play a key role: the higher the ratios, the better the IRC performance. Rather than associating the BER with the DIR and SINR variables, we take the approach of first computing a single approximate SINR from all the user-signal and interfering signal powers. The SINR can then be associated with the burst-wise BER, and a SINR to BER mapping can be derived. One possibility is to use the classical SINR for the IRC filter in the presence of narrowband signals [6]. Another option is to combine the per-antenna DIR and CIR values to produce a single [SINR, DIR] pair [7]. Both choices provide sufficient accuracy in terms of verification procedures.

5.3.4.4 OSC/VAMOS Uplink Receivers

The OSC/VAMOS concept allows two users to simultaneously share a frequency and a time slot. For single antenna mobiles, the base station and the two users form a 2 x 2 multiple-input multiple-output (MIMO) system. On the side of the mobiles, the data transmission cannot be coordinated, as opposed to a MIMO system with a single user employing two receive antennas. In order to separate the data stream from both users, the base station employs an iterative procedure known as successive interference cancellation (SIC), which consists of three steps:

1. *Sorting*: determine which user has the larger SINR. Here, the IRC-style SINR can be used by considering that user 1 is interference to user 2, and vice versa. This user is denoted as the dominant user. Since there is one SINR per antenna per user, the maximum SINR across antennas is used for classification.
2. *Cancellation*: process the data for the dominant user with the IRC receiver and decode its data stream. Rebuild the received signal from the decoded bits by using the channel estimate which is required by the IRC filter, and subtract the rebuilt signal from the composite signal. The quality of this cancellation depends on how close the rebuilt signal is to the true received signal of the user that was decoded first.
3. *Use the signal after the cancellation step*, to decode the other user.

The receiver could, of course, iterate this sequence by subtracting the second user from the original composite signal and executing steps 1–3. However, this increases the computational complexity of the algorithm and the increments in performance benefits become smaller with each iteration.

From the first stage mapping point, of view, the BER for both users is computed as follows:

1. From the signal and interfering powers on the receiver antennas, compute IRC-style SINRs as described in Section 5.3.4.3. Here, the signal power of user 1 is added to the list of interfering powers of user 2, and vice versa. We denote the user with larger SINR as the dominant user.
2. Use a SINR to BER mapping, which is specific to the dominant user. The mapping is selected on the basis of the maximum DIR across the user's antennas. This assumes that a set of mappings is available for different DIR values, and the value closest to the maximum DIR is used.

3. Assume that the cancellation of the dominant user signal is perfect, and compute a new SINR for the other user. Use a SINR to BER mapping, which is specific to the non-dominant user. The mapping is selected from a set of curves, based on the maximum DIR, similar to the previous point.
4. A burst-wise BER value for each user has been determined.

The channel estimation quality depends on the orthogonality properties of the training sequences used by the users. However, it turns out that mappings built for different sequence pairs can be approximated by simple shifts of the mappings built for the best performing pair. This simplifies the L2S slightly, because the system simulator only needs to apply a sequence-pair dependent shift to the computed SINRs, instead of using different mappings for the dominant and non-dominant user.

An example of the mappings for the dominant and non-dominant users, for different maximum DIR values is shown in Figure 5.3. This illustrates the performance of the training sequence pair (0,8).

5.3.4.5   Second Stage Mapping: From BERs to FER/BLER

The second stage of the L2S takes the burst-wise BERs and gives an expected BLER, which then determines the throughput of a link and ultimately the network performance. Modeling the performance of channel coding schemes is in general not easy. However, the simplified

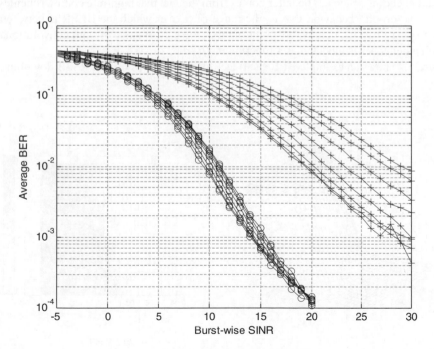

Figure 5.3   Link to system mappings for the dominant and non-dominant users, selection of different DIR values.

model described in this section is widely used in network performance simulations, and is also considered to be accurate enough for these purposes.

The second stage mapping predicts the BLER for a group of consecutive bursts, according to the interleaving period of the channel coding scheme. This period is typically 4, 2 or 1 burst. The BLER is approximated by a function of two variables: the average and standard deviation of the burst-wise BERs, which we denote by μ and σ, respectively. There are two types of (μ, σ) pairs that are of interest: those where the BLER is between zero and one, and those where the surface is either zero or one.

Many channel coding schemes benefit from channel variations within the interleaving period, for example, from frequency hopping. This is seen in the BLER surface by following a line with constant μ, and observing that the BLER decreases as σ increases. This gain also depends on the level of μ considered. On the other hand, one can identify a range for μ, outside which the surface is either zero or one, regardless the value of σ. This range is of interest, as the system simulator does not need to evaluate the BLER surface in these cases.

The use of frequency hopping has an impact on the range of σ values that can be observed in the link and system-level simulations. For systems without frequency hopping, the BER variability within the interleaving period is largely determined by the mobile speed. When frequency hopping is used, however, the speed does not have a big impact in the distribution of σ. It must be noted that μ and σ are not independent variables, because they are both computed upon the same burst-wise BERs. Thus, at small values of μ, large values of σ are not possible. This implies that the accuracy of the BLER mapping is more critical in certain zones, depending on the usage of frequency hopping, the mobile speeds and the error-correction capabilities of the channel coding scheme. The latter comes from the fact that high-rate coding schemes tend to show an "on-off" behavior, that is, the range of μ over which the BLER is between zero and one is very narrow. Finally, we point out that the performance of both convolutional and turbo channel codes is modeled as a function of (μ, σ).

Figure 5.4 shows an example BLER surface, for MCS DBS8 of EGPRS2B downlink.

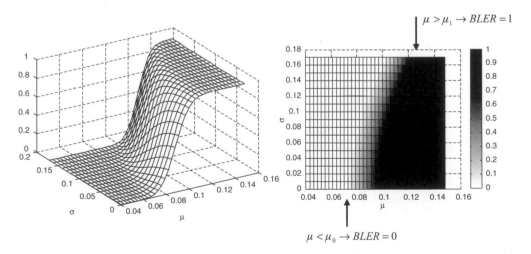

Figure 5.4 Link-to-system Second Stage Mapping, BER to FER/BLER conversion.

5.3.4.6   Mapping Chain Verification Procedures

One key aspect of the link performance modeling is its accuracy. The L2S is ultimately responsible for the network simulation results, because the throughput estimates of each link
are computed by the two-stage mapping procedure described in the previous sections. The
verification considered throughout this book assesses how close the mapping chain approximates the link-level BLER/FER results, for a given interference profile (e.g., a DTS-2 type of
scenario, or a custom interference profile derived from network simulations). The comparison
is done as follows:

1. Group consecutive simulated bursts according to the interleaving depth of the MCS under
   consideration.
2. Use the mapping chain to determine whether the RLC blocks of the group were successfully
   decoded.
   a. Use the first-stage mapping to get burst-wise BERs.
   b. Compute the average and standard deviation of the BERs of each group.
   c. For each group, compute a BLER estimate with the second-stage mapping. Use this
      estimate to draw binary random variables and assign decoding success/failure to the
      RLC blocks of the group.
3. Compute the L2S-predicted BLER as the ratio of blocks marked as failure in 2.C) to the
   total number of simulated blocks.
4. Repeat steps 1–3 for each point in the link-level BLER curve, and compare the predicted
   BLER with the actual BLER given by the link-level simulator.

The accuracy of the L2S has a different impact on the system-level simulator results, depending
on the BLER levels at which the whole chain is not accurate. For example, if the network is
designed to operate at 10% BLER, errors in the mapping chain for BLER levels below 1% may
have a negligible impact in the system-level results. Similarly, inaccuracies at BLER levels over
20 or 30% may only produce slight distortions in the throughput percentiles representing the
cell edges. The L2S mappings employed throughout this book typically show an accuracy of
1 dB or better, at 10% BLER. The accuracy of the L2S tends to worsen with larger constellation
sizes. For example, MCS for EGPRS2 using QPSK tend to be more accurate than MCS using
32QAM, especially at lower BLER levels.

An example of the verification procedure result is shown in Figure 5.5.

5.4   Key Performance Indicators

5.4.1   *Key Performance Indicators for Link Level Analysis*

Commonly agreed Key Performance Indicators (KPI) are fundamental when specifying and
comparing the performance between transmitter and receiver. The aim of the standardization is
to specify the expected minimum performance requirement, but not to specify the methodology
how to achieve the target performance. This way, each mobile and network vendor can develop
high performing platforms with a proprietary cost-efficient architecture, and ensure the network
operation and compliant performance are defined by the 3GPP standard.

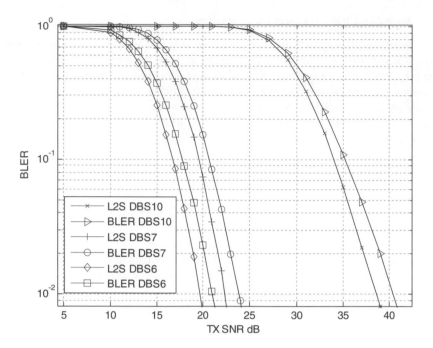

Figure 5.5 Link-to-system mapping verification.

5.4.1.1 Bit Error Ratio and Received Signal Quality

The bit error ratio (BER) in GSM systems represents the likelihood of a misinterpreted bit between the transmitter and receiver in a communications link due to noise and interference. That is, BER represents the signal quality over the air interface. BER is a function of energy quantity per transmitted bit to noise power spectral density ratio of the desired signal, Eb/N0. BER can be measured simply by dividing the number of erroneous bits received with the total number of transmitted bits. For a communications link, the simplest form of performance measure is typically done in an Additive White-Gaussian-Noise channel (AWGN). For any practical wireless application the realistic radio propagation channels with different mobile station velocities need to be evaluated in terms of BER. These prototype propagation channels, such as Typical Urban, Hilly Terrain and Rural Area, are specified in [8].

In GSM, both the mobiles and the base transceiver stations estimate BER and report the result to the Base Station Controller (BSC) where BER is used as a measure to control the traffic channel parameters and management, such as handover and power control decisions. BER is stored in BSC also for statistical purposes, where the network average performance can be monitored and optimized over different traffic profiles such as busy hour traffic and night time. Along with the performance requirements, also the BER measurement reporting accuracy and reliability are standardized. Received Signal Quality (RXQual) represents the average bit error rate over the measurement period in signal quality reporting over a slow associated common control channel frame of 480 ms (SACCH).

5.4.1.2 Frame Erasure Ratio

For GSM, the Frame Erasure Ratio (FER) is the ratio of discarded speech frames compared to all the received speech frames. A speech frame is generally discarded if after the decoding and error correction process a speech frame is considered to be corrupted. It is better to reject a suspicious or an erroneous frame rather than pass it onto the speech decoding, since a bad frame usually degrades the subjective speech quality.

The FER is a measure of how successfully the speech frame was received after the error correction process. Therefore it is a better indication of the subjective speech quality compared to the RXQUAL, which gives an estimate of the link quality in terms of BER, but does not indicate the distribution of errors over the measurement period.

Different speech codecs will have a slightly different FER to MOS correlation (Mean Opinion Score), since the higher the speech codec bit rate is, the more sensitive it becomes to frame erasures, thus degrading the quality experienced.

5.4.1.3 Block Error Ratio

Similarly to FER, Block Error Ratio (BLER) is a ratio of discarded data service blocks compared to all the received data blocks. A data block is generally discarded if the block-wise decoding of data header fails or if any of the received bits is found to be erroneous after the decoding and error correction process.

5.4.2 Key Performance Indicators for System Level Analysis

In system level simulations the chosen relevant Key Performance Indicators (KPI) are analyzed to understand how the studied features are performing. The same KPIs are also used in the field trial, thus making it possible to compare the simulation with the real live network performance.

There are various KPIs defined for voice and data services. Table 5.4 and Table 5.5 contain information on the most common KPIs that are being used in the latest standardization contributions. Detailed descriptions of the KPIs are found in [1].

5.5 EFL Methodology[1]

The spectral efficiency (SE) of a deployed cellular network is very hard to quantify in practice. This is due to the fact that cellular networks have an operating point in terms of local traffic load, which varies due to busy hour, site-to-site configuration and different geographical conditions, for example. The performance measurements collected during a trial, will identify the performance of the network associated with the traffic load at each operating point. However, this information is not sufficient to measure the Spectral Efficiency of the measured network, because it is usually not possible to quantify the maximum capacity without driving the network to a maximum load condition. This is the point where the average load during the measurement period exceeds the maximum achievable capacity.

---

[1]Reproduced from Timo Halonen, Javier Romero, Juan Melero, *GSM, GPRS and EDGE Performance Evolution Towards 3G/UMTS, Second Edition*, John Wiley & Sons, Ltd.

Table 5.4 Voice service KPIs

KPI	Abbreviation	Definition
Blocked call rate (%)	BCR	Percentage of the blocked calls
Dropped call rate (%)	DCR	Percentage of the dropped calls
Bad quality calls (%)	BQC	Percentage of the calls with quality below chosen threshold
Bad quality samples (%)	BQS	Percentage of the bad samples that are defined in SMART as sample of 1.92s with FER > 4.2%
Quality of Service (%)	QoS	100% - BCR DCR - BQC
Spectral efficiency (Erl/MHz/sector)	SE	Describes how many Erlangs can be achieved per MHz per sector
Hardware efficiency (Erl/TRX)	HE	Describes how many Erlangs can be achieved per TRX unit

5.5.1   Network Performance Characterization

The Effective Frequency Load (EFL) quantifies how much of each available frequency can be loaded in the system [9]. It is independent of the frequency reuse and deployed TRX configuration.

When EFL methodology is deployed in the system under evaluation, it is possible to estimate the load associated with any benchmark quality. Maximum network capacity for a given benchmark quality can be calculated, or how much the network quality will degrade as the predicted load increases. EFL is directly proportional to the traffic and inversely proportional to the available bandwidth.

In real networks the traffic load is fixed but the number of frequencies deployed can be modified in order to increase the EFL. During trials the network performance can be characterized by modifying the number of frequencies used.

Figure 5.6 displays the characterization of a sample network. As an example both the BCCH and frequency hopping layers have been characterized to indicate a much differentiated performance. Typically there is hardly any degradation on BCCH layer performance, because BCCH TRX is always transmitting at full nominal power to meet the cell coverage, even if the temporary load decreases or increases. On the other hand, the hopping layer experiences degradation as load increases, as expected. The overall network performance using the allocated frequency

Table 5.5 Data service KPIs

KPI	Abbreviation	Definition
Blocked TBFs (%)	BCR	Percentage of the blocked TBFs
Dropped TBFs (%)	DCR	Percentage of the dropped TBFs
Throughput (kbps)	TP	Mean gross or net session data throughput
Delay (s/kbit)		Mean gross or net normalized LLC frame delay

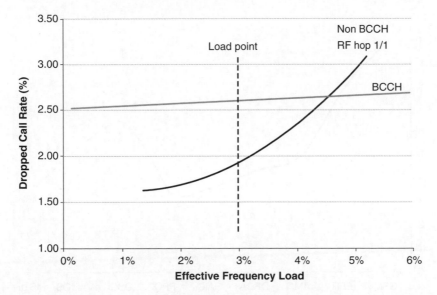

Figure 5.6 Characterization of network performance. Reproduced from Timo Halonen, Javier Romero, Juan Melero, *GSM, GPRS and EDGE Performance Evolution Towards 3G/UMTS, Second Edition*, John Wiley & Sons, Ltd.

resource will be the combination of both layers weighted by the traffic distribution per layer. Different ELF points in the 1/1 hopping curve mean that each EFL point represents a different hopping configuration, while in the BCCH layer the different traffic load operating points are mainly due to different channel allocation strategies.

5.5.1.1 Network Capacity

GSM voice spectral efficiency is used to quantify the number of voice users a network (or site or cell) is able to carry for a given spectrum. A spectrum is a limited resource for GSM, where a large number of subscribers continuously load the network. Moreover, in many countries portions of the same spectrum are allocated to new systems such as UMTS900 and E-UTRAN/LTE. Re-farming of GSM spectrum to new systems has started and therefore it would be beneficial to optimize GSM to use a narrower spectrum with the same target quality and level of service. GSM voice performance is a very important asset to maintain since most of the subscriber voice traffic will be carried in the GSM networks.

In this book, network capacity is measured using the EFL formula (Equation 5.1) [1]:

$$EFL\,(\%) = \frac{Erl_{BH}}{Tot\#\,freq} \frac{1}{Ave\#\left(\frac{TSL}{TRX}\right)} \tag{5.1}$$

where

ErlBH = Busy Hour carried traffic (Erlangs)
Tot#freq = Total number of available frequencies in the system
Ave#(TSL/TRX) = Average number of TSLs per TRX available for traffic.

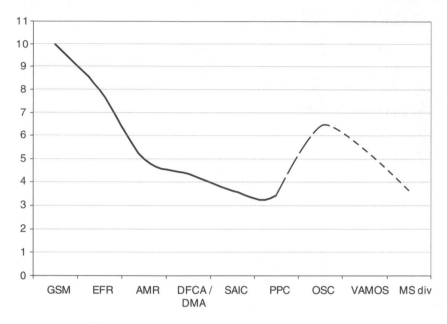

Figure 5.7 Relative frequency reuse for hopping layer.

EFL can be translated into Erl/MHz/sector using the following formula (Equation 5.2):

$$Spectral\_efficiency\,(Erl/MHz/sector = 40^*EFL\,(\%) \qquad (5.2)$$

where

   Ave#(TSL/TRX) has been considered to be 8 time slots.

EFL can easily be calculated from the system-level simulation or from the network field trial. EFL provides a direct measure of how loaded the frequency resources are. In this book, EFL has been used as a measure of voice traffic load and Spectral Efficiency.

   One way to improve the network SE is to use a tighter frequency reuse factor. As Figure 5.7 shows, the GSM network utilizing AMR voice and enhanced radio resource management algorithms can be operated at tighter frequency reuses. Tighter frequency reuse factor enables more TRXs per cell when the available spectrum is limited.

   Starting from the introduction of the initial GSM technology, the relative frequency reuse has been reduced three times to a lower number by applying standardized features and supporting proprietary features in the network. As shown in Figure 5.7, GSM voice evolution OSC/VAMOS allows usage of looser frequency reuses since the hardware efficiency is doubled, for example the number of voice channels per TRX is doubled. Even further improvements would be possible utilizing GSM Evolution features standardized in 3GPP Release 7. For example, mobile station diversity with OSC/VAMOS could be leading GSM networks to even higher spectrum efficiency and tighter frequency reuses.

5.6   Further Reading

There are several documents on the 3GPP standards, which support not only the performance requirements but also the network level design and technology deployment. Documents closely related to link performance engineering are given in the 3GPP TR 45 "Radio Aspects" series, which is useful reading when working with the radio performance engineering details. For system-level information, the 3GPP TR 43.030 "Technical Realization" series contains several documents where the technical deployment scenarios are explained. This is especially useful for radio network planning aspects.

For a detailed background and description of the subjects discussed in this section, the reader is recommended to read [1].

References

[1] T. Halonen, J. Romero and J. Melero, *GSM, GPRS and EDGE Performance*, Chichester: John Wiley & Sons, Ltd, 2002, pp. 162–90.

[2] 3GPP GP-081178, "Working Assumptions for MUROS", Nokia Siemens Networks, TSG GERAN #39, August 2008.

[3] E. Malkamaki, F. de Ryck, C. Mourot and A. Urie, "A Method for Combining Radio Link Simulations and System Simulations for a Slow Frequency Hopped Cellular System," in *Proc. IEEE VTC*, Spring, Stockholm, June 1994.

[4] H. Olofsson, M. Almgren, C. Johansson, M. Höök and F. Kronestedt, "Improved Interface between Link Level and System Level Simulations Applied to GSM," in *IEEE ICUPC*, San Diego, October 1997.

[5] S. Hämäläinen, P. Slanina, M. Hartman, A. Lappeteläinen, H. Holma and O. Salonaho, "A Novel Interface between Link and System Level Simulations", in *Proc. ACTS Mobile Telecommunications Summit*, Aalborg, Denmark, October 1997, pp. 599–604.

[6] J. Winters, "Optimum Combining in Digital Mobile Radio with Cochannel Interference", *IEEE Trans. Veh. Technol.*, 33(3): 144–55, August 1984.

[7] K. Ivanov, C.F. Ball, R. Mullner, H. Winkler, R. Perl and K. Kremnitzer, "GERAN Evolution: Voice Capacity Boosted by Downlink Dual Antenna Interference Cancellation", in *Proc. IEEE PIMRC 2007*, Athens, Sept. 2007.

[8] ETSI TC-SMG, GSM 05.05, European Digital Cellular Telecommunication System (Phase 2) – Radio Transmission and Reception, March 1995.

[9] J. Wigard, P. Mogensen, P.H. Michaelsen, J. Melero and T. Halonen, "Comparison of Networks with Different Frequency Reuses, DTX and Power Control using the Effective Frequency Load Approach", *VTC2000*, Tokyo, Japan.

6

EGPRS2 and Downlink Dual Carrier Performance

Mikko Säily, Kolio Ivanov, Khairul Hasan, Michal Hronec,
Carsten Ball, Robert Müllner, Renato Iida, Hubert Winkler,
Rafael Paiva, Kurt Kremnitzer, Rauli Järvelä, Alexandre Loureiro,
Fernando Tavares and Guillaume Sébire

6.1 Introduction

The GSM/EDGE Radio Access Network (GERAN) is today's backbone of mobile communications, providing worldwide access and roaming for voice and packet data services. Current deployments of GPRS and especially the widespread rollout of EDGE in existing GPRS networks have opened the door to worldwide mobile internet services [1]. Commercially available mobile stations with a typical downlink multi-slot capability of 4 to 5 time slots support application peak data rates of up to 75–90 kbps with GPRS and 225–275 kbps with EDGE [2]. The aim of the upcoming GERAN data evolution features is to significantly increase capacity and spectrum efficiency along with a boost in user throughput and significantly reduced overall latency [3].

The principal GERAN data evolution feature is the EGPRS2 concept, a comprehensive feature package including the introduction of higher order modulation schemes (such as QPSK, 16-QAM and 32-QAM) along with 20% increased symbol rate, spectrally wide transmit pulse shaping filter and turbo coding. Other essential features are mobile station receive diversity, reduced latency by improved multiplexing schemes and fast Ack/Nack reporting [4, 5]. A detailed study of the EGPRS2 performance deployed in the GSM uplink can be found in [6].

3GPP Release 7 also introduced Downlink Dual Carrier (DLDC), where the user's signal bandwidth can be increased on the downlink, potentially increasing the throughput twofold or even more. EGPRS2 in combination with DLDC, in particular, will break through the current 4–5 TS MS limit to up theoretical 16 time slots per connection, and open the possibility of four times higher data rates compared to conventional single carrier EDGE and resulting data

GSM/EDGE: Evolution and Performance   Edited by Mikko Säily, Guillaume Sébire and Eddie Riddington
© 2011 John Wiley & Sons, Ltd

rates up to 1.89 Mbps in downlink. GSM evolution is essential for data service continuity with HSPA and LTE, since today 68% of EDGE networks and the majority of EDGE devices support HSPA.

This chapter is organized as follows. Section 6.2 briefly discusses the GSM data performance evolution, and the performance evaluation of the new EGPRS2 physical layer is presented in Section 6.3. System level performance evaluations for data services are discussed for EGPRS2 and Downlink Dual Carrier in Sections 6.4 and 6.5 respectively. Dual Transfer Mode for simultaneous voice and data traffic is studied in Section 6.6. Finally, Section 6.7 gives a summary of the data performance enhancements for GSM Evolution.

6.2   Overview of GSM Data Performance Evolution

General packet radio service (GPRS) introduced the convenience of "always online" direct internet connections adding packet switching to GSM with a packet-based air interface on top of the current circuit switched mode of operation. These standards and their evolution have enabled the first sophisticated end-user services of mobile internet connectivity and personal multimedia.

In 3GPP Release 99, Enhanced Data rates for GSM Evolution (EDGE) was introduced to boost network capacity and the data rates of GPRS threefold. EDGE is especially attractive to GSM 900, GSM 1800 and GSM 1900 operators that do not have a license for UMTS, but still wish to offer competitive personal multimedia applications utilizing the existing band allocation. EDGE was designed to be backward-compatible with GSM with the support of GMSK modulation while allowing improved data transmission rates and a data throughput peak rate over 400 kbps per carrier. EDGE networks deployments started in 2000–01. EDGE has become a huge success and has outgrown the number of fixed internet connections, which typically use local area network connections and DSL. Today EDGE is supported by many chip vendors and all major mobile vendors.

In 3GPP Release 7, GSM data evolution also called EDGE Evolution, Evolved EDGE or EDGE II, is an upgraded version of GSM technology that was ratified by the 3GPP. Release 7 enhances GSM data performance through many techniques, which can be deployed in the network independently or as a combination of features. These new features apply some of the techniques employed also in HSPA and LTE to increase the data throughput while lowering the transmission latency.

In Release 7, a key part of the evolution is the utilization of two radio frequency carriers, a technique called Downlink Dual-Carrier (DLDC). This is designed to overcome the inherent limitation of the narrow channel bandwidth of a single GSM carrier. DLDC enables higher effective channel bandwidth for user data. These carriers can be allocated independently along with the existing frequency hopping carrier allocation, thus allowing compatibility with network frequency plans and fast deployments in existing networks. For example, DLDC with ten time slots per connection enables peak rates above 1 Mbit/s and typical average data rates of 400 kbps or more can be expected.

Another important step for GSM data evolution in the 3GPP Release 7 was the introduction of EGPRS2, which specified higher symbol rate (HSR) and higher order modulations for the downlink and uplink directions. Higher symbol rate with spectrally wide transmit pulse shaping filter means higher bandwidth, which is now increased to 325 kHz exceeding the normal

270 kHz symbol rate by 20%. Higher symbol rate is combined with three new modulations QPSK, 16-QAM and 32-QAM. As a result, EGPRS2 provides 118.4 kbps peak rates per time slot, which is double compared to EDGE maximum data rate of 59.2 kbps per time slot. When combining the DLDC with EGPRS2-B, up to 16 time slots can be used with a higher symbol rate, thus providing the GSM data evolution maximum peak rate of 1.89 Mbps.

GSM data evolution brings not only improvements in the highest peak rate but also in the average bit rates which is especially beneficial to users who do not have access to the latest technical steps of LTE or HSPA due to lack of network coverage or lack of support in the user equipment. The launch of GSM data evolution means a dramatic increase in average data rates. All the user equipment that is capable of GSM Rel-7 evolution services is also backwards compatible with the EDGE systems. In many cases, GSM data evolution can be implemented as a software upgrade of infrastructure hardware in existing networks.

The transferred volume of wireless data is growing fast and is expected to reach 100 times the data transmitted today during the next stages of evolution in the wireless systems. Technologies such as WLAN, HSPA and LTE increase the usage of data services and this will increase the volume of GSM data transmission as well. Therefore the spectrum resource allocated to GSM should be maintained and continuously improved.

6.3 EGPRS2 Link Performance

6.3.1 Overview

In this section the performance of radio link is studied in downlink and uplink. Performance is usually characterized in sensitivity and interference-limited conditions to reflect the nature of a cellular network with limited coverage and spectrum resource. For data services the performance is usually evaluated in terms of throughput per time slot of connection. In many cases the throughput gain is based on the link-level improvement, such as EGPRS2, where both the symbol rate and modulation have been improved to double the data throughput per time slot compared to EGPRS.

GERAN data services consist of a layered protocol structure, and the EGPRS2 concept has an impact only to the radio interface protocols below the Logical Link Control (LLC). EGPRS2 can seamlessly co-exist with other GSM data. Therefore, the EGPRS2 concept does not require any new base station sites or new spectrum, and has no impact on the existing cell or frequency plans. The EGPRS2 services come with two Network and Mobile Station (NW/MS) support levels, namely Level A and Level B (which may be deployed depending on network capabilities) and two transmit pulse shaping options for the uplink are defined for EGPRS2 support Level B, where either the Linearized GMSK pulse or a new spectrally wide pulse shape can be selected by the network.

6.3.2 Simulation Models

The performance of the EGPRS2 concept has been studied by means of classic link-level simulations for both downlink and uplink, and for both support levels A and B, using the available transmit pulse shaping options. The link simulation chain used for this performance evaluation was discussed in Chapter 5. This is a one-way link between one transmitter and one receiver. For this performance study, the block error rate (BLER), the bit error rate before channel decoding (RawBER) and the throughput per time slot are evaluated.

In downlink, a single antenna mobile station is assumed while in the uplink, a dual antenna base station employing interference cancellation is assumed (antenna diversity is used in the vast majority of base stations). Channel coding is implemented according to 3GPP Release 7. The multi-path radio propagation channel is Typical Urban channel model, with a mobile speed of 3 km/h with ideal frequency hopping, which is also reflected in the naming convention of the simulation scenarios, for example TU3 ideal FH.

Frequency hopping has been either disabled, or assumed to be ideal. Ideal frequency hopping means that tne radio channel does not have a memory and consecutive radio bursts do not correlate.

Noise- and interference-limited scenarios are simulated with Additive White Gaussian Noise and single co-channel interference (C/I) scenario respectively.

Table 6.1 gives an overview of the modulation and coding schemes available in GERAN for packet switched data transmission along with the RLC maximum data rate per time slot.

Table 6.1 also shows the code rate used in forward error correction (FEC) for different EGPRS and EGPRS2 modulation and coding schemes. The code rate is defined as the ratio between number of bits given as the input to the channel encoder to the number of bits coming out of the encoding and puncturing or rate matching process. The encoding process involves addition of systematic redundancy bits and, therefore, the lower value of FEC code rate indicates more redundant bits, stronger coding and better protection against impairments in radio channel. A value of 1.0 implies no coding gain.

EGPRS service offers a peak throughput of 59.2 kbps (MCS-9), while EGPRS2 offers 118.4 kbps – double that of EGPRS. EGPRS2 level A, however, offers peak throughput of 98.4 kbps in the downlink (DAS-12) and 76.8 kbps in the uplink (UAS-11). Owing to the weak channel coding, these coding schemes (i.e. MCS-9, DAS-12, UAS-11, DBS-12, and UBS-12) can achieve peak throughputs only in good radio conditions. However, EGPRS2 offers significantly better throughput than EGPRS even in moderate and poor channel conditions. This is explained further in following sub-sections.

6.3.3   EGPRS Reference Link Performance

EGPRS performance has been evaluated as a reference performance level for EGPRS2. Figures 6.1 to 6.4 provide the link-level performance of EGPRS coding schemes MCS-1 to MCS-9. Some of these coding schemes are also reused in EGPRS2 service, and hence the link performances of these schemes are the same between EGPRS and EGPRS2. The plots show BLER and RawBER performance at different input signal levels in the case of noise limited scenario (sensitivity) and at different C/I ratios in the case of an interference-limited scenario.

The reference RawBER performance of 8PSK modulation scheme is clearly worse than that of GMSK modulation scheme, indicating the expected coverage with GMSK modulated coding schemes should be the best of the EGPRS coding schemes. However, owing to strong channel coding used in MCS-5 and MCS-6 (both 8PSK), the performance of these channels is better than MCS-4 and sometimes than MCS-3 which have relatively weak channel coding.

Table 6.1 GPRS, EGPRS and EGPRS2 modulation and coding schemes and RLC user data rates

Service	Coding scheme	Modulation scheme	RLC blocks per radio block	FEC code rate	RLC Data rate per time slot [kbps]
GPRS	CS-1	GMSK	1	0.45	8.0
	CS-2		1	0.65	12.0
	CS-3		1	0.75	14.4
	CS-4		1	n/a	20.0
EGPRS, EGPRS2	MCS-1	GMSK	1	0.51	8.8
	MCS-2		1	0.64	11.2
	MCS-3		1	0.83	14.8
	MCS-4		1	0.98	17.6
EGPRS, *EGPRS2-A uplink	*MCS-5	8-PSK	1	0.37	22.4
	*MCS-6		1	0.49	29.6
	MCS-7		2	0.75	44.8
	MCS-8		2	0.91	54.4
	MCS-9		2	0.99	59.2
EGPRS2-A downlink	DAS-5	8-PSK	1	0.37	22.4
	DAS-6		1	0.45	27.2
	DAS-7		1	0.54	32.8
	DAS-8	16-QAM	2	0.56	44.8
	DAS-9		2	0.67	54.4
	DAS-10	32-QAM	2	0.63	65.6
	DAS-11		3	0.80	81.6
	DAS-12		3	0.96	98.4
EGPRS2-B downlink	DBS-5	QPSK	1	0.48	22.4
	DBS-6		1	0.63	29.6
	DBS-7	16-QAM	2	0.46	44.8
	DBS-8		2	0.61	59.2
	DBS-9		3	0.70	67.2
	DBS-10	32-QAM	3	0.73	88.8
	DBS-11		4	0.90	108.8
	DBS-12		4	0.98	118.4
EGPRS2-A uplink	UAS-7	16-QAM	2	0.54	44.8
	UAS-8		2	0.61	51.2
	UAS-9		2	0.71	59.2
	UAS-10		3	0.83	67.2
	UAS-11		3	0.94	76.8
EGPRS2-B uplink	UBS-5	QPSK	1	0.47	22.4
	UBS-6		1	0.61	29.6
	UBS-7	16-QAM	2	0.45	44.8
	UBS-8		2	0.59	59.2
	UBS-9		3	0.69	67.2
	UBS-10	32-QAM	3	0.70	88.8
	UBS-11		4	0.88	108.8
	UBS-12		4	0.95	118.4

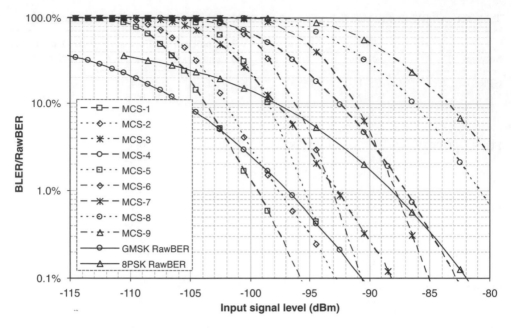

Figure 6.1   EGPRS downlink performance (sensitivity, TU3 ideal FH).

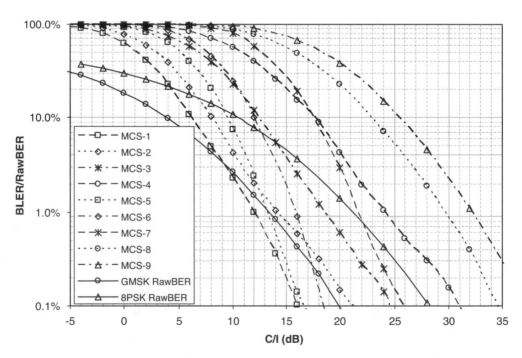

Figure 6.2   EGPRS downlink performance (co-channel interference, TU3 ideal FH).

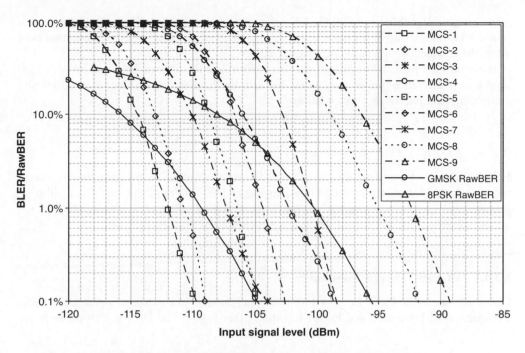

Figure 6.3 EGPRS uplink performance (sensitivity TU3 ideal FH).

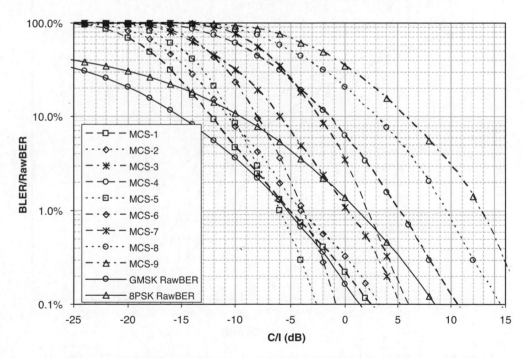

Figure 6.4 EGPRS uplink performance (co-channel interference, TU3 ideal FH).

At a higher signal level or C/I, MCS-7 even outperforms MCS-4. This clearly shows that mobiles not supporting 8PSK modulation will not achieve expected EGPRS throughput even when the channel conditions are below average.

It is worth noting that the link performance is significantly different between downlink and uplink in interference-limited conditions, because the base station receiver uses a diversity combining with interference cancellation.

6.3.4   EGPRS2 Link Performance

EGPRS2 downlink performance was evaluated using the same link-level simulation, except the concept level details were modified according to EGPRS2 requirements. The results are grouped into Level A and Level B and presented in the following sub-sections. Only BLER and RawBER of different coding schemes are presented as link-level performance.

6.3.4.1   EGPRS2 Level A Downlink

Figure 6.5 shows EGPRS2 Level A link-level performance in co-channel interference case. Comparison of Figure 6.2 and Figure 6.5 shows that DAS-5 and DAS-6 performances are

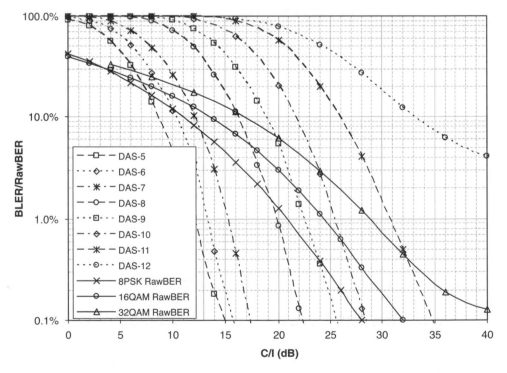

Figure 6.5   EGPRS2 Level A downlink performance (co-channel interference, TU3 ideal FH).

almost equal to MCS-5 and MCS-6 due to the same modulation and almost the same channel coding specification. DAS-7 performs significantly better than MCS-7 despite using the same modulation because the former is protected by the FEC code rate of 0.54 while the latter is protected by the code rate of 0.75.

Weaker coding schemes, that is, DAS-10, 11 and 12, show a reasonable performance only at high C/I levels. DAS-12, in particular, experiences an error floor which indicates that the peak throughput of this coding scheme is difficult to achieve in this propagation scenario.

6.3.4.2   EGPRS2 Level B Downlink

Figures 6.6 and 6.7 show the EGPRS2 Level B downlink performance in an interference-limited scenario. Figure 6.6 shows performance with the Linearized GMSK pulse shape filter in the transmitter. This is the narrow-band legacy filter specified in 3GPP TS 45.004. Figure 6.7 shows the performance when using the spectrally wide pulse shape filter (Chapter 4, WIDER) designed for higher symbol rate transmission and reception in the downlink.

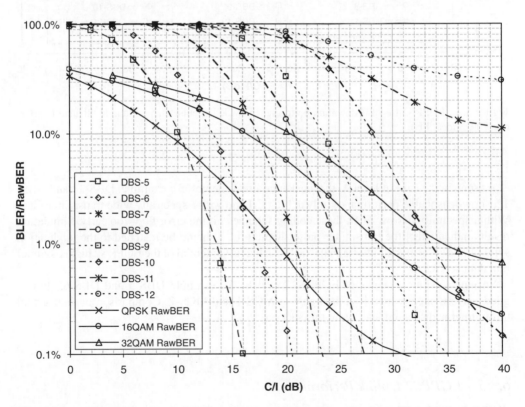

Figure 6.6   EGPRS2 Level B downlink performance (narrow Tx pulse shape filter, co-channel interference, TU3 ideal FH).

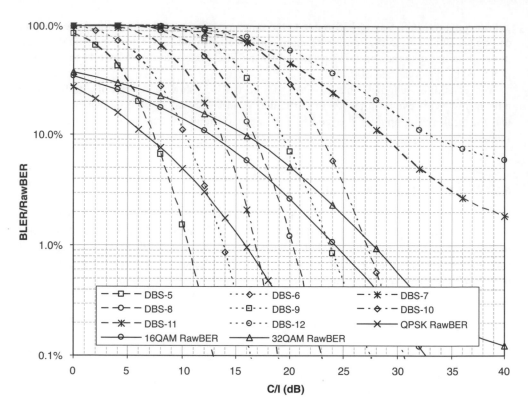

Figure 6.7 EGPRS2 Level B downlink performance (wide Tx pulse shape filter, co-channel interference, TU3 ideal FH).

As seen in the case of level A, the weaker coding schemes, for example, DBS-11 and DBS-12, show reasonable performance only at high C/I and with a spectrally wide transmitter filter. With a spectrally narrow filter, these coding schemes hit the error floor too early. Simulation results indicate that the transmitter pulse shaping filter should be matched to the modulation symbol rate of the transmitted signal to achieve the full potential of the EGPRS2 higher symbol rate data service.

Stronger coding schemes of EGPRS2 Level B, for example DBS-5 and DBS-6, show a good performance, which is comparable to MCS-1 and MCS-2. Comparison of their actual throughput is discussed later on in this section.

6.3.5 EGPRS2 Uplink Performance

EGPRS2 uplink performance was also evaluated using link-level simulation. The results are grouped into Level A and Level B and presented in the following sub-sections. Only BLER and RawBER of different coding schemes are presented as link-level performance.

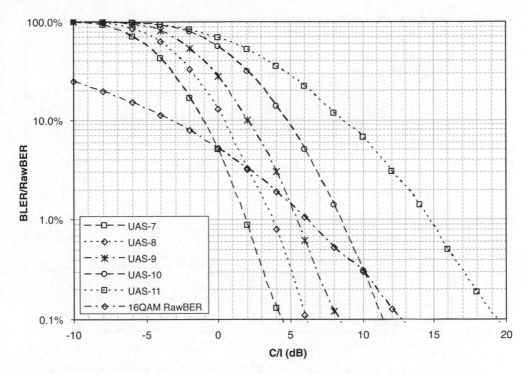

Figure 6.8   EGPRS2 Level A uplink performance (co-channel interference, TU3 ideal FH).

6.3.5.1   EGPRS2 Level A Uplink

In the uplink, EGPRS2 Level A supports MCS-1 to MCS-6 using legacy modulation schemes and UAS-7 to UAS-11 using the new modulation scheme, namely 16-QAM. Figure 6.8 shows the link-level performance of only UAS-7 to UAS-11 coding schemes in co-channel interference scenario. Earlier sub-section 6.3.3 on the EGPRS reference link performance has already profiled MCS-1 to 6 performance.

When comparing Figure 6.4 and Figure 6.8, it is clear that UAS coding schemes start to contribute to the link performance where MCS-1 to 6 coding schemes reach the limit of their performance.

6.3.5.2   EGPRS2 Level B Uplink

EGPRS2 Level B performance for uplink was evaluated using link-level simulation in co-channel interference scenario for both spectrally narrow and wide pulse transmitter shaping filters. Figure 6.9 shows the performance with the linearized GMSK pulse shape filter in the transmitter. This is the same pulse shaping filter as used in EGPRS, and is also referred to as the spectrally narrow pulse shape. Its performance evaluation indicates about the same limitations as in the downlink when using LGMSK filter, which is not matched to the higher symbol rate.

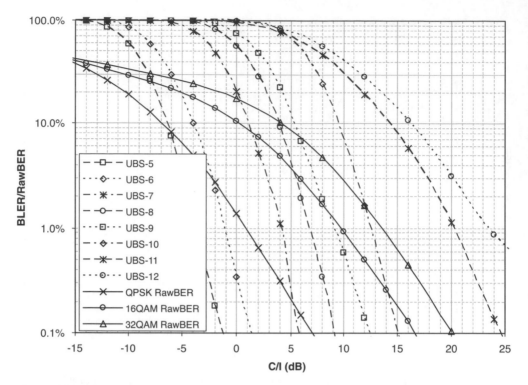

Figure 6.9 EGPRS2 Level B uplink performance (narrow Tx pulse shaping filter, co-channel interference, TU3 ideal FH).

Figure 6.10, on the other hand, shows the link performance with the spectrally wide pulse shape filter as specified in 3GPP TS 45.004 for uplink only. The weaker coding schemes in the uplink, for example, UBS-11 and UBS-12, show quite good performance at high C/I and it is possible to achieve a performance quite close to the peak throughput in realistic network conditions. The robust coding schemes perform well in average network conditions and show doubled throughput throughout the whole practical C/I range.

6.3.6   EGPRS2 Pulse Shaping Filters

In 3GPP TS 45.004, two transmit pulse shaping filter types are specified for EGPRS2 uplink: (1) a spectrally wide pulse, optimized for throughput performance at the higher symbol rate; and (2) the existing linearized GMSK pulse, for scenarios where the wide pulse shape might not be suitable. For example, the narrow pulse can be used in scenarios where the edge channels in a network operator's bandwidth allocation are next to another network allocation and there is no guard channel, or EGPRS2 data layer interferes with another service layer of the same operator's network.

Figures 6.11 a) and b) show a performance comparison of the wide and narrow Tx pulse shape filters in uplink in sensitivity and co-channel interference scenarios. Performance is

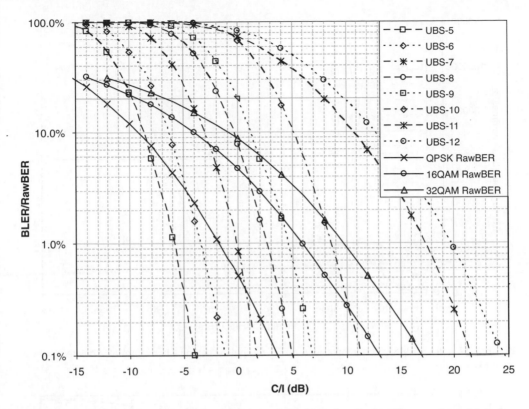

Figure 6.10   EGPRS2 Level B uplink performance (wide Tx pulse shaping filter, co-channel interference, TU3 ideal FH).

given as throughput for selected coding schemes at different input signal levels (noise limited) and C/I levels (co-channel interference). Sensitivity performance is important for coverage and Figure 6.11(a) shows about 5 dB loss due to narrow pulse shaping for the high bit rate coding schemes. At a moderate input level, the wide pulse provides a throughput improvement of about 40 kbps compared to the narrow pulse and more than double the throughput at high signal levels.

Interference-limited networks are co-channel limited in most cases, thus a higher adjacent channel interference by a wider transmit spectrum is partly compensated for by the lower co-channel interference from the pulse shape. At the network level, the expected shorter transmit time will further reduce the impact of higher adjacent channel interference. Figure 6.11(b) shows that the narrow Tx filter will result in about 4 dB loss in the case of UBS-12 in an interference-limited scenario.

No decision had been made at the time of writing about the specification of a spectrally wide pulse shape for the downlink, which is currently being evaluated as part of a study in 3GPP [7]. Figure 6.12 shows that, with the narrow Tx filter, the level B peak throughput (contributed by DBS-11) is similar to the level A peak throughput (contributed by DAS-12). Note that, due to high level of block errors, the throughput of DBS-12 with the narrow Tx filter is lower than that of DBS-11. This clearly underlines the importance of wide pulse shape filter in downlink.

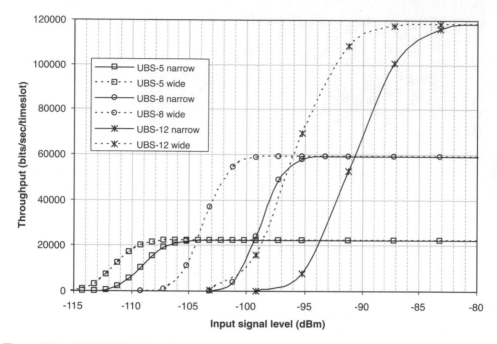

Figure 6.11 (a) EGPRS2 Throughput with optimized pulse shaping filter in uplink: sensitivity, TU3 ideal FH.

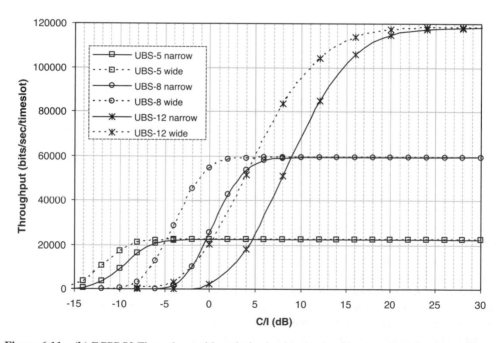

Figure 6.11 (b) EGPRS2 Throughput with optimized pulse shaping filter in uplink: Co-channel interference, TU3 ideal FH.

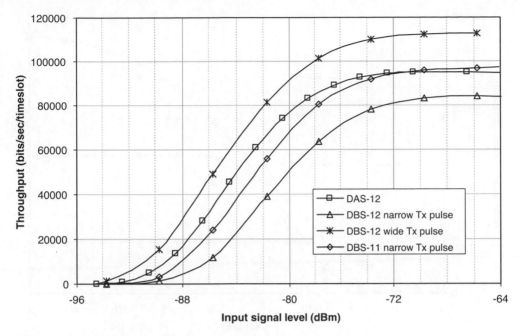

Figure 6.12   EGPRS2 Level B throughput performance of wide and narrow pulse shaping filter in downlink compared to Level A.

6.3.7   Incremental Redundancy in EGPRS2

Incremental redundancy is a crucial feature in EGPRS and EGPRS2 to improve the data throughput with retransmissions. Each coding scheme in EGPRS2 has multiple puncturing patterns, which support the efficient combining of soft bits from repeated transmissions of same user data. Each retransmission adds some time diversity and increases the coded robustness of the user signal.

Figure 6.13 shows an example of the incremental redundancy performance in EGPRS2-B downlink. The throughput improvement is mainly seen in cases of moderate or poor radio conditions and in weaker coding schemes (e.g., DBS-12). In stronger coding schemes (e.g. DBS-5) or in very good channel conditions, the delay associated with retransmissions offsets the gain achieved by improved channel decoding.

6.3.8   Effect of Frequency Hopping

Frequency hopping has a mixed impact on link-level performance as seen in Figure 6.14 for EGPRS2-B and Figure 6.15 for EGPRS. There is a gain from hopping for the stronger coding schemes in the low C/I regions. In other cases, frequency hopping is giving no gain or degrades performance. This behavior is expected in EGPRS and same conclusions apply to EGPRS2 for the equivalent code rates. For the high bit rate codecs with least robust coding, the increased variance of the radio channel makes errors more probable, while channel coding is insufficient to correct them.

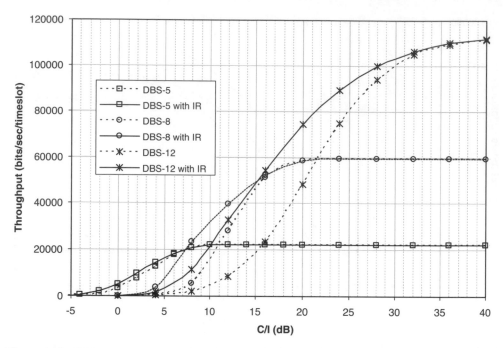

Figure 6.13 EGPRS2 Level B incremental redundancy in downlink (wide Tx pulse, TU3 ideal FH, co-channel interference).

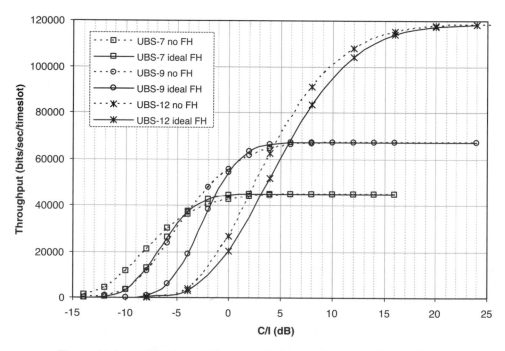

Figure 6.14 EGPRS2 Level B frequency hopping performance in uplink (TU3).

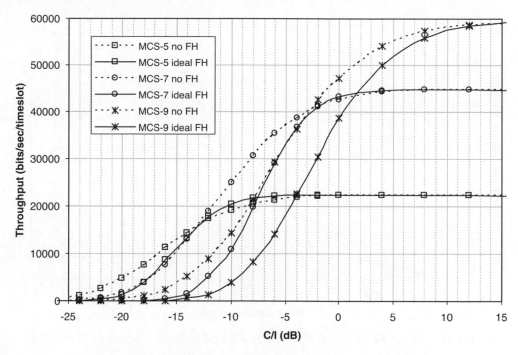

Figure 6.15 EGPRS frequency hopping performance in uplink (TU3).

From the performance point of view, EGPRS2-B with frequency hopping enabled can provide better or equal C/I limited performance throughout the whole practical C/I range compared to EGPRS. At 15 dB C/I where EGPRS achieves its peak rate of 59.2 kbps, EGPRS2-B throughput is 115 kbps, that is close to the peak throughput of 118.4 kbps.

6.3.9 EGPRS vs. EGPRS2 RLC Throughput

Figures 6.16 and 6.17 show the comparison of overall throughput of EGPRS and EGPRS2. These plots represent the throughput envelope vs. input level of all the coding schemes of respective services in EGPRS, EGPRS2-A and EGPRS2-B. These envelopes represent the ideal RLC throughput of each service. For this analysis, EGPRS service takes into account MCS-1 to MCS-9 while EGPRS2-A and EGPRS2-B consider only the new modulation and coding schemes. It can be seen that the EGPRS2 service offers significant throughput gains over EGPRS. At the very low signal levels (e.g. less than −104 dBm in downlink and less than −109 dBm in uplink), EGPRS2-B is equal to EGPRS because GMSK modulation has been specified as part of EGPRS2.

EGPRS2 Level A is not able to provide significantly better throughput than EGPRS unless the radio conditions are fairly good. In other words, throughput improvement by EGPRS2 Level A compared to EGPRS service cannot be sustained over the whole cell coverage area.

EGPRS2 Level B service provides better coverage and throughput compared to EGPRS throughout the whole practical signal level range. At high signal levels, EGPRS2 Level B

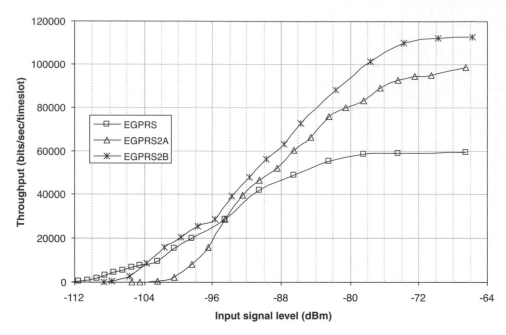

Figure 6.16 EGPRS, EGPRS2-A and EGPRS2-B/WB RLC throughput comparison in downlink.

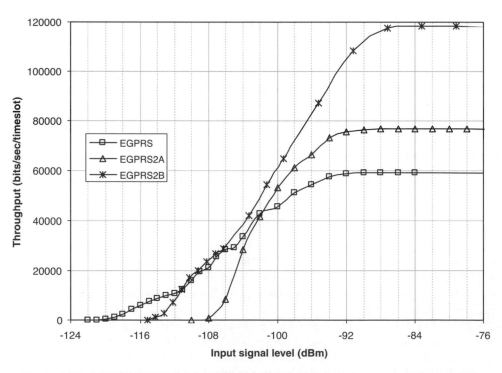

Figure 6.17 EGPRS, EGPRS2-A and EGPRS2-B/WB RLC throughput comparison in uplink.

provides almost twice the throughput of EGPRS, thanks to higher symbol rate, higher order modulation and the wide transmit pulse shaping filter.

6.3.10   Conclusion

This chapter presents the GSM/EDGE data evolution concept EGPRS2, which has been standardized in 3GPP Release 7. The key aspects of the link-level transmission techniques were described, and link-level performance results were provided for EGPRS and EGPRS2. It was shown that EGPRS2 level A outperforms EGPRS with a 30% higher data peak rate. However, EGPRS2-A does not improve the coverage of the packet data services. With EGPRS2-B, the peak rate is 100% higher and the link simulations with the wide transmit pulse showed about 15 kbps per time slot better data rates in average network conditions which increased to twice the peak rate in good conditions. EGPRS2-B codecs are able to improve the packet data service throughout the whole cell, whereas EGPRS2-A has some improvements at good network conditions but does not provide gain, when the network conditions are below average.

The use of different traffic layers to deploy voice and data services can decrease the overall impact to legacy services [6]. Reduced transmission times associated with the higher throughputs of EGPRS2-B can compensate for the increased ACI power, in interference-limited networks.

6.4   EGPRS2 System Performance

In this section, EGPRS2 performance is evaluated and discussed based on results obtained from a dynamic system-level network simulator. First, the simulation assumptions are described, such as scenarios, network layout and set-up, simulated services, RRM algorithms and the key performance indicators. Finally, the simulation results are presented and the conclusions are drawn.

6.4.1   Simulation Assumptions

EGPRS2 system performance evaluations were performed to compare the performance of EGPRS, EGPRS2 level A and EGPRS2 level B (using wide pulse, for example WIDER, Chapter 4, Data Evolution) by running data-only simulations. Also the aim was to see the difference in connection throughput and evaluate the impact to speech users of EGPRS2 level B with different pulse shapes. For these simulations, both data and speech connections were simulated simultaneously within the same network.

6.4.1.1   Scenarios

The simulations were performed in a homogeneous hexagonal grid network containing 75 cells. Three cells were grouped in a site. Time slots were synchronized in the whole network, TDMA frames were aligned within a site and randomly aligned between different sites.

Table 6.2 Common scenario parameters

Parameter	Value	Comment
Frequency band	900 MHz	
Sector antenna pattern	65° beam width	
Propagation model	UMTS 30.03	
BCCH layer	4/12 frequency reuse	
MS speed	3 km/h	
Max BTS output power	20 W (~43 dBm)	
Max MS output power	2 W (~33 dBm)	
Power control	uplink only	
Incremental redundancy	disabled	
Link adaptation	enabled	MCS adaptation

For the data-only simulations three scenarios were defined: two interference-limited scenarios with different resources and frequency hopping arrangement (A, B) and one noise-limited non-hopping scenario (C).

For the pulse shape comparison, the scenarios specified in the 3GPP study into the spectrally wide pulse shapes [7] (referred to as the WIDER study) were utilized. In this case the WIDER-1 scenario was split into two separate scenarios to investigate the performance in BCCH and TCH layers independently. The scenario parameters are summarized in Tables 6.2 to 6.6.

6.4.1.2   RRM Description

In scenarios A and C, there was only one BCCH transceiver per BTS, with the first time slot reserved for BCCH and the remaining time slots dedicated to packet data traffic, as shown in Figure 6.18.

In scenario B, there were three transceivers per BTS: one BCCH and two hopping. The BCCH transceiver was blocked for traffic and all the data traffic was carried by the two hopping transceivers, see Figure 6.19.

The WIDER scenarios had the time slot resources configured according to Figure 6.20.

Power control was used in uplink direction only. In downlink, full transmit power was used with a back-off which depended on the current burst's modulation. The back-off values are listed in Table 6.7.

All packet data services used modulation and coding scheme (MCS) adaptation, where the adaptation algorithm selected the MCS providing maximum throughput, based on past link

Table 6.3 Scenario A

Parameter	Value	Comment
Cell radius	500 m	
Simulated bandwidth	2.4 MHz	1 BCCH TRX per BTS
BCCH layer	4/12 reuse, no hopping	

Table 6.4 Scenario B

Parameter	Value	Comment
Cell radius	500 m	
Simulated bandwidth	4.4 MHz	
BCCH layer	4/12 reuse, no hopping	2.4 MHz + 0.2 MHz guard band
hopping layer	1/1 random RF hopping	1.8 MHz, 9 ARFN in MA list

Table 6.5 Scenario C

Parameter	Value	Comment
Cell radius	4 km	
Simulated bandwidth	2.4 MHz	1 BCCH TRX per BTS
BCCH layer	4/12 reuse, no hopping	

Table 6.6 WIDER scenarios

Parameter	WIDER-1 BCCH	WIDER-1 TCH	WIDER-2	WIDER-3
Frequency band	900 MHz	900 MHz	900 MHz	900 MHz
Cell radius	500 m	500 m	500 m	500 m
Bandwidth	2.4 MHz	4.4 MHz	11.6 MHz	8.0 MHz
Guard band	0 MHz	0.2 MHz	0.2 MHz	0.2 MHz
Number of channels (excl guard band)	12	21	57	39
Number of TRX	1	3	6	4
BCCH frequency reuse	4/12	4/12	4/12	4/12
TCH frequency reuse	n/a	1/1	3/9	3/9
Frequency hopping	none	synthesized	baseband	baseband
Length of MA	0	9	5	4[1]
BCCH or TCH under interest	BCCH	TCH	BCCH and TCH	BCCH and TCH

[1] includes BCCH carrier

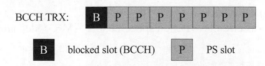

BCCH TRX: | B | P | P | P | P | P | P | P |

B blocked slot (BCCH)   P PS slot

Figure 6.18 Time slot resources in scenarios A and C.

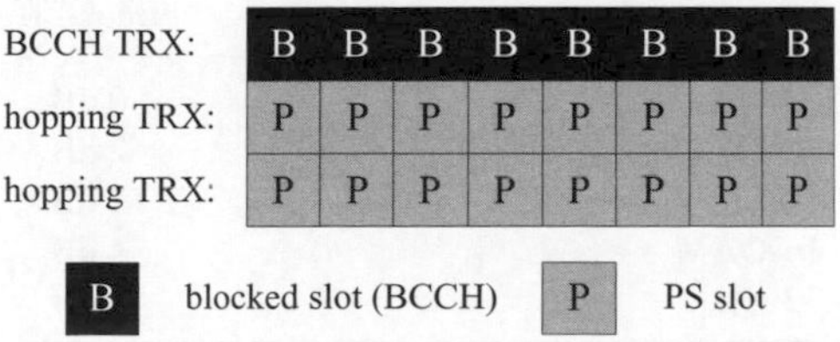

BCCH TRX: | B | B | B | B | B | B | B | B |
hopping TRX: | P | P | P | P | P | P | P | P |
hopping TRX: | P | P | P | P | P | P | P | P |

B blocked slot (BCCH)   P PS slot

Figure 6.19 Time slot resources in scenario B.

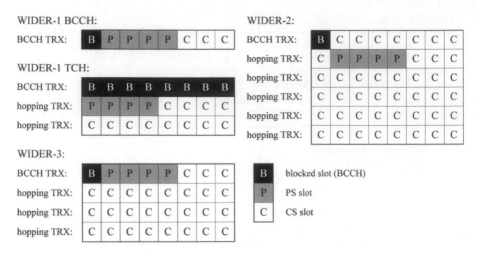

Figure 6.20 Time slot resources in WIDER scenarios.

quality measurements. The terminals were allowed to use all MCS corresponding to EGPRS, EGPRS-2 A or EGPRS-2 B respectively. Incremental redundancy was not used.

6.4.1.3 Simulated Services

Data connections consisted of FTP single file downloads using terminals with EGPRS, EGPRS2-A and EGPRS2-B with narrow and spectrally wide transmit pulse shaping filter support (LGMSK pulse and pulse candidates 2 and 3 from WIDER). Mobile terminals supported up to four time slots in downlink. The traffic generation model was very simple: one block of data in one direction representing one file. There was neither FTP control signaling nor TCP signaling modeled.

This service represented the worst case for measuring throughput under various loads. The highest load point was when the network reached saturation, that is when the rate of newly arriving data calls roughly equaled the rate of ending calls. The observed performance indicators are summarized in Section 6.4.1.4, connection throughput being the most important performance indicator in this case.

The second service was circuit-switched speech using AMR codec adaptation on a full-rate channel.

Table 6.7 Back-off values applied in downlink

Burst modulation	Back-off
GMSK	0 dB
QPSK	2 dB
8-PSK	2 dB
16-QAM	4 dB
32-QAM	4 dB

Table 6.8 FTP service parameters

Parameter	Value	Comment
Traffic generator	FTP	no control signaling
Number of files/call	1	
File size	1 MB	
RLC mode	acknowledged	
Load	network saturation and less	
Penetration of wide pulse usage in EGPRS2-B	100%	
Multislot capability	4 DL + 1 UL	

Parameters of the simulated services are summarized in Tables 6.8 and 6.9.

6.4.1.4   Key Performance Indicators

In the pure data simulations, an FTP service was used by all terminals with performance indicators of mean connection throughput and throughput versus distance from BTS.

Connection throughput was calculated as the amount of transmitted data divided by the time needed for error-free transmission. Mean connection throughput is the arithmetic mean of all simulated connections.

Throughput versus distance from BTS shows the average throughput relative to the BTS to which the terminal is associated. The throughput samples were taken in one second intervals. The indicator demonstrated the level of throughput gain at the cell edge or gains in coverage (i.e. the gain in coverage given a fixed throughput reference such as EGPRS). Cell edge was here defined as distance from BTS, beyond which only 10% of the samples remained.

In addition to the above, a throughput map was collected showing the average throughput versus network location. Also here the sample duration was one second.

For the mixed data and speech simulations, the most important performance indicators were 10th, 50th and 90th-percentile connection throughputs and proportion of speech users experiencing 1% frame erasure rate (FER) or less.

6.4.2   Simulation Results

The simulation results are shown and discussed in separate sections for each simulated service.

Table 6.9 AMR speech service parameters

Parameter	Value	Comment
Traffic generator	speech	
AMR codec modes	12.2, 7.4, 5.90, 4.75 kbit/s	same in DL and UL
Mean call length	50 s	neg. exponential distribution
Load	2% blocking	

6.4.2.1   FTP in Downlink

This section shows results of the FTP service evaluated in the downlink using the scenarios described in Section 6.4.1.1. The FTP service was configured according to the description in Section 6.4.1.3. The FTP service in downlink was used to compare the throughput for EGPRS, EGPRS2-A and EGPRS2-B. For statistical confidence the simulations consist of 10,000 connections per data point in the graph.

Figures 6.21, 6.23 and 6.25 (b) show the mean throughput for EGPRS, EGPRS2 A and B in scenarios A, B, and C respectively. In these figures, a significant throughput gain is seen

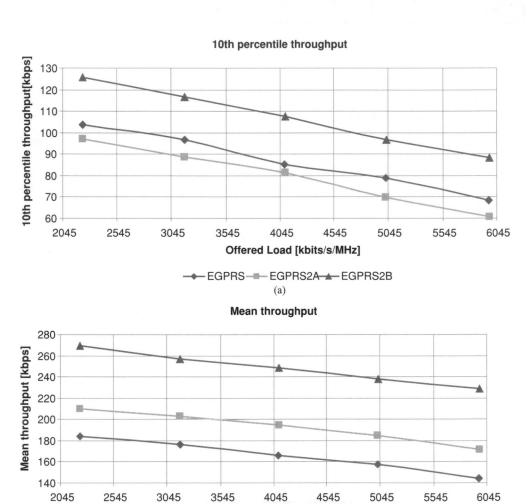

Figure 6.21   Scenario A simulation results (a) 10th percentile throughput; (b) Mean connection throughput.

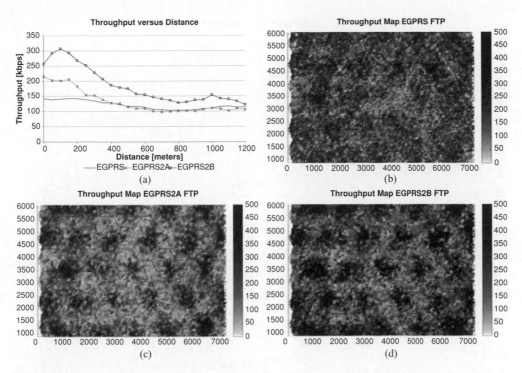

Figure 6.22   Throughput statistics for Scenario A using FTP service in downlink. (a) Throughput versus distance statistics, and throughput map for (b) EGPRS, (c) EGPRS2-A and (d) EGPRS2-B.

with EGPRS2 relative to EGPRS. Positive gains are observed which ranged from 7–55% for EGPRS2-A and 40–82% for EGPRS2-B.

Figures 6.21, 6.23 and 6.25 (a) show the minimum guaranteed throughput achieved in 90% of the samples. In all the figures, a significant gain in the 10th percentile throughput can be seen with EGPRS2-B relative to EGPRS and EGPRS2-A. While EGPRS2-A shows a loss in the 10th percentile throughput for scenario A, in scenario B it shows a significant improvement in throughput compared to EGPRS.

When considering the network level performance results for BCCH layer in Figure 6.22 (a), the throughput with EGPRS2-A and -B increases as the distance reduces. With EGPRS2-A, the throughput decreases at a faster rate with distance than EGPRS2-B.

Figure 6.24 (a) shows a more constant throughput behavior with distance. Figure 6.26 (a) shows throughput over a higher distance range than Figure 6.22 (a) and Figure 6.24 (a).

When comparing scenarios A and C (where the only difference is the higher site-to-site distance in scenario C), it is seen that EGPRS2-B throughput for a higher site-to-site distance is comparable to EGPRS throughput in scenario A. The results indicate that PS spectrum resources can be reduced or eliminated in certain BTSs to save costs without degradation of performance.

Table 6.10 shows the average throughput gain of EGPRS2 (A and B) over EGPRS for different offered load points. FTP throughput indicates that EGPRS2-B network performance

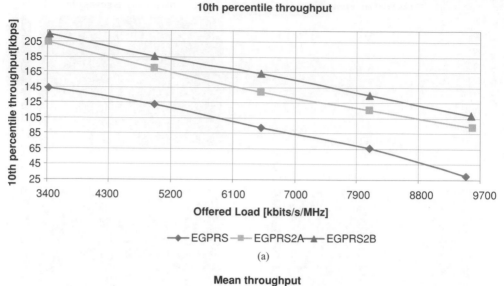

(a)

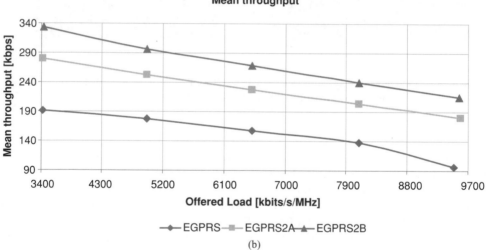

(b)

Figure 6.23 Scenario B simulation results: (a) 10th percentile throughput, (b) Mean connection throughput.

is higher than EGPRS2-A in all scenarios. It can be expected that even higher gains can be obtained from EGPRS2-B with further optimization of the link adaptation algorithm.

6.4.2.2   FTP in Uplink

This section shows results for the FTP service evaluated in the uplink direction using the scenarios described in Section 6.4.1.1. The FTP service was configured with file sizes of 1 MByte using 4 time slots. FTP service in uplink was used to compare the throughput for

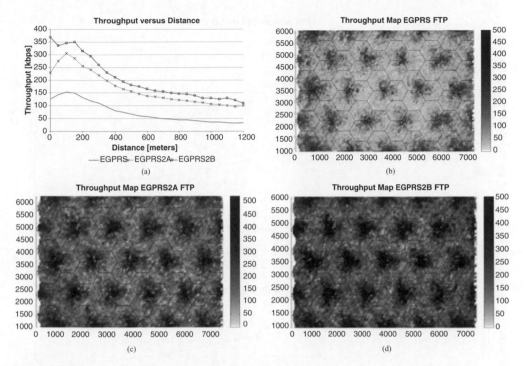

Figure 6.24 Throughput statistics for Scenario B using FTP service in downlink. (a) Throughput versus distance statistics, and throughput map for (b) EGPRS; (c) EGPRS2-A and (d) EGPRS2-B.

EGPRS, EGPRS2-A and EGPRS2-B (Figures 6.27–6.32). For statistical confidence, simulations consist of 10,000 connections per data point in the graph.

Table 6.11 shows the average throughput gain of EGPRS2 (A and B) over EGPRS at different offered load points. It can be seen that the EGPRS2-B performance is higher than EGPRS2-A in all scenarios. Note that even higher gains can be expected for EGPRS2-A and -B with further optimizations of the link adaptation algorithm.

6.4.2.3 FTP Uplink and Downlink Comparison

By comparing Figure 6.22 (a) with Figure 6.28 (a) and Figure 6.24 (a) with Figure 6.30 (a), it can be seen that throughput increases in the uplink with distance from the BTS. This behavior is explained by the IRC receiver and power control in uplink.

Figure 6.33 shows the CIR over distance for Scenario A and B. The performance gain of the uplink compared to the downlink is evident in both scenarios.

6.4.2.4 Optimized Pulse Impact in Mixed Voice and Data Simulations

This section shows the impact of two spectrally wider Tx pulse candidates W1 (pulse candidate #2) and W2 (pulse candidate #3) on the speech quality and on the connection throughput. More

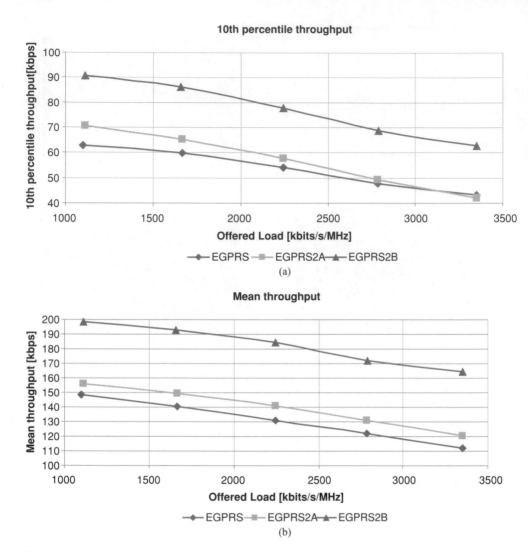

Figure 6.25 Scenario C simulation results (a) 10th percentile throughput; (b) Mean connection through-put.

details about these pulse shapes can be found in Chapter 4 and in [7]. The reference simulation uses the Linearized GMSK transmit pulse (narrow pulse, N) and a data traffic load set to saturation, that is the highest rate of arriving data calls which the network still managed to serve and match with the rate of ending calls.[1] The offered load in the simulations with wide pulse shapes W1 and W2 was roughly the same as in the reference simulation.

---

[1] The data calls were never blocked due to insufficient resources. Rather, once the data terminals arrived in the network, they were waiting idle (if there were currently insufficient resources available).

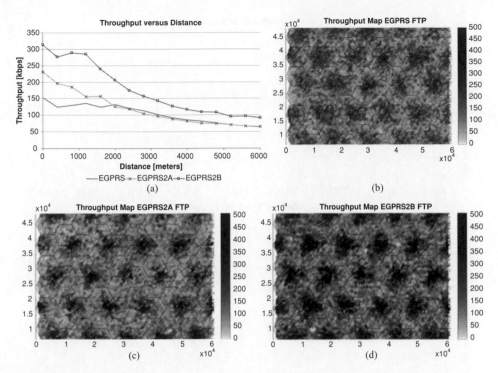

Figure 6.26 Throughput statistics for Scenario C using FTP service in downlink: (a) Throughput versus distance statistics, and throughput map for (b) EGPRS; (c) EGPRS2-A; and (d) EGPRS2-B.

Figure 6.34 shows the impact of wide pulse used in the data connections to the voice quality of the speech connections. In general, the impact was rather negligible and there was no significant difference between wide pulse shape candidates 1 and 2. The biggest impact is observed in the WIDER-1 BCCH scenario, as a result of the large proportion of resources dedicated to data traffic.

However, the wide pulse has a significant impact on the connection throughput as shown in Figure 6.35 and Figure 6.36. The median throughput gains with the pulse shape candidate W1 ranges from 36–44%, while pulse shape candidate W2 provides slightly less gains – between 31% and 39%, depending on the scenario.

Table 6.10 Summary of the achieved throughput relative gains for FTP services using EGPRS2 over EGPRS in downlink (%)

	10th percentile throughput		Mean connection throughput	
	EGPRS 2A	EGPRS2 B	EGPRS 2A	EGPRS2 B
Scenario A	−9.23	25.43	17.57	52.70
Scenario B	43.81	57.12	45.21	71.64
Scenario C	5.6	44.27	6.94	40.28

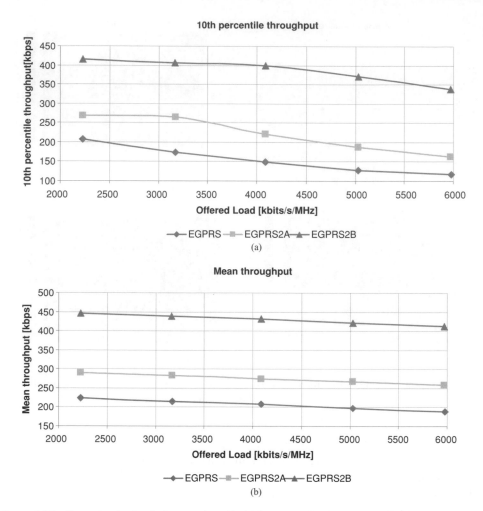

Figure 6.27 Scenario A simulation results: (a) 10th percentile throughput; (b) Mean connection throughput.

6.4.2.5 Capacity Gain with Wide Pulse

This section shows results from the same scenarios as shown in Section 6.4.2.4 but from a capacity point of view. In this case, the network is close to the saturation point in all three scenarios: reference pulse (N, narrow), wide pulse candidate W1 and wide pulse candidate W2.

Figure 6.37 shows again the impact of the wide pulse on the voice quality. Comparing with results with Figure 6.34 it can be concluded that the impact is about the same – rather negligible.

Achieved connection throughput in those scenarios is shown in Figure 6.38 and the corresponding gains in Figure 6.39.

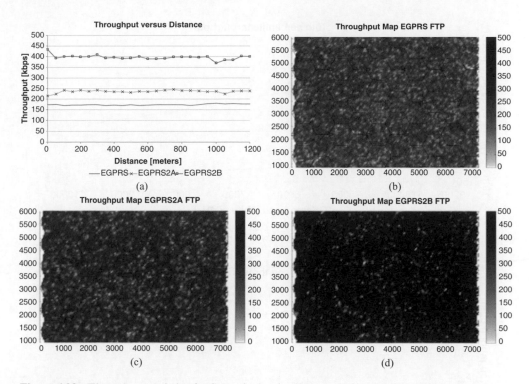

Figure 6.28 Throughput statistics for Scenario A using FTP service in uplink. (a) Throughput versus distance statistics, and throughput map for (b) EGPRS; (c) EGPRS2-A; and (d) EGPRS2-B.

6.4.2.6 Conclusions with Wide Pulse

Wide pulse candidate W1 gives median throughput gains from 16–27% and candidate pulse W2 from 13–23%, depending on the scenario. Median throughput fell by about 11% due to the increased traffic load (and hence interference in the network). Nevertheless, the wide shaping candidates still outperformed the reference narrow pulse.

In addition, there were also capacity gains, in terms of increased offered load: W1 provides minimum 18% gain and W2 minimum 16% gain, which were measured from scenario WIDER-2.

6.5 Downlink Dual Carrier Performance[2]

The introduction of the data evolution feature package to legacy GSM/EDGE deployments offers operators a significant boost in network capacity and mobile data users UMTS/HSDPA such as high speed packet data services along with competitive latency. Intelligent radio

---

[2] This section is based on a revised conference paper. "Paving the path for high data rates by GERAN Evolution EDGE2 with Dual Carrier", by K. Ivanov, C.F. Ball, R. Müllner and H. Winkler, in *IEEE PIMRC*, 2008, Cannes. © 2008 IEEE.

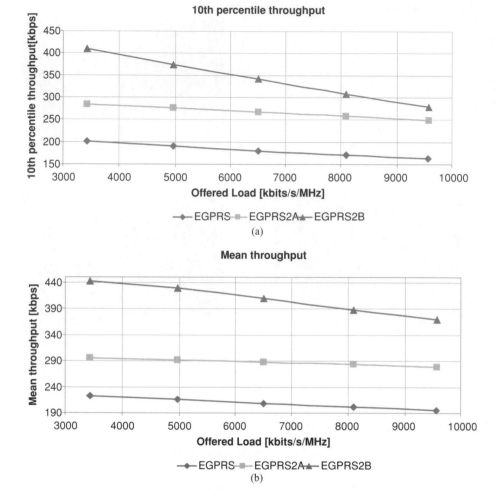

Figure 6.29 Scenario B simulation results (a) 10th percentile throughput; (b) Mean connection throughput.

resource management supports novel downlink dual-carrier (DLDC) capable mobile stations (MS) by the dynamic configuration of GPRS/EGPRS packet data channels (PDCHs) on multiple non-BCCH carriers. Multislot capable Mobile Stations with 4–5 time slots (TS) provide 225–275 kbps with EDGE. Along with 3GPP Release 7, the EGPRS2 feature provides enhanced PDCH data rates up to 118.4 kbps per TS and 1.89 Mbps in multislot connections using the maximum 16 time slots.

System-level simulation results for the end-to-end performance of EGPRS/EGPRS2 over TCP/IP, assuming conventional four time slots up to a future potential 12 time slots, for capable mobiles show up to 680/1360 kbps peak data rates.

In this section, FTP application throughput has been investigated with respect to both download file size and important TCP settings such as, for example receiver window size. The

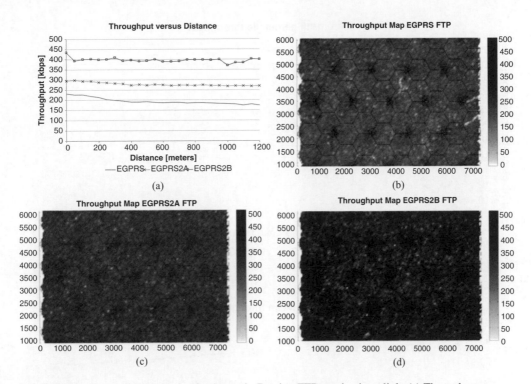

Figure 6.30 Throughput statistics for Scenario B using FTP service in uplink. (a) Throughput versus distance statistics, and throughput map for (b) EGPRS; (c) EGPRS2-A; and (d) EGPRS2-B.

DC performance of legacy EGPRS has been compared with that of EGPRS2 both under ideal radio conditions and in regular hexagonal cellular deployments depending on system load.

The simultaneous allocation of 4 DL TS on single TRX and up to 12 downlink TS on two TRX has been assumed. Preserving the present EGPRS coding schemes the focus is on the end-to-end performance under ideal radio conditions (single cell, not limited coverage scenario) as well as in a real network interference-limited environment. In addition, the new EGPRS2 concept has been investigated under the same conditions to evaluate the resulting performance gain in terms of user throughput and network capacity.

As a consequence, on the network side, both an intelligent radio resource management (RRM) as well as an efficient radio link quality control (LQC) strategy has to be implemented for dynamically handling MS DC allocations on BCCH and non-BCCH transceivers (TRX) with multiple reuse planning (MRP) characterized by variable radio conditions [8].

The effects of TCP/IP as today's dominant transport layer protocol over the internet on the application throughput in wireless networks have been thoroughly investigated [8–10]. Valuable recommendations concerning the setting of the TCP receiver window size on the client side have been given. Furthermore, the dependency of the application throughput on the FTP download file size has been derived. FTP application throughput results under varying system load are presented for slow-moving MS in cellular hexagonal deployments with relaxed 4/12 frequency reuse.

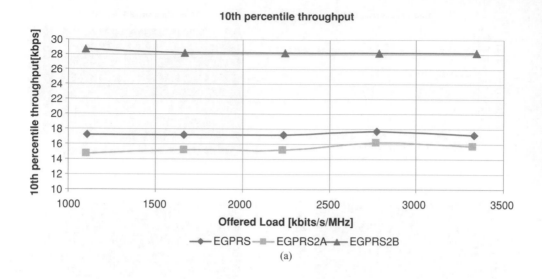

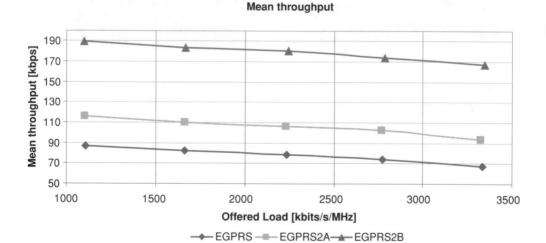

Figure 6.31 Scenario C simulation results (a) 10th percentile throughput; (b) Mean connection throughput.

6.5.1 *Network Simulation Model Including GERAN DLDC Architecture*

The network simulation model shown in Figure 6.40 includes all GERAN network elements considering latency, queuing, transmission delay and all relevant call processing features. The following layers of the protocol stack have been implemented [11, 12]:

1. The physical layer covers GPRS, EGPRS and EGPRS2 link adaptation (LA) as well as incremental redundancy (IR). The physical link is modeled by the block erasure rate (BLER) vs.

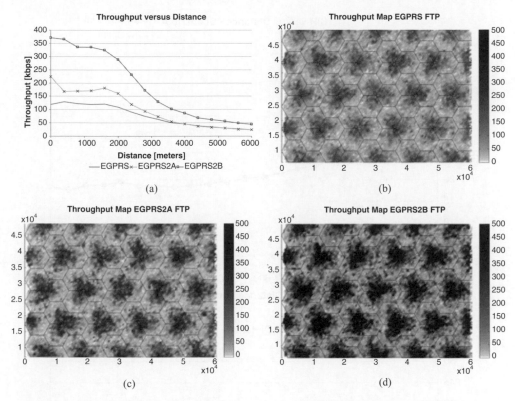

Figure 6.32 Throughput statistics for Scenario C using FTP service in uplink. (a) Throughput versus distance statistics, and throughput map for (b) EGPRS; (c) EGPRS2-A; and (d) EGPRS2-B.

carrier-to-interferer-ratio (CIR) mapping obtained from link-level simulations performed for TU3 (no FH) in the case of EGPRS and TU3 (ideal FH) in the case of EGPRS2 Level-B (EGPRS2-B). A decent MS receiver performance has been assumed, excluding advanced features like MSRD, and single/dual antenna interference cancellation (SAIC/DAIC). Table 6.11 gives an overview of the modulation and coding schemes available in GERAN for DL packet switched data transmission along with the RLC maximum data rate per TS.

Table 6.11 Summary of the achieved relative throughput gains for FTP services using EGPRS2 over EGPRS in uplink (%)

	10th percentile throughput		Mean connection throughput	
	EGPRS 2A	EGPRS2 B	EGPRS 2A	EGPRS2 B
Scenario A	44.17	173.98	34.97	112.99
Scenario B	44.30	97.49	35.13	98.00
Scenario C	−10.72	63.25	36.07	131.73

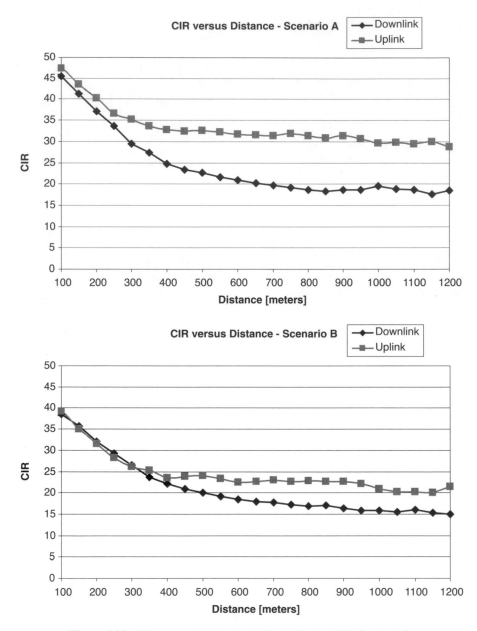

Figure 6.33   CIR versus distance in (a) Scenario A and (b) Scenario B.

2. Radio Link Control/Medium Access Control (RLC/MAC) layer: the selective ARQ protocol for RLC has been completely implemented. For EGPRS, the round-trip time (RTT) on RLC level (i.e. signaling delay from the MS to the packet control unit (PCU) and vice versa) was adjusted to 100 ms, the Relative Reserved Block Period (RRBP) to 40 ms [13]. For EGPRS2, both RTT and RRBP have been reduced, to 80 ms and 20 ms, respectively.

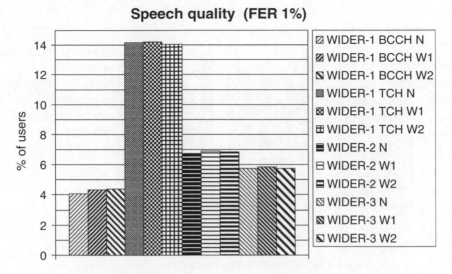

Figure 6.34   Impact of wide pulse to speech quality.

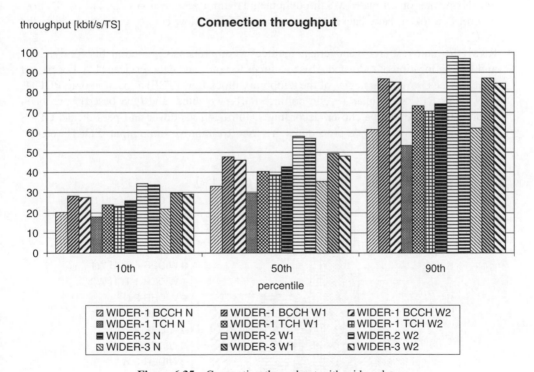

Figure 6.35   Connection throughput with wide pulse.

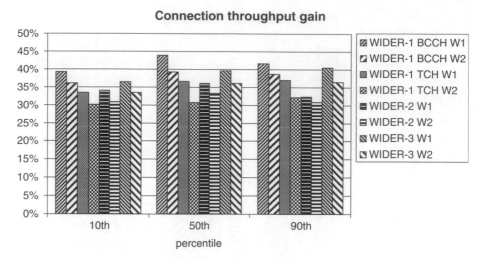

Figure 6.36 Connection throughput gains with wide pulse.

3. Logical Link Control (LLC) layer: the mobile specific LLC flow control function in the SGSN operates on estimated MS throughput and memory congestion state. The SGSN flow control is a token leaky bucket algorithm and receives flow control commands from the PCU.
4. RRM: the radio resource management includes a comprehensive functionality for dynamic and fixed allocation of radio and Abis resources to voice and data services [14]. For packet data services, several strategies of the temporary block flow (TBF) allocation onto PDCHs as well as RLC schedulers (cyclic polling, fairly weighted, and QoS-based) have been implemented and can be chosen accordingly. Intra-cell handover and periodical EGPRS TS downgrade and upgrade procedures are used to improve throughput. PDCHs can be

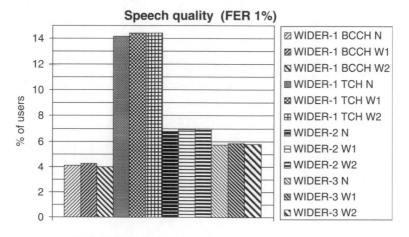

Figure 6.37 Impact of wide pulse at saturated load.

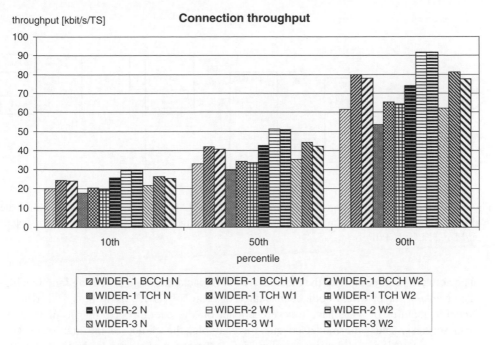

Figure 6.38 Connection throughput with wide pulse at saturated load.

configured arbitrarily on BCCH and/or non-BCCH carriers in different reuse patterns. Shared (on demand) PDCHs might suffer from voice service soft preemption. The novel DC approach allows dynamic configuration of the mobile's PDCH allocation simultaneously on two TRXs. As an example, a mobile might utilize in downlink 12 PDCHs (6 PDCHs on TRX-2 and 6 PDCHs on TRX-3) as well as 1 PDCHs in uplink on TRX-2.

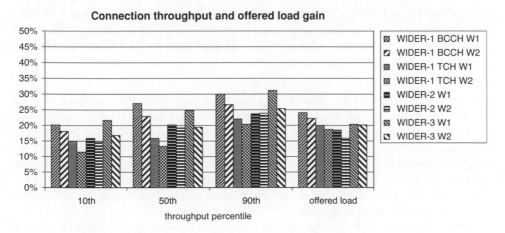

Figure 6.39 Connection throughput and offered load gains with wide pulse.

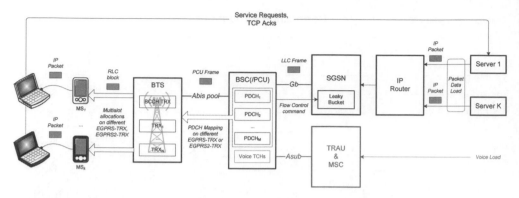

Figure 6.40   Network simulation model according to the GERAN evolution architecture featuring the dual-carrier approach and including all end-to-end network entities and relevant protocol stacks such as RLC/MAC, LLC, TCP/IP.

5. The network layer comprises the transmission of IP packets as well as routing functionality.
6. The transport layer offers both User Datagram Protocol (UDP) as well as TCP (Reno). Specific features of TCP have a severe impact on the overall performance of wireless data services. Thus, the model covers, for example, the choice of the maximum TCP segment size (MSS), advertising window size of the receiver/client (AWND), congestion window management at the sender/server and TCP slow start. The TCP round-trip time is continuously measured and filtered to update the retransmission timeout (RTO). RTO expiry causes TCP retransmissions and a new slow start. In addition, the effects of duplicate acknowledgments (DUPACKs) combined with fast recovery and fast retransmit are part of the model. Hence the complete TCP Reno implementation of the transport layer has been incorporated in the simulator.
7. The application layer consists of a variety of traffic models for WAP, HTTP, email, FTP, SMS, MMS and streaming services. Because of the open architecture of the simulator, new traffic models or traces of real sessions can easily be imported [15, 16]. User's behavior is modeled by probability distributions of the number and size of downloads per Internet session and reading times between separate downloads. Nevertheless the network simulation results presented in this study are exclusively performed for the FTP download service with deterministic file volume of 500 kByte.

Figure 6.40 shows the network elements and interfaces included in the simulation model as well as the path through the network for an IP packet (from the server to the client) on a download request. When a mobile leaves the idle state, a packet data protocol (PDP) context is generated at the SGSN. The mobile makes an access to the GPRS/EDGE network and submits a download request via the mobile network and the internet to a server. The server divides the requested data volume into TCP segments, adds a TCP/IP header and sends them as IP packets via a router to the SGSN. Furthermore, the server initializes the TCP flow control parameters, for example to perform the slow start.

The SGSN creates LLC frames out of the IP packets and transmits them over the Gb interface to the PCU, if a permission has been obtained from the leaky bucket flow control, otherwise the

LLC frames are queued. Packet queuing on the Gb interface due to congestion is considered. Meanwhile the PCU allocates radio resources (PDCHs) and the necessary bandwidth on the dynamic Abis interface. The LLC frames wait in a queue in the PCU to be segmented into RLC blocks. The RLC blocks are scheduled and transmitted over the air interface to the mobile. The PCU polls the mobile for a bitmap to indicate the correctly and erroneously received RLC blocks. The latter are retransmitted.

During the TBF lifetime the PCU performs a periodic TS upgrade/downgrade and LA. As soon as the MS has correctly received all the RLC blocks belonging to the same LLC frame, it reassembles the LLC frame and sends it to the connected laptop/PC client. In the client, the corresponding IP packet and hence the TCP segment are reassembled.

On the receipt of a TCP segment with the expected sequence number, the client sends an acknowledgement to the server. Delayed acknowledgement is considered. For segments out of sequence, the TCP layer of the client transmits DUPACKs. Depending on the state of the TCP parameters, the server invokes on receipt of the TCP acknowledgements and their sequence numbers the appropriate TCP algorithm, for example flow control and congestion window management, RTO handler and retransmission management, as well as fast recovery/retransmissions. As soon as the client has received all the TCP segments of the application data volume, the network resources of the packet call are released. The client/user might send additional download requests after a certain idle period. Otherwise the GPRS/EDGE session is finished and the PDP context is deleted in the SGSN.

6.5.2   Simulation Results for Ideal Radio Conditions

The GSM/EDGE standard specifies nine modulation and coding schemes MCS-1–MCS-9 utilizing both GMSK and 8-PSK and providing RLC data rates of up to 59.2 kbps per PDCH. The GERAN evolution concept EGPRS2-B supporting higher order modulation and coding schemes with higher symbol rate and turbo coding allows for 118.4 kbps per PDCH, that is twice higher than that provided by legacy EDGE. A BLER-based LA algorithm selects the most appropriate MCS/DBS according to the radio conditions optimizing the overall throughput [17, 18]. Hence under ideal radio conditions (zero BLER) the highest MCS-9/DBS-12 with EGPRS/EGPRS2-B will be selected all the time. Furthermore the support of the extended UL TBF feature has been assumed.

Figure 6.41 shows the simulated FTP 2 MByte end-to-end mean application throughput for a single TS MS (one PDCH allocated) up to 12 TS MS depending on the TCP receiver window size ranging from 16 kByte (Windows XP default) up to 48 kByte. Application throughput means that upper layer effects such as TCP slow start as well as overhead including TCP/IP and LLC headers have been considered, the latter reducing the peak data rates by up to 5%.

The Windows XP default TCP window size of 16 kByte is absolutely sufficient for 1 up to 4 TS allocations. Target throughput of 56 kbps for single TS, 112 kbps for 2 TS, 169 kbps for 3 TS and 225 kbps for 4 TS MS has been achieved.

Obviously a 16 kByte receiver window size is completely insufficient for the DC approach and has to be properly adjusted. A throughput degradation of roughly 10% from approximately 340 kbps down to 300 kbps is clearly visible for 6 TS MS. A TCP receiver window size of 24 kByte is required for 6 and 8 TS MS to obtain peak throughput of 340 and 450 kbps, respectively. A 10 TS MS needs a TCP window size of 32 kByte to achieve data rates of

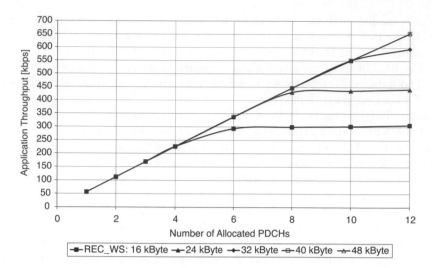

Figure 6.41   EGPRS MCS-9 FTP 2 MByte mean application throughput of 1 up to 12 TS MS depending on the TCP receiver window size (Rel. 6 performance with extended UL-TBF assumed).

550 kbps. For a potential throughput of 650 kbps to be feasible with 12 TS MS, a TCP receiver window size of at least 40 kByte is recommended.

The results shown in Figure 6.42 for EGPRS2-B reveal that the TCP receiver window size has to be adjusted to 64 kByte to support data rates of up to 1.25 Mbps achievable with 12 TS MS on an EGPRS2-B DC. It is worth noting that for a certain application throughput the TCP

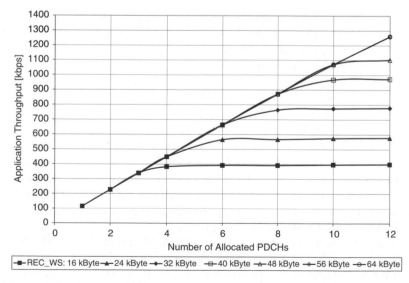

Figure 6.42   EGPRS2-B DBS-12 FTP 2 MByte mean application throughput of 1 up to 12 TS MS depending on the TCP receiver window size. (30% EGPRS2 latency improvements assumed in addition).

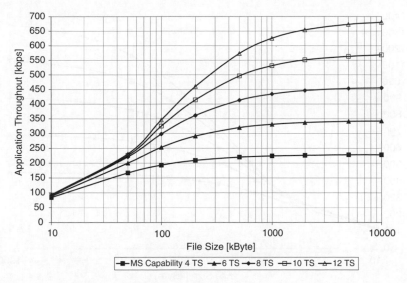

Figure 6.43   EGPRS MCS-9 FTP application throughput for 4 TS up to 12 TS MS depending on download file size (48 kByte TCP receiver window size and Rel. 6 performance with extended UL-TBF assumed).

receiver window size required with EGPRS2-B is significantly less than that required with EGPRS due to the latency reduction features to be introduced with EGPRS2-B significantly improving the TCP/IP round-trip time (ping reduction from currently 160 ms down to less than 100 ms expected).

The impact of FTP download file size on the achievable peak application throughput for different MS TS capabilities ranging from 4 TS up to 12 TS is illustrated in Figure 6.43 for EGPRS and in Figure 6.44 for EGPRS2-B respectively. The file size varies from rather small 10 kByte to a quite large one of 10 MByte. For a small file size, no major difference in throughput has been observed with different MS multi-slot capabilities. Due to the adverse TCP slow start effect, the user data rate is heavily degraded down to approximately 100 kbps for both EGPRS and EGPRS2-B. With increasing file size the application throughput grows very rapidly up to a certain saturation level depending on the MS multi-slot capability, for example 225 kbps for a 4 TS MS in EGPRS and 450 kbps in EGPRS2-B. Apparently the higher the MS multi-slot class, the larger the file size required for throughput saturation, for example a file size of 1 MByte is sufficient to obtain the target throughput of 225/450 kbps for 4 TS MS, however, 5 MByte are required for 8 TS MS to achieve 450/900 kbps and 10 MByte for 12 TS MS at 680/1360 kbps with EGPRS/EGPRS2-B, respectively.

6.5.3   System Level Simulation Results for Regular Hexagonal Cell Deployment

System level simulations have been performed for a dual-carrier deployment scenario assuming 8 TS MS and FTP download service with a constant download volume of 500 kByte (not a full

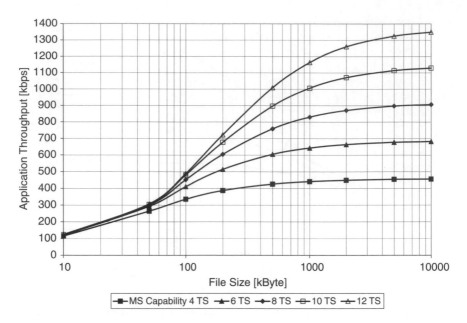

Figure 6.44 EGPRS2-B DBS-12 FTP application throughput for 4 TS up to 12 TS MS depending on file size with 64 kByte TCP receiver window size. (30% EGPRS2 latency improvements assumed in addition).

buffer!) in a regular hexagonal cells interference-limited network with a 4/4/4 configuration (i.e. a 3-sector site with four TRX per sector) and 700 m cell radius. In each cell (sector) 8 reserved PDCHs have been configured on two non-BCCH TRXs planned in a relaxed 4/12 frequency reuse. LA has been enabled in both scenarios EGPRS and EGPRS2-B, while IR has been enabled only in EGPRS.

Figure 6.45 depicts the mean application throughput along with the 10th and 90th user percentiles for varying system load measured in terms of mean user busy hour (BH) data rate. In the investigated scenarios an offered load of for example 500 bps translates to 18.6 kbps per TS.

It should be pointed out that the EGPRS dual-carrier mean user throughput of 350 kbps to 360 kbps achieved with 8 TS MS under very low load conditions (up to 100 bps offered load) is as high as in 3G-UMTS Rel.99 networks. Some 10% of the EGPRS users enjoy the top data rates of roughly 400 kbps and 90% of the subscribers achieve data rates higher than 250 kbps. As the offered load increases, the perceived user throughput gradually decreases due to increased interference level in the network and resource sharing between users. An excellent mean user throughput of 200 to 250 kbps has been obtained at medium load (400 to 600 bps), and even in a fully loaded system (800 bps) mean user data rates well above 100 kbps are feasible. Further increase of the data load pushes the EGPRS network into congestion. The worst 10% of the users get practically nothing from the service (less than 32 kbps).

EGPRS2-B outperforms EDGE in terms of both user throughput and capacity over the entire system load range (Figure 6.46). The gain in user throughput achieved by the introduction of the higher order modulation and coding schemes with turbo coding DBS-5 through DBS-12, increased 1.2 symbol rate as well as the latency reduction features in GERAN evolution

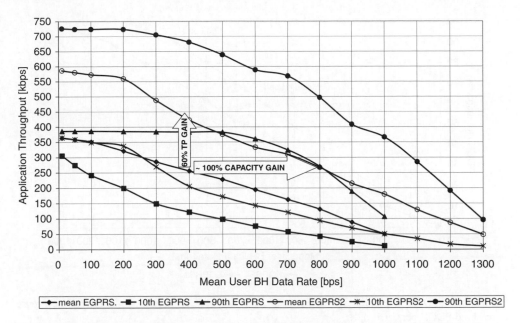

Figure 6.45 EGPRS FTP 500 kByte application throughput vs. system load for 8 TS MS on 2 non-BCCH carriers in 4/12 frequency reuse (4 reserved PDCH per carrier).

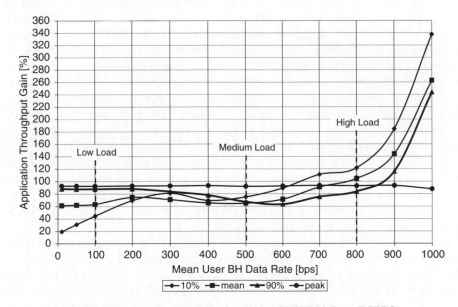

Figure 6.46 Application throughput gain of EGPRS2-B vs. EGPRS.

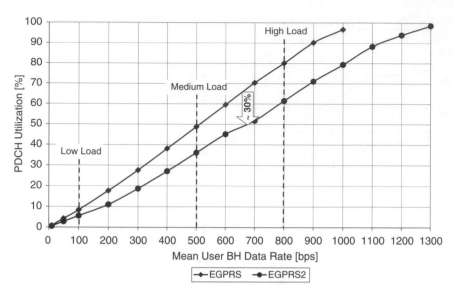

Figure 6.47 PDCH utilization vs. system load on 2 non-BCCH carriers in 4/12 frequency reuse (4 reserved PDCH per carrier).

EGPRS2 has been evaluated as a function of the offered system load. Obviously the gain in peak data rates of roughly factor 2 is independent of the load. The gain in both mean user throughput and that of the best 10% users varies in the range of 60–90% for an offered load up to 700 bps. At system load beyond 800 bps the EGPRS scenario runs into congestion causing the exponential gain growth. For an offered load higher than 650 bps (24 kbps/TS), EGPRS2-B provides more than double throughput for the worst 10% users.

Furthermore, the improvement in throughput performance reduces the effective load in the network since the sojourn time of each EGPRS2-B user gets shorter. This improves the capacity of the system. Figure 6.47 clearly indicates the reduced PDCH-utilization with EGPRS2-B as the offered load increases. At medium to high offered load (300 to 700 bps), the PDCH-utilization measured in the EGPRS scenario has been reduced by nearly 30% in the EGPRS2-B scenario. The spared resources along with the enhanced link-level performance and latency reduction features in EGPRS2-B translate to a roughly 100% capacity gain as indicated in Figure 6.46. While the EGPRS scenario runs into an overload situation at 900 to 1000 bps offered load, the PDCH-utilization of 70–80% observed in the EGPRS2-B scenario still allows excellent mean user throughput of about 200 kbps.

The cumulative distribution functions (CDF) of the application throughput at low (100 bps), medium (500 bps) and high system load (800 bps) are presented in Figure 6.48.

The distributions clearly demonstrate the optimum exploitation of the good radio conditions (good CIR) on the non-BCCH carriers in 4/12 reuse. Especially at low load EGPRS2-B users can overwhelmingly profit from the excellent CIR which is typically higher than 25 dB in 90% of the cell area. While the application throughput for about 70% of the EGPRS users is limited to the peak value of 380 kbps, 50% of the EGPRS2-B users enjoy 240 kbps higher throughput (higher than 620 kbps). The best 10% EGPRS2-B users perceive almost double

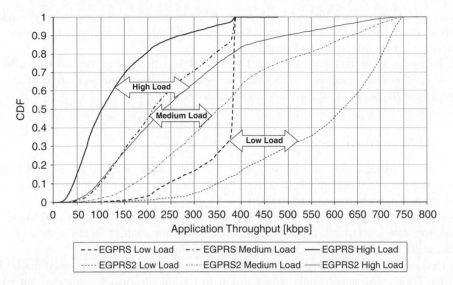

Figure 6.48 CDFs of the application throughput at low (100 bps), medium (500 bps) and high system load (800 bps).

data rates (higher than 720 kbps). Furthermore, it is worth mentioning that the user throughput perception with EGPRS2-B at high system load is at least as good as with EGPRS at medium load revealing the significantly improved spectral efficiency of EGPRS2-B. In addition, the best 10% EGPRS2-B users achieve data rates higher than 500 kbps.

6.5.4 Summary

The novel DC approach as well as the currently standardized EGPRS2-B GERAN evolution feature package have been investigated by means of system level simulations under both ideal and realistic radio conditions. Preserving the modulation and coding schemes currently used with EGPRS, a DC implementation on two non-BCCH carriers planned in 4/12 frequency reuse demonstrates a substantial performance gain over today's single carrier approach. A 3G-UMTS Rel-99 like mean user throughput can be obtained in EGPRS DC networks.

Mean user throughput of 360 kbps has been measured for FTP 500 kByte download with 8 TS capable MS at low system load. TCP receiver window size and download file size have a much stronger impact on the end-to-end performance of EGPRS2-B compared to that of EGPRS. As an example, a TCP receiver window size of 24 kByte is required for 8 TS MS to achieve the peak data rate of approximately 450 kbps in EGPRS while the receiver window size has to be adjusted to 40 kByte for an EGPRS2-B MS with 8 TS to support a peak data rate of 900 kbps. Following this recommendation and assuming a sufficiently large download file size target throughput of up to 680/1360 kbps could be achieved under ideal radio conditions (single cell with not coverage-limited scenario) with the DC approach using EGPRS/EGPRS2-B capable MS with 12 TS.

In a cellular interference-limited deployment (4/12 frequency reuse) at medium system load, EGPRS2-B outperforms EGPRS providing more than 60% gain in mean user throughput and an increase in network capacity of about factor 2.

GERAN evolution including DC and EGPRS2-B is a promising method of enhancing GERAN packet data service to UMTS/HSDPA so that in the near future subscribers can enjoy seamlessly high data rates in multi RAT mobile networks.

6.6   DTM performance

Dual Transfer Mode (DTM) is a GSM feature in 3GPP standard Release 99. It allows simultaneous transfer of voice and data in different channels. This feature is optional and requires DTM support in both terminal and network.

When DTM is enabled in the network and the DTM terminal is setting up a DTM call, the connection state is set to dual transfer mode. This means that connection has reserved CS and PS channels. The transmission states are described in Figure 6.49.

In Release 99, the dual transfer mode can be entered only from the dedicated mode. This means a DTM MS cannot establish a CS call in the middle of a PS session, but has to first release the assigned PS resources before entering dedicated mode. When the MS has entered dedicated mode, the MS or the network are able to start the DTM call establishment procedure. The procedure assigns PS resources during an existing CS connection and the MS finally enters into dual transfer mode [19]. In Release 5, DTM enhancements were brought in, allowing direct transitions between packet transfer mode and dual transfer mode as illustrated in Figure 6.49.

DTM introduces a new challenge for the network. There are limitations, for example in channel allocation due to terminal capabilities. This section discusses important network

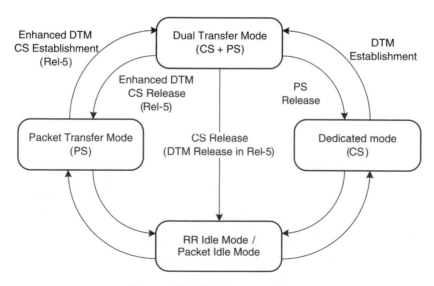

Figure 6.49   DTM state transitions.

features that improve the overall performance of the network where DTM is implemented. Various features have been developed to make DTM deployment successful. In the early research phase, new features are developed into the dynamic system level simulator SMART and major Key Performance Indicators (KPIs) are evaluated to understand the benefits of the feature. The discussion section finally gives conclusions on how essential it is to deploy RRM algorithm improvements along with the DTM.

The DTM call establishment process takes some time. The reason is that the CS connection needs to be moved to the PS territory. This requires the territory downgrade procedure that lasts 400 ms and intra-cell handover which takes additional 320 ms. From the user's point of view, the total delay for initiating data transfer in dedicated mode is around 900 ms. Once the DTM call has been established, PS data transfer runs as a normal (E)GPRS session. The major difference is that for each different multi-slot class (MSC), 1+1 time slots are reserved for CS call.

6.6.1 Fragmentation Issue

This section presents the PS territory fragmentation issue due to the DTM call allocation. PS and CS resources are divided into PS and CS territories. The territory mechanism uses a set of BSS parameters to move the border dynamically based on the CS traffic load. Figure 6.50 gives an example of the dynamic resizing of the PS territory.

When a DTM call is established in the PS territory, it causes fragmentation of PS resources if the PS configured time slots in the territory are no longer subsequent to each other due to the DTM/CS resource reservation. The fragmentation makes multi-slot PS allocation for other users more difficult, because the time slots in the multi-slot allocation have to be subsequent to each other.

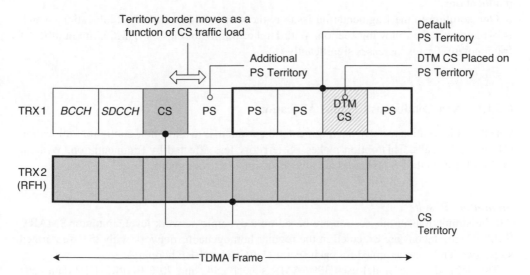

Figure 6.50 Dynamic resizing of the PS territory.

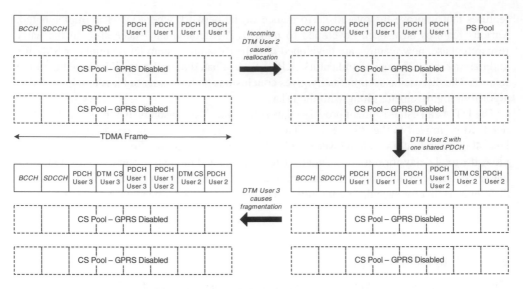

Figure 6.51   DTM channel allocation procedure.

An example of the allocation procedure is demonstrated in Figure 6.51. In the first step, four consecutive time slots were assigned in the regular PS connection (user 1). Then a DTM call (user 2) is allocated to the PS territory. This causes reallocation of the first regular PS connection. Once the DTM call is allocated, there is one shared time slot. Another allocation of DTM call (user 3) causes fragmentation. PS traffic is allocated to the neighboring time slots. A third DTM connection would get at most two consecutive time slots and would cause many reallocations.

One solution to the fragmentation issue is the non-contiguous channel allocation, which is explained in the following section. With Higher multi-slot classes (HMC), the number of fragmentation cases increases significantly.

6.6.1.1   Non-Contiguous Channel Allocation

One possible solution to solve the fragmentation problem is to allow non-contiguous allocation (NCA) of PS calls. This option makes PS territory less affected by fragmentation. Without DTM allocation the PS resources are always assigned as contiguous.

Simulation Results

All the simulations in this section are based on a dynamic network level simulator SMART. SMART simulations are executed in the regular homogeneous network with 25 3-sectorized sites. Two TRXs are assumed for each sector (1 BCCH + 1 RF hopping).

The defined traffic model uses 30% AMR speech calls and 70% EGPRS FTP data calls with a file size of 120 kB. The DTM penetration rate is set to 30% and the DTM user's Class is assumed to be Class 11 [19].

Territory configuration is set as follows: default PS territory size is 5 and maximum PS territory size is 14. The maximum number of DL TS is set to 2 for DTM and 5 for non DTM users (higher multi-slot class – HMC).

Figure 6.52 (a) shows the mean number of time slots per user due to the impact of territory fragmentation. A network without DTM service is used as a reference case. On average, there are 1.5 time slots less for 30% of DTM penetration compared to a network without DTM. By allowing non-contiguous PS allocation we observe around 1 time slot more per connection.

The distribution of used time slots is actually more critical than the average values as shown in Table 6.12.

Figure 6.52 (b) shows the average throughput per user for DTM users. When non-continuous allocation is enabled, the frequency of higher throughputs is increased. The fragmentation impact in the throughput, delay and other KPIs can be decreased with non-continuous allocation. Non-contiguous channel allocation can be considered a must for higher DTM penetrations.

6.6.2   DTM Multiplexing

In the first network simulations of DTM, only FR was supported in DTM/CS. The aim of this section is to show the impact of supporting also half-rate speech (HR) in DTM/CS. Lower and upper bounds for network capacity with DTM support obtained by simulations are presented in the following section.

Using HR with DTM/CS provides higher flexibility on sharing remaining PS resources. Figure 6.53 shows examples of multiplexing of resources.

6.6.2.1   Simulation Results

The simulations in this section are based on a regular grid with 75 sectors and with 2 TRXs (1 BCCH + 1 RF hopping) per sector. The traffic model assumes 30% speech AMR and 70% EGPRS FTP with 120 kB file size. A penetration rate of 30% DTM Class 11 users is assumed. The dedicated/default/maximum PS territory sizes are set to 2/4/6 respectively. The maximum number of DL TSL is 2 for DTM users and 4 for non-DTM users.

Figure 6.54 shows the TBF blocking gain due to DTM/CS multiplexing when forcing the CS part of the DTM call to half-rate mode. The reference case is based on a network with DTM disabled. Analyzing the TBF blocking impact on DTM users, the forced use of HR in DTM mode provides a gain of about 20% compared to HR not supported in DTM, considering a 1% TBF blocking threshold.

6.6.3   Intra-Cell Handover Failures

Transition from dedicated mode to dual transfer mode is necessary before DTM call establishment. Transition implies an intra-cell handover. The intra-cell handover is triggered by the DTM assignment and expiry of packet flow timer (after PS release). A CS call needs to be moved from CS territory into PS territory to start a DTM call. A PS call needs to release PS resources first before starting the DTM call. Either a CS or PS call starts the DTM call, it implies resource reallocation. The DTM assignment increases the intra-cell handover and these extra handovers can cause the drop of the call due to handover signaling failure.

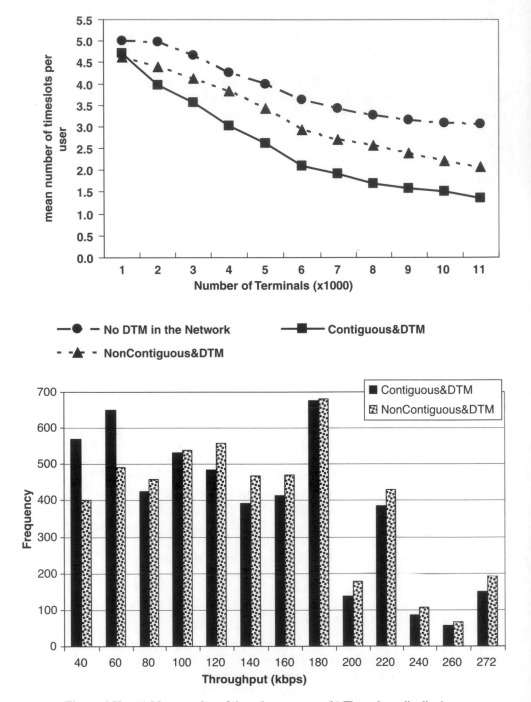

Figure 6.52 (a) Mean number of time slots per user; (b) Throughput distributions.

Table 6.12 Distribution of used time slots

GPRS traffic	No DTM (%)	DTM (%)	DTM with NCA (%)
with 1 TS	0.00	79.5	19.3
with 2 TS	0.00	8.7	60.4
with 3 TS	97.4	9.0	15.1
with 4 TS	0.01	2.1	4.4
with 5 TS	2.59	0.7	0.8

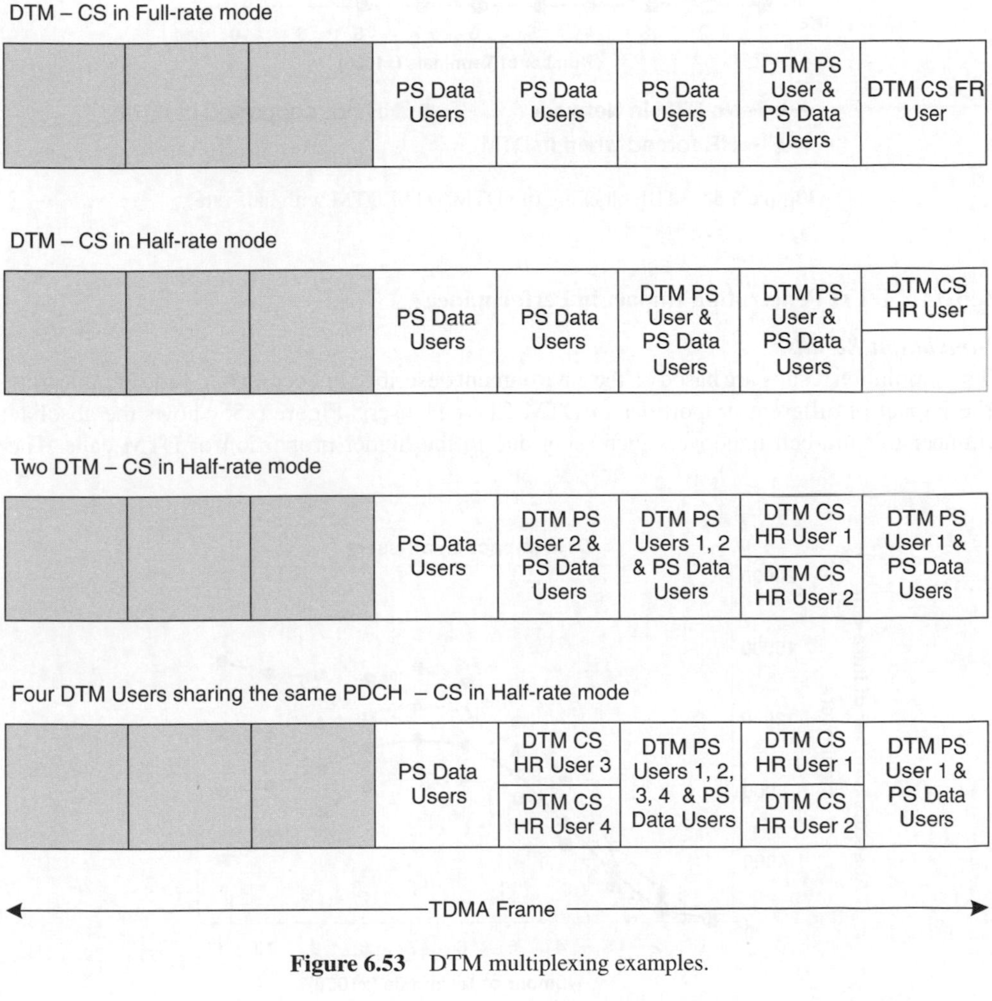
DTM – CS in Full-rate mode
PS Data Users
PS Data Users
PS Data Users
DTM PS User & PS Data Users
DTM CS FR User
DTM – CS in Half-rate mode
PS Data Users
PS Data Users
DTM PS User & PS Data Users
DTM PS User & PS Data Users
DTM CS HR User
Two DTM – CS in Half-rate mode
PS Data Users
DTM PS User 2 & PS Data Users
DTM PS Users 1, 2 & PS Data Users
DTM CS HR User 1
DTM CS HR User 2
DTM PS User 1 & PS Data Users
Four DTM Users sharing the same PDCH – CS in Half-rate mode
PS Data Users
DTM CS HR User 3
DTM CS HR User 4
DTM PS Users 1, 2, 3, 4 & PS Data Users
DTM CS HR User 1
DTM CS HR User 2
DTM PS User 1 & PS Data Users
TDMA Frame

Figure 6.53 DTM multiplexing examples.

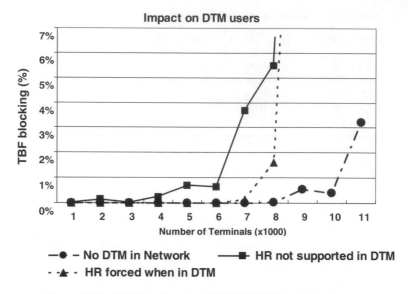

Figure 6.54   TBF blocking (no DTM/DTM/DTM with half-rate).

6.6.3.1   DTM Penetration Impact in Performance

Simulation Results
The simulation results are based on the environment described in Section 6.6.3 and demonstrate the impact of different proportions of DTM Class 11 users. Figure 6.55 shows the absolute number of intra-cell handovers increasing due to the higher proportion of DTM calls. The

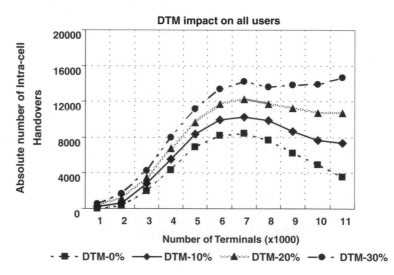

Figure 6.55   Absolute number of intra-cell handovers' impact due to increased DTM penetration.

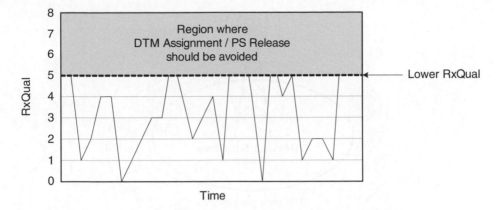

Figure 6.56   RXQUAL region not allowed to DTM assignment.

numbers contain intra-cell handovers of DTM and non-DTM users. Handover signaling failures play an important role, but they are not analyzed here.

6.6.3.2   Solutions to Intra-Cell Handover Failures Caused by DTM

In this section features for improving the intra-cell handover performance are studied and discussed: the introduction of a RXQUAL threshold and FACCH repetitions. Two features are studied in this section: RXQUAL threshold and FACCH repetition.

RXQUAL Threshold for Intra-Cell Handover Caused by DTM
The basic principle of this feature is to use an RXQUAL threshold for intra-cell handovers caused by DTM, similar to the switch from AMR-HR operation to AMR-FR mode triggered by radio conditions. In DTM assignment, there are good reasons for blocking TBF if the connection has a substantially high RXQUAL representing poor radio conditions (Figure 6.56). And after PS release, it is recommended to reset the packet flow timer (PFT) until RXQUAL is good enough. Figure 6.57 shows the range layers where the intra-cell handover due to DTM assignment will not be allowed and where a DTM connection with RXQUAL higher than 5 will have the TBF blocked.

FACCH Repetition for DCR HO Caused by DTM
FACCH signaling is bidirectional (uplink and downlink) and is used when a handover is required. FACCH failures imply an increase of Dropped Call Rate (DCR) during handover process. It may also be employed in call establishments. The solution is to use the FACCH repetition/boosting to increase the FACCH performance and reduce the handover failures. Figure 6.58 shows that FACCH FER performance without repetition creates a bottleneck in handover since performance of signaling channels is worse than that of the lower rate AMR codecs.

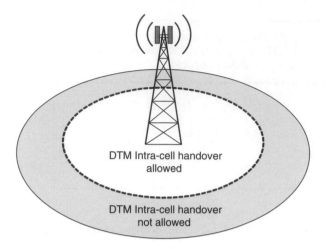

Figure 6.57 Range layer not allowed for intra-cell handover due to DTM assignment.

6.6.4 Conclusion

Deployment of DTM in the network is challenging with respect to the overall performance.
DTM affects both voice and data services and further improvements are needed to maintain
network performance on the desired level.

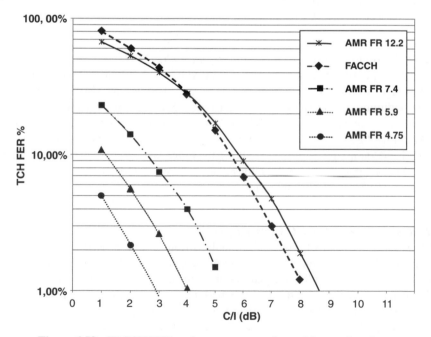

Figure 6.58 FACCH FER performance comparison with speech codecs.

DTM increases the blocking (voice and data) and the DTM/CS HR mode has to be used. It increases the flexibility of dedicated PS territory and hence smooths the impact of DTM deployment.

DTM also increases the number of intra-cell handovers and these additional handovers will cause a higher rate of dropped calls due to handover signaling failure. Essential features to solve this issue are the enabling of RXQUAL threshold and the application of FACCH repetition. Both features clearly improve the overall performance by reducing the handover failures.

Non-contiguous allocation, half-rate for DTM/CS and FACCH enhancements are essential for DTM to be successfully deployed. It is clearly seen that any features that improve speech service performance will also improve overall DTM performance. For example, dynamic allocation of frequency resources, for example DFCA, deployment improves the capacity of DTM/CS and leaves more flexibility to DTM/PS to improve performance by higher data throughput.

DTM gets even more interesting when LTE is deployed. Future mobile stations will support GSM and LTE technologies. Future dual transfer mode may consist of GSM speech service and LTE data service. Smooth interaction between GSM and LTE is essential to optimize overall network performance. Spectrum migration plays an important role and multi-radio DTM would provide more flexibility to an operator.

6.7 GSM Data Evolution Performance Summary

GSM/EDGE data evolution consists of complementary techniques that taken together could achieve quadrupling data rates compared to average data rates available in existing EDGE networks. These data evolution features can be employed to allow a flexible bandwidth extension with EGPRS2 and Downlink Dual Carrier, which both provide doubling of the throughput compared to traditional EDGE services. EGPRS2 supports higher order modulation schemes, such as 32-QAM and together with the higher symbol rate, spectrally wide optimized transmit pulse shaping filter and turbo codes boost peak data rates to 118.4 kbps per time slot. Downlink Dual Carrier, on the other hand, doubles the bandwidth and hence the number of assigned time slots can be increased to 16 thus enabling the theoretical peak data rates of up to 1.89 Mbps with EGPRS2-B.

Table 6.13 summarizes the data evolution performance characteristics per allocated time slot using the new EGPRS2 modulations. Table 6.13 also lists Mobile Station Receive Diversity (MSRD) as one potential technology to increase the data throughput, although MSRD is not evaluated here.

The 3GPP Release 7 improvements increase peak bit rates, and more importantly, the average bit rates when the data service coverage area and spectrum efficiency are improved. Data rates offered by EGPRS2 and DLDC complement HSPA speeds allowing multi-mode users to experience a new level of service continuity, where the boundaries between 2G and 3G do not limit the usage of latest packet data application. It is worth noting that almost all HSPA and LTE users are also GSM/EDGE users.

One of the key design criteria in 3GPP Release 7 is that the deployment of evolution features has minimal or even no hardware impact on legacy network equipment. Therefore it is possible to introduce the specified data enhancements only with software upgrades, without any new

Table 6.13 GSM Data evolution performance summary

EDGE evolution	Peak data rate[1]	Average throughput[2]	Cell edge throughput[3]
EGPRS	59.4 kbps	45/50 kbps	20/21 kbps
MSRD (DL only)	0% gain	40..100% gain	80..120% gain
DLDC (DL only)	100% gain	100% gain	100% gain
EGPRS2-A (DL/UL)	98.4/76.8 kbps	52/62 kbps	10/2 kbps
EGPRS2-B (DL/UL)	118.4/118.4 kbps	62/72 kbps	26/26 kbps

Notes: [1]Theoretical maximum per timeslot.
[2]Average (50th percentile) input levels for a coverage limited network. In downlink, 2 W (MS)
−88 dBm median input level. In uplink −98 dBm median input level.
[3]Cell edge (95th percentile) input levels for a coverage limited network. In downlink, 2 W (MS)
−98 dBm input level at cell edge. In uplink −108 dBm input level at cell edge.

hardware roll-out. Mobile Stations will require modifications and there are already trials ongoing to demonstrate the expected capabilities of the GSM/EDGE data evolution features. Key enablers of Release 7 data evolution are the huge subscriber base of GSM/EDGE and the cost-efficient four-fold enhancement of the EDGE performance using EGPRS2 and Downlink Dual Carrier when compared to average EDGE data speeds.

References

[1] M. Taferer and E. Bonek, *Wireless Internet Access over GSM and UMTS*, Berlin: Springer, 2002.
[2] C.F. Ball, K. Ivanov, R. Müllner, and P. Stöckl, "Impact of Configuration and Parameter Settings on GPRS/EDGE Latency and Throughput", in *IEEE Global Mobile Congress* (GMC), 2004.
[3] 3GPP TSG GERAN, "Feasibility Study for Evolved GSM/EDGE Radio Access Network (GERAN)", 3GPP TR 45.912, Ver.7.1.0.
[4] 3GPP TSG GERAN WG1, "Multiplexing and Multiple Access on the Radio Path", 3GPP TS 45.002.
[5] 3GPP TSG GERAN WG1, "Channel Coding", 3GPP TS 45.003.
[6] M. Säily, E. Zacarías, J. Hulkkonen, O. Piirainen and K. Niemelä, "EGPRS2 Uplink Performance for GERAN Evolution", in *Proceedings of IEEE VTC*, Spring, 2008.
[7] 3GPP TR 45.913 v1.3.0, "Optimized Transmit Pulse Shape for Downlink Enhanced General Packet Radio Service (EGPRS2-B)", November 2009.
[8] C.F. Ball, K. Ivanov and F. Treml, "Contrasting GPRS and EDGE over TCP/IP on BCCH and Non-BCCH Carriers", in *Proceedings of IEEE VTC*, Fall , 2003.
[9] R. Sanchez, J. Martinez, J. Romero and R. Järvelä, "TCP/IP Performance over EGPRS Network", in *Proceedings of IEEE 56th VTC*, 2002, pp. 1120–1124.
[10] M. Meyer, "TCP Performance over GPRS", in*Proceedings of IEEE Wireless Communications and Networking Conference*, 1999.
[11] 3GPP TS 43.064 v7.9.0, "General Packet Radio Service (GPRS); Overall Description of the GPRS Radio Interface".
[12] 3GPP TS 44.060 v7.13.0, "General Packet Radio Service (GPRS); Mobile Station – Base Station System Interface; Radio Link Control/Medium Access Control (RLC/MAC) Protocol".
[13] C.F. Ball, K. Ivanov, L. Bugl and P. Stöckl, "Analysis and Optimization of the (E)GPRS RLC Protocol by Simulations and Measurements", in *Proceedings of IEEE PIMRC*, 2004.
[14] C.F. Ball, K. Ivanov, R. Müllner and F. Treml, "Performance Analysis of Dynamic TDM-Transport for GSM Voice and GPRS/EDGE Packet Data Services", in *Proceedings of IEEE VTC*, Fall , 2003.
[15] C.F. Ball, C. Masseroni and R. Trivisonno, "Multi RAB-Based and Multimedia Services over GERAN Mobile Networks", in *Proceedings of IEEE VTC*, Fall, 2005.

[16] C.F. Ball, C. Masseroni and R. Trivisonno, "Introducing 3G Like Conversational Services in GERAN Packet Data Networks", in *Proceedings of IEEE VTC*, Spring, 2005.
[17] C.F. Ball, K. Ivanov, P. Stöckl, C. Masseroni, S. Parolari and R. Trivisonno, "Link Quality Control Benefits from a Combined Incremental Redundancy and Link Adaptation in EDGE Networks", *IEEE VTC*, Spring, 2004.
[18] C.F. Ball, K. Ivanov, L. Bugl and P. Stöckl, "Optimizing GPRS/EDGE End-to-End Performance by Link Adaptation and RLC Protocol Enhancements", in *Proceedings of IEEE Global Mobile Congress – GMC*, 2004.
[19] 3GPP TS 43.055 v8.1.0, "Dual Transfer Mode (DTM)", February 2009.

7

Control Channel Performance

Eddie Riddington and Khairul Hasan

7.1 Introduction

7.1.1 Overview

Since the beginning of GSM, a significant amount of effort has been spent by the ETSI and 3GPP standardization bodies to improve the quality, robustness and capacity of the speech channel leading to a number of advancements in speech codec technology. In this respect, a milestone was reached with the Adaptive Multi-Rate (AMR) codec. AMR introduces a plurality of codec *modes* from which the network adapts the robustness and the speech quality of the codec to suit the condition of the radio channel.

This makes AMR the codec of choice in bandwidth efficient networks where the need for increased robustness against interference is matched by the need to maintain a reasonable speech quality performance. While AMR represents a significant step forward towards increased network efficiency, the performance of the control channels associated with the speech traffic channel represent a major bottleneck since they were designed with less robust traffic channels in mind such as the Full-Rate, Enhanced Full-Rate and Half-Rate channels. To realize the full potential of AMR, a need was identified within the 3GPP standardization body to improve the robustness of the associated control channels so that their performance relative to the robust modes of AMR could be ensured.

In this chapter on control channel performance, we describe the improvements to the associated control channels that were standardized as part of 3GPP Release 6. It is complementary to Chapter 12 on Capacity Enhancements for GSM, where a number of non-standardized enhancements to the associated control channels are described and which can be considered as complementary to the standardized methods. First, in Section 7.1.2, we give an introduction to the AMR codec where we highlight the benefits of this codec over previous codecs both in terms of speech quality performance and robustness to interference. Then in Section 7.1.3, we give an introduction to the associated control channels: the FACCH and SACCH, where we describe their role in maintaining the link between the network and the mobile. Next,

GSM/EDGE: Evolution and Performance   Edited by Mikko Säily, Guillaume Sébire and Eddie Riddington
© 2011 John Wiley & Sons, Ltd

in Section 7.2 we give an introduction to the Repeated SACCH, the standardized method to improve the robustness of the SACCH. A case study relating to radio link failure is also given to demonstrate the benefits of the technique. Finally, in Section 7.3, we give an introduction to the Repeated Downlink FACCH, the standardized method to improve the robustness of the FACCH. The benefit of the technique is demonstrated in a case study relating to handover.

7.1.2   Introduction to Adaptive Multi-Rate (AMR) Coding

Speech service in GSM began with the Full-Rate (FR) codec based on Regular Pulse Excitation-Long Term Prediction (RPE-LTP) coding and operating at an average bit rate of 13 kbps. Later, the Enhanced Full-Rate (EFR) codec based on Algebraic Code Excited Linear Prediction (ACELP) coding was introduced to improve speech quality while operating at a slightly lower bit rate (12.2 kbps). To increase the network capacity of voice traffic, ETSI also standardized the Half-Rate (HR) codec which is based on Vector Sum Excited Linear Prediction (VSELP) coding and operates at 5.6 kbps.

The bits from these speech codecs are transmitted over GSM logical channels TCH/FS, TCH/EFS and TCH/HS respectively after having undergone channel coding for protection against impairments in the wireless channel. Channel coding adds redundancy bits and matches the source bit rate to the bit rate of the GSM channel (which is 22.8 kbps for the TCH/FS and TCH/EFS and 11.4 kbps for TCH/HS). More details about the channel coding for these logical channels can be found in 3GPP TS 45.003 [1].

While these speech codecs operate only at constant bit rates, the Adaptive Multi-Rate (AMR) codec supports a range of bit rates that can be adapted depending on the condition of the channel. AMR utilizes the ACELP algorithm employed in GSM EFR and supports eight codec modes with bit rates 12.2, 10.2, 7.95, 7.4, 6.7, 5.9, 5.15 and 4.75 kbps. All eight modes are supported by the TCH/AFS full-rate traffic channel while the last six are supported by the TCH/AHS half-rate traffic channel. AMR delivers optimum speech quality and increased system capacity through fast in-band signaling and link adaptation. It adapts to the radio and traffic conditions by selecting the optimum traffic channel mode (full-rate or half-rate) and codec mode.

Channel coding for the AMR codec modes is based on recursive systematic convolutional coding with puncturing to obtain the required GSM channel bit rates (22.8 kbps for full-rate and 11.4 kbps for half-rate).

In the basic AMR operation (depicted in Figure 7.1), both the mobile station (MS) and the base transceiver station (BTS) perform channel quality estimation on the received signal. On the basis of these channel quality measurements, a codec mode command (CMC) on the downlink or codec mode request (CMR) in the uplink is sent over the radio interface in in-band messages and the receiving end uses this information to choose the best codec mode for the prevailing channel conditions. A codec mode indication (CMI) is also sent to indicate the current mode of operation at the sending side. The basic principle for codec mode selection is that the mode chosen in the uplink may differ from the one used on the downlink direction, but the channel mode (full-rate or half-rate) must be the same.

The benefit of the in-band method is that it does not require a separate signaling channel for message transfer. By sending the messages and the indicators together with the speech, the link adaptation can be made faster, leading to improvements in system performance.

The network controls the uplink and downlink codec modes and channel modes, while the MS must obey the codec mode commands coming from the network. To determine which

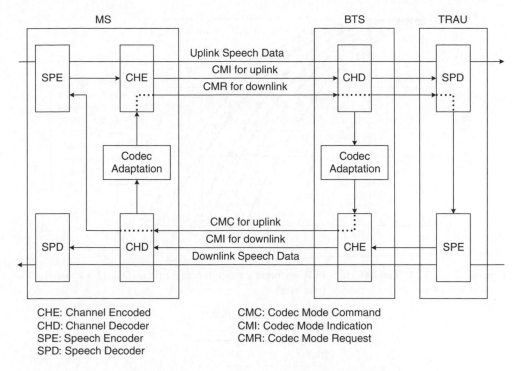

CHE: Channel Encoded
CHD: Channel Decoder
SPE: Speech Encoder
SPD: Speech Decoder

CMC: Codec Mode Command
CMI: Codec Mode Indication
CMR: Codec Mode Request

Figure 7.1   Schematic showing AMR operation [2].

codec mode to use, the network may use any complementary information in addition to the codec mode request coming from the MS.

The performance of the channel encoder (CHE) for the TCH/AFS channel is depicted in Figure 7.2 as a function of carrier-to-interference ratio (C/I), where performance is given in terms of frame erasure rate (FER) and where a frame erasure is counted by an error in the class 1a bits (the most sensitive bits of the speech frame). In many cases, the speech quality begins to degrade when the FER exceeds 1%. Hence when we compare the most robust mode of AMR with the TCH/FS channel reference (also depicted in Figure 7.2), we see a benefit of about 5 dB (corresponding to the relative level of interference that can be tolerated).

The other side of the AMR coin are the bit rates supported by the Speech Encoder (SPE), as these together with the compression algorithm determine the quality of the speech. In AMR, the supported bit rates range from 12.2–4.75 kbps, which compare to fixed bit rates of 13 kbps for FR, 12.2 kbps for EFR and 5.6 kbps for HR. Their performance in terms of speech quality can be compared after performing listening tests and rating the listening quality according to the Mean Opinion Score (MOS) in Table 7.1 [3]. For example, Figure 7.3 depicts the delta MOS for each of the AMR codec modes as a function of FER.

The performance of the CHE and SPE combined is depicted in Figure 7.4.[1] To maximize speech quality performance and channel robustness, the AMR link adaptation algorithm needs

---

[1] The relatively high C/I values in this figure reflect the performance of the receiver assumed in 3GPP TS 26.975 [2].

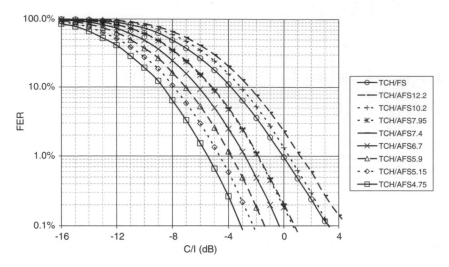

Figure 7.2 TCH/AFS and TCH/FS FER performance as a function of C/I (DL, TU3 channel, ideal frequency hopping, single co-channel interferer).

Table 7.1 Mean Opinion Score for listening quality

MOS	Quality
5	Excellent
4	Good
3	Fair
2	Poor
1	Bad

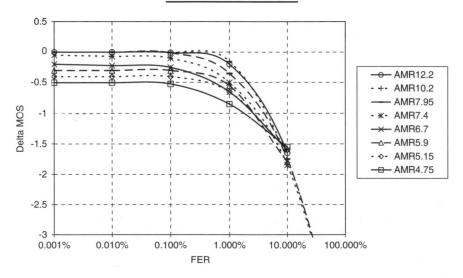

Figure 7.3 AMR MOS degradation as a function of FER [2].

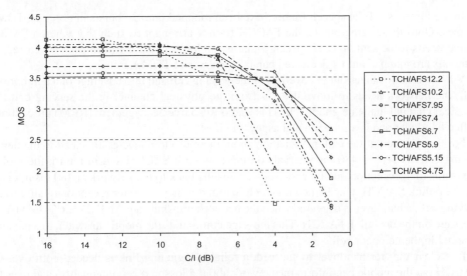

Figure 7.4 AMR MOS as a function of C/I [2].

to adapt the codec mode such that the MOS follows the envelope in Figure 7.4 as a function of the C/I.

7.1.3 Introduction to the Associated Control Channels

From the beginning of a call when a traffic channel (TCH) is allocated, until the end of the call when the TCH is released, the network and mobile need to send and receive radio resource messages that are vital in maintaining the link between the two entities. To transmit these messages, a signaling link is provided over two control channels that are associated with the speech traffic channel (TCH): the Fast Associated Control Channel (FACCH) and the Slow Associated Control Channel (SACCH). These associated control channels are time-multiplexed on the same physical channel as the TCH as depicted in Figure 7.5 (for the full-rate TCH channel).

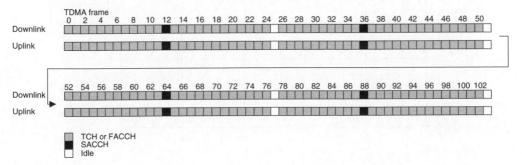

Figure 7.5 Time organization of a TCH + FACCH + SACCH channel combination (mapped onto a full rate physical channel).

In the figure, six TCH speech frames have been mapped onto 24 bursts in every 26 TDMA frames. Onto these same bursts, the FACCH frames are mapped such that when a FACCH frame needs to be sent, a TCH speech frame is 'stolen'. SACCH frames are sent in regular intervals are mapped onto 4 dedicated bursts in every 104 TDMA frames.

When a mobile moves out of coverage of the serving cell and into coverage of a neighboring cell, the network needs to move the call from the physical channel in the serving cell to a physical channel in the neighbor cell. This Radio Resource Management (RRM) procedure is called handover and is a fundamental aspect of GSM.

For the duration of the call, information about the radio frequencies allocated to the neighboring cells is sent at 480 ms intervals on the downlink SACCH and in return, the mobile measures their reception quality and sends the results back to the network at 480 ms intervals on the uplink SACCH. Once a neighbor cell is found to have a better reception quality to the serving cell, a handover is triggered when the network transmits a HANDOVER COMMAND message on the downlink FACCH. This message commands the mobile to switch to a physical channel in the neighbor cell.

It is clear that for handover to succeed, a reliable signaling link is needed both over the SACCH (so the mobile can inform the network about a loss in reception quality) and over the FACCH (so the network can order the mobile switch to a physical channel in a neighbor cell).

Another Radio Resource Management procedure that is a necessary part of GSM is connection release, when a radio resource allocated to a call needs to be terminated. This is normally performed when either party ends the call, but is also needed when the radio link fails and it has not been possible to re-establish the connection. The criteria for radio link failure is based on the success rate of the messages sent over the SACCH. Any deficiencies in the SACCH performance may translate into a prematures release of a call (i.e. increased proportion of dropped calls due to radio link failure).

When we consider the FACCH and SACCH without enhancements, their performance is as depicted in Figure 7.6.

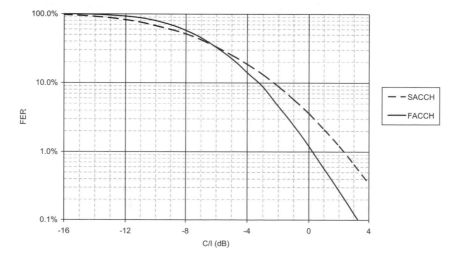

Figure 7.6 FACCH and SACCH link level performance on downlink (TU3 channel, ideal frequency hopping, single co-channel interferer).

7.2   Repeated SACCH

7.2.1   Introduction

The method adopted in 3GPP to improve the robustness of the SACCH is called Repeated SACCH and is based on the principle that when repeated transmissions are combined in the receiver, a decoding gain is achieved thanks to the time diversity between the transmissions. In the case of Repeated SACCH, duplicate SACCH frames are sent by making use of information redundancy that is already inherent in the SACCH transmissions. These are then exploited at the receiving end to produce a second attempt at decoding the frame. The Repeated SACCH was first proposed in 2005 [4] and was standardized to become part of Release 6 of the 3GPP specifications in 2005.

7.2.2   The Repeated SACCH Procedure

The Repeated SACCH procedure, which is described in 3GPP TS 44.006 [5], is depicted in Figure 7.7. For SACCH reception on the downlink, the mobile attempts to decode a SACCH frame, first without attempting to combine it with the previously received SACCH frame. If the decoding fails, the mobile notifies the network (by setting the SACCH Repetition Request indicator bit on the uplink SACCH to "SACCH Repetition Required") and makes a second attempt, this time after combining it with the previously received SACCH frame. The process is then repeated for the next downlink SACCH frame. For SACCH transmission in the uplink, the mobile first checks if Repeated SACCH is enabled in the uplink (by checking if the SACCH Repetition Order indicator bit in the last successfully decoded downlink SACCH frame was set to "SACCH Repetition Order") and that the previously transmitted SACCH frame is not already a duplicate. The mobile then transmits a duplicate frame instead of the originally scheduled frame. The process is then repeated for the next uplink SACCH frame.

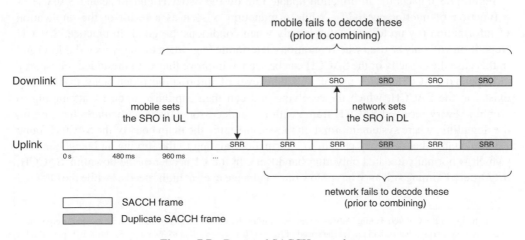

Figure 7.7   Repeated SACCH procedure.

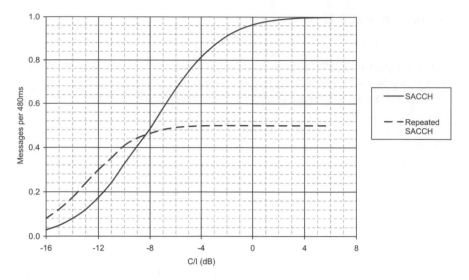

Figure 7.8   Throughput performance of SACCH and Repeated SACCH on downlink.

7.2.3   The Repeated SACCH Concept

To understand the Repeated SACCH concept, it helps to recognize that one of the functions of the SACCH is to enable serving cell measurements during the time that speech frames are not being transmitted (i.e. when the user is muted during discontinuous transmission operation). For this reason, measurement result messages have to be transmitted continuously on the uplink SACCH and system information messages have to be transmitted continuously on the downlink SACCH (unless something else needs to be sent, such as an SMS message).With this principle in mind, it can be understood that the SACCH serves not only to provide an acceptable refresh rate of the information carried by the SACCH but also to allow the mobile and the network to continuously perform its link quality measurements.

Figure 7.8 depicts the information update rate on the SACCH and Repeated SACCH as a function of interference (C/I). While a reduction is seen as a result of the duplication of information (by up to 50%), this is only when conditions are good. In practice, SACCH Repetition will only be used when conditions deteriorate (e.g. when the network C/I is low) and in this case the capacity of the SACCH can be seen to improve thanks to the added robustness.

Repeated SACCH makes use of a certain level of information redundancy that already exists on the SACCH, which the receiving end can then capitalize upon by attempting to combine every received SACCH frame with its predecessor. This is particularly the case on the downlink, where system information messages sent in the main body of the SACCH frame seldom change because they pertain to information relating to the serving and neighbor cells (which is normally updated only after handover). In the L1 header of the downlink SACCH, the Ordered Timing Advance has a slow rate of change even at high speeds,[2] while the Ordered

---

[2] A change in the Ordered Timing Advance will occur after the mobile has traveled a distance which is equivalent to one symbol period, that is 0.55 km (at the speed of light). This is equivalent to 39.6 s (or 82.4 SACCH periods) for a mobile traveling at 50 km/h.

MS Power Level changes at a rate typically determined by uplink quality measurements which are often averaged over a number of SACCH periods to improve their accuracy. Clearly, the criterion that enables Repeated SACCH in uplink and downlink will be critical if the gains are to be to maximized and any unnecessary repetitions minimized. While this criterion has been left for each network vendor to specify, the SACCH Repetition Request indicator bit may be employed, which has been introduced in the uplink SACCH to indicate the occurrence of SACCH frame failures on the downlink.

7.2.4   Case Study: Radio Link Failure

In this section, we evaluate the impact of Repeated SACCH on the probability of radio link failure (RLF). As mentioned earlier, an orderly release of the network's resources needs to be performed when the radio link fails and it is not possible to re-establish the connection. The criteria that triggers a radio link failure is based on SACCH frame decoding failures and a parameter broadcast on the BCCH called RADIO LINK TIMEOUT. Specifically, a radio link failure occurs when the *radio link counter* expires, where this counter is initialized by and never exceeds the value of RADIO LINK TIMEOUT and is decremented by one with every unsuccessfully decoded SACCH frame and incremented by two with every successfully decoded SACCH frame [7].

In 3GPP TS 45.008 [7], it states that a radio link failure should not occur until the call has degraded to beyond the point at which the user would have manually released the call. "This ensures that, for example, a call on the edge of a radio coverage area, although of bad quality, can usually be completed if the subscriber wishes." Hence the criteria that triggers a radio link failure should be robust enough in "bad" quality speech.

To calculate the probability of a radio link failure, the radio link counter can be modelled as a state machine following a Markov process with a limiting border at one side (the value of RADIO LINK TIMEOUT) and an absorbing border on other side (when the counter is zero corresponding to a radio link failure).

The RADIO LINK TIMEOUT value determines the number of states, and the decoding state of each SACCH frame determines the transitions between the states. The probability of a state is accumulated with each SACCH transmission and hence the probability of a radio link failure can be determined by calculating the probability at the absorbing state after a specified number of SACCH transmissions.

If we assume the average call duration is 3 minutes and the probability of a SACCH frame failure is given by the FER shown in Figure 7.9 (for the SACCH and Repeated SACCH) and when considering RADIO LINK TIMEOUT values of 16, 32 and 64, then the probability of a RLT occurring can be calculated, as shown by the curves in Figure 7.9. Given that the probabilistic model above assumes uncorrelated SACCH frame failures, these curves should be taken as an upper bound (because the SACCH frame failures will in practice be correlated due to the multipath and shadow fading effects of the channel).

In Table 7.1, the listening quality for speech has been classified into a Mean Opinion Score (MOS) where a score of "1" is given for bad quality and a score of "2" for poor quality. In this case study, we have assumed the boundary between these levels (i.e. "1.5") as the point at which a user would choose not to maintain the call under any conditions. If a premature

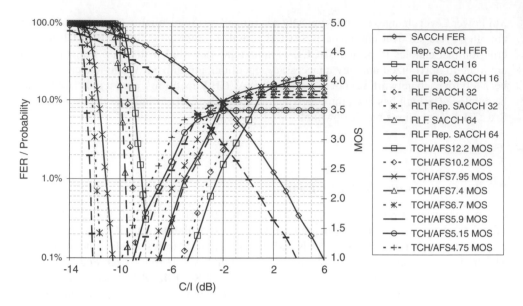

Figure 7.9   Probability of Radio Link Failure verses Mean Opinion Score of AMR.

call drop is therefore to be avoided, then the probability of a radio link failure occurring when speech quality is above this "1.5" level should be zero.

Finally, the MOS verses C/I curves that are needed to compare with the curves depicting the probability of a radio link failure can be obtained from the FER verses C/I curves in Figure 7.2 (which assume the same channel conditions as the FER verses C/I curves for the SACCH and Repeated SACCH in Figure 7.9) and by mapping the FER to MOS by using the MOS verses FER curves in Figure 7.3.

In Figure 7.9, at the point at which the MOS for the most robust AMR codec mode (TCH/AFS4.75) is "1.5", the probability of a radio link failure for RADIO LINK TIMEOUT values of 16, 32 and 64 respectively are 19%, 1.3% and 0.0% for the SACCH and 0.0% in all cases with Repeated SACCH. Clearly the performance of the SACCH is too weak for RADIO LINK TIMEOUT values of 16 and 32 if a zero probability of radio link failure is targeted. In this case study, increasing the RADIO LINK TIMEOUT value to 64 could be used to prevent unnecessary radio link failures, but RADIO LINK TIMEOUT values has high as this are not normally used as they can lead to a late release of a call (where a prolonged radio link failure will affect a user's satisfaction and may result in higher interference levels).

7.3   Repeated Downlink FACCH

7.3.1   Introduction

The method that was adopted to improve the robustness of the FACCH is called Repeated Downlink FACCH and, like the Repeated SACCH, makes use of time diversity between repeated transmissions to increase the likelihood of a successful decode. In the case of Repeated Downlink FACCH, the duplicate FACCH frame is sent by making use of the speech codec's tolerance to having the speech frames "stolen" for the purpose of FACCH transmissions and which can be exploited by the receiving end to provide a second attempt at a successful decoding. Repeated FACCH was first proposed in 2001 [8] and Repeated Downlink FACCH was standardized to become part of Release 6 of the 3GPP specifications in 2005.

7.3.2 The Repeated Downlink FACCH Procedure

The Repeated Downlink FACCH procedure is described in 3GPP TS 44.006 [5] and is depicted in Figure 7.10. The network may transmit a duplicate FACCH frame 8 TDMA frames after sending a FACCH frame (or 9 TDMA frames, if a SACCH or idle frame occurs in-between). This maximizes the time diversity on the full-rate channel and minimizes any FACCH signaling latency on the half-rate channel. If the mobile receives both frames but is unable to decode either of them independently, then a further attempt is made this time after combining them.

The data link layer sends information on the FACCH within LAPDm command and response frames (the latter frame type being specific to acknowledged information transfer) and both frame types are applicable with Repeated Downlink FACCH. Repeated Downlink FACCH is also applicable to legacy mobiles as well as those indicating support for Repeated ACCH.

With a legacy mobile, the use of Repeated Downlink FACCH is restricted to command frames only and in this case a diversity gain is achieved when the legacy mobile attempts to decode both FACCH frames independently. If a mobile explicitly indicates support for Repeated ACCH (a capability indicator in the MS Classmark 3 information element [10]), then the use of Repeated Downlink FACCH can extend to response frames as well. This is because the capability indicator implicity ensures that the mobile has allocated enough time to

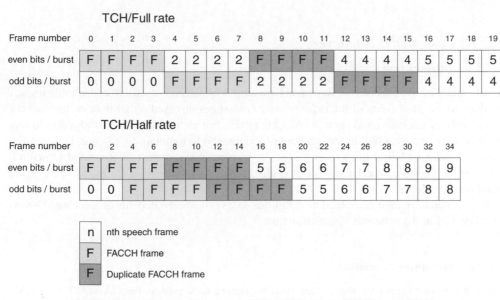

Figure 7.10 Timing of the repeated FACCH frame on TCH/F and TCH/H.

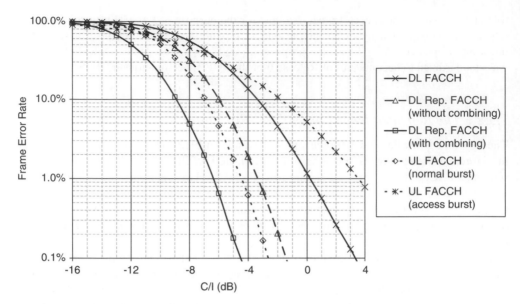

Figure 7.11   FACCH link-level performance (TU3 channel, ideal frequency hopping, single co-channel interferer).

receive a duplicated response frame. Additional diversity gains are obtained in this case with the support of combining. Figure 7.11 shows the layer 1 performance of the downlink FACCH and Repeated Downlink FACCH. For legacy mobiles, the gain is about 3 dB and for mobiles indicating support for Repeated Downlink FACCH, the gain is about 6 dB (when measured at 5% FER).

7.3.3   Case Study: Handover

7.3.3.1   Introduction

To illustrate the effectiveness of Repeated Downlink FACCH, it is useful to consider its impact on the success probability of the radio resource messages involved rather than on the success probability of the individual layer 1 FACCH frames. For example, in the introduction to this chapter, handover was highlighted as being fundamental to maintaining the link between the network and a mobile. In this case study, we investigate the impact of Repeated Downlink FACCH on the success probability of the radio resource messages that are used as part of the handover procedure.

We then analyze their combined probabilities to investigate the benefits of Repeated Downlink FACCH to the handover completion time.

7.3.3.2   Handover Procedure

The relevant messages sent over the air interface during handover are the HANDOVER COMMAND message, the HANDOVER ACCESS message, the PHYSICAL INFORMATION

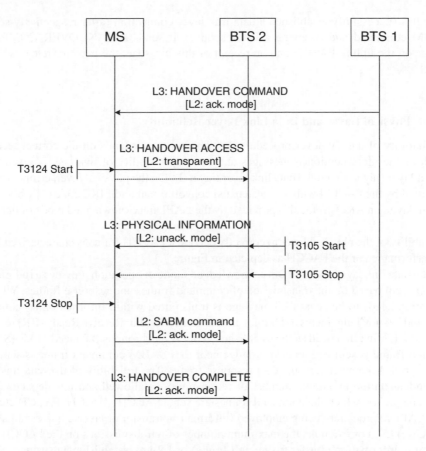

Figure 7.12   Signaling diagram for handover.

Note: The transmission reliability of messages that are not sent in acknowledged mode has been provided by blind retransmissions (shown as dashed lines) that are governed by timers or counters.

message, the SABM command frame and the HANDOVER COMPLETE message [6]. Their transmission sequence is depicted in the signaling diagram in Figure 7.12.

Handover is initiated by the transmission of a HANDOVER COMMAND message on the downlink FACCH. This message instructs the mobile to change channel to a neighboring cell. When the mobile receives a HANDOVER COMMAND message and changes channel, it starts timer T3124 and begins repeating the HANDOVER ACCESS message in access bursts on the uplink FACCH. HANDOVER ACCESS messages allow the network to calculate the timing advance information needed to synchronize the mobile to the network. Once the network has calculated this information, it sends a Timing Advance command back to the mobile in the PHYSICAL INFORMATION message on the downlink FACCH and starts timer T3105. On expiry of T3015, the network retransmits the PHYSICAL INFORMATION message and resets the timer. This is then repeated for a maximum of Ny1 times or until a layer 2 or TCH frame is received. When the mobile receives the PHYSICAL INFORMATION message, it stops transmitting HANDOVER ACCESS messages and stops timer T3124. Finally, when the mobile has successfully established a data link layer connection (after respectively sending a SABM command and receiving a UA response), it sends a HANDOVER COMPLETE message on the uplink FACCH and on receipt of this message, the network releases the old channel.

7.3.3.3 Physical Layer and Data Link Layer Reliability

The reliability of the handover procedure described above depends on the correct reception of each of the aforementioned messages and the responsibility of this task is given to the physical layer (layer 1) and the data link layer (layer 2). On the physical layer, error correction is provided by the FACCH with a concatenated convolutional and FIRE code [1] while on the data link layer, an ARQ protocol is provided by the LAPDm acknowledged mode of operation [9].

The ability of the physical layer to correctly decode a FACCH frame is characterized by the FER performance of the FACCH as depicted in Figure 7.11.

On the data link layer, the LAPDm acknowledged mode of operation provides a rudimentary ARQ protocol based on the transmission of command frames and response frames. When L3 information needs to be sent, a L3 message is transmitted within Information (I) command frames and acknowledgements to I frames are transmitted within Receive Ready (RR) response frames. If a L2 link first needs to be established, a Set Asynchronous Balanced Mode (SABM) command frame is sent and an acknowledgement to a SABM command frame is sent in an Unnumbered Acknowledgement (UA) response frame. The reliability of the data link layer is dependent on the maximum number of retransmissions allowed and this depends on the logical channel on which the command frames are sent (SACCH, SDCCH, FACCH etc.) and on the LAPDm procedure being employed (information transfer or layer 2 link establishment and release) [5]. For example, the maximum number of retransmissions on the FACCH when performing information transfer (using an I frame) or L2 link establishment (using a SABM command frame) is given in Table 7.2.

7.3.3.4 Maximum Transmission Times on the Data Link Layer

The maximum time permitted before a response frame shall be transmitted (after receiving a command frame) or before a command frame shall be retransmitted (after failing to receive a response frame) is also dependent on the logical channel used. For example, the maximum transmission times for the FACCH/Full-rate and FACCH/Half-rate channels are given in Table 7.3. When assuming these maximum times, then the "round-trip" time for a single

Table 7.2 Maximum number of retransmissions on the data link layer on the FACCH

Channel	Maximum number of I frame retransmissions (TDMA frames)	Maximum number of SABM frame retransmissions (TDMA frames)
FACCH/Full rate	34	5
FACCH/Half rate	29	5

Table 7.3 Maximum transmission times for a response frame and a command frame retransmission

	Maximum response delay (TDMA frames)[†]	Maximum retransmission delay(TDMA frames)[‡]
FACCH/Full rate	9	39
FACCH/Half rate	10	44

Notes: [†]delay corresponds to the first burst carrying the response frame and is relative to the last burst carrying the command frame.
[‡]delay corresponds to the first burst carrying the retransmission and is relative to the first burst carrying the initial transmission.
TDMA frame $\approx$ 4.615 ms.

acknowledged LAPDm transmission can be calculated. For example, on the FACCH/Full-rate channel it is 25 TDMA frames or 115 ms. If Repeated Downlink FACCH is used, this is increased by 9 TDMA frames to 157 ms. Further, when we assume the maximum time allowed before a retransmission has to be sent, then the accumulated transmission time is extended by 39 TDMA frames or 180 ms per retransmission.

7.3.3.5 Success Probability Fundamentals

When a LAPDm frame is transmitted within a FACCH frame, the probability of the frame being received in error as a function of C/I can be determined by the FER performance of the FACCH. For example, the FER performance of the FACCH/Full-rate in uplink (using normal and access bursts) and in downlink (with Repeated Downlink FACCH enabled and disabled both for legacy mobiles and mobiles indicating support for the Repeated Downlink FACCH) is depicted in Figure 7.11.

Note that the FER performance shown reflects the receiver's performance against a single interferer, while in practice a channel will consist of many interferers.

Let p denote the probability of a frame received in error (e.g. as determined by the FER for a given C/I). Then the probability of a successful transmission is $1 - p$ and the probability of a successful transmission in one or more of N attempts is

$$1 - p^N \tag{7.1}$$

This equation can be used to determine the success probability of a L3 message transmitted over the data link layer when in unacknowledged mode, where N is the number of transmissions and when assuming uncorrelated transmissions.

The probability of a successful transmission both on the downlink and in the uplink (for example, when a transmission in one direction needs to be acknowledged with a transmission in the other direction), is $(1 - p_d)(1 - p_u)$ and the probability of a successful transmission both on the downlink and in the uplink in one or more of N attempts is

$$1 - (1 - (1 - p_d)(1 - p_u))^N \tag{7.2}$$

This equation can be used to determine the success probability of a L3 message transmitted over the data link layer in acknowledged mode, where N is the number of transmissions and where p_d and p_u denote the probability of a frame received in error on the downlink and uplink respectively.

When the size of a L3 message exceeds the payload size of a L2 frame, the L3 message first needs to be segmented into K L2 frames. In which case, the success probability of K L2 frames after N transmissions (where $N \geq K$) is

$$1 - \sum_{i=0}^{K-1} \binom{N}{i} ((1 - p_d)(1 - p_u))^i (1 - (1 - p_d)(1 - p_u))^{N-i} \tag{7.3}$$

7.3.3.6   HANDOVER COMMAND Message

One of the purposes of the HANDOVER COMMAND message is to convey the physical channel in the neighbor cell that the mobile must switch to. This includes the mobile's allocation of carrier frequencies in the case of frequency hopping. Depending on the number of hopping frequencies, the HANDOVER COMMAND message may need segmenting over two L2 frames. Transmission of the HANDOVER COMMAND message is in acknowledged mode, hence to calculate the probability of a successful transmission we use Equation (7.3), where $K = 2$, p_d and p_u is the FER of the FACCH on the downlink and uplink respectively (given in Figure 7.12 for normal bursts) and N is 1. . .35 (where the maximum number of retransmissions i.e. the maximum number of transmissions minus one is given in Table 7.2 for information transfer on the FACCH/Full-rate). The maximum transmission delay will be 115 + $n \times 180$ ms for an acknowledged LAPDm command frame on the FACCH/Full-rate and this increases to $157 + n \times 180$ ms with Repeated Downlink FACCH, where n is the number of retransmissions. When taking into account the above assumptions, it is possible to calculate the success probability and transmission delay of the HANDOVER COMMAND message on the FACCH/Full-rate. This is shown in Figure 7.13 for FACCH and Repeated Downlink FACCH.

7.3.3.7   HANDOVER ACCESS Message

On receipt of a HANDOVER COMMAND message, the mobile starts repeatedly transmitting the HANDOVER ACCESS message in access bursts in successive TDMA frames on the uplink FACCH(resulting in a transmission delay of 5ms per HANDOVER ACCESS message). On the network side, their time of arrival provides the information needed to synchronize the mobile to the neighbor cell. Between non-synchronized cells, the HANDOVER ACCESS message is sent in consecutive TDMA frames until the PHYSICAL INFORMATION message is received or the timer T3124 expires. On a traffic channel, T3124 is fixed at 320 ms, hence to calculate the probability of a successful HANDOVER ACCESS message we use Equation (7.1), where N is 69 and p is the FER of access bursts sent on the uplink FACCH (given in Figure 7.12). When taking into account the above assumptions, it is possible to calculate the success probability and transmission delay of the HANDOVER ACCESS message on the FACCH. This is shown in Figure 7.14.

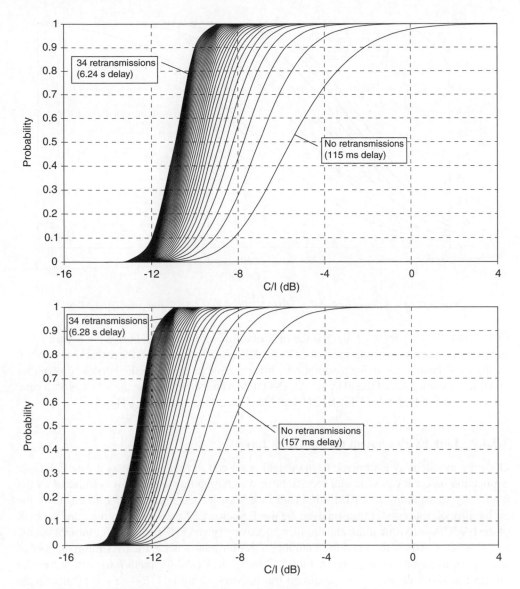

Figure 7.13   Success probability of HANDOVER COMMAND and associated delay for FACCH (top) and DL Repeated FACCH (bottom).

7.3.3.8   PHYSICAL INFORMATION Message

The timing advance information sent to the mobile to allow it to synchronize to a neighbor cell is transmitted by the network in successive PHYSICAL INFORMATION messages. These messages are sent in intervals determined by the timer T3105 until either a SABM command frame or a speech frame is received, or until a maximum of Ny1 retransmissions have been sent. Ny1 is a network dependent parameter that ranges from 3 to 64, hence we use Equation

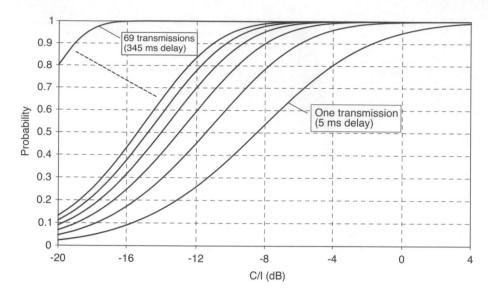

Figure 7.14   Success probability of HANDOVER ACCESS and associated delay.

(7.1), where N is $1...\text{Ny}1+1$ and p is the FER of the downlink FACCH (given in Figure 7.11). In this case study we assume the values 3 and 80ms for Ny1 and T3105 respectively. When taking into account the above assumptions, it is possible to calculate the success probability and transmission delay of the PHYSICAL INFORMATION message on the FACCH/Full-rate. This is shown in Figure 7.15 for FACCH and Repeated Downlink FACCH.

7.3.3.9   Link Establishment on Data Link Layer

Once the mobile is synchronized to the neighboring cell, it first establishes a data link layer connection before it can send and receive layer 3 signaling messages. This is initiated by the mobile by transmitting a SABM command in acknowledged mode. Hence to calculate the probability of a successful transmission we use Equation (7.2), where p_d and p_u are the FER of the FACCH on the downlink and uplink respectively (given in Figure 7.11 for normal bursts) and N is $1...6$ (where the maximum number of retransmissions i.e. the maximum number of transmissions minus one is given in Table 7.2 for L2 link establishment information transfer on the FACCH/Full-rate). The maximum transmission delay is $115 + n \times 180$ ms for an acknowledged LAPDm command frame on the FACCH/Full-rate and this increases to $157 + n \times 180$ ms with Repeated Downlink FACCH, where n is the number of retransmissions. When taking into account the above assumptions, it is possible to calculate the success probability and transmission delay of the SABM command on a FACCH/Full-rate. This is shown in Figure 7.16 for FACCH and Repeated Downlink FACCH.

7.3.3.10   HANDOVER COMPLETE Message

The mobile informs the network about a successful handover by sending a HANDOVER COMPLETE message. Transmission of the HANDOVER COMPLETE message is in

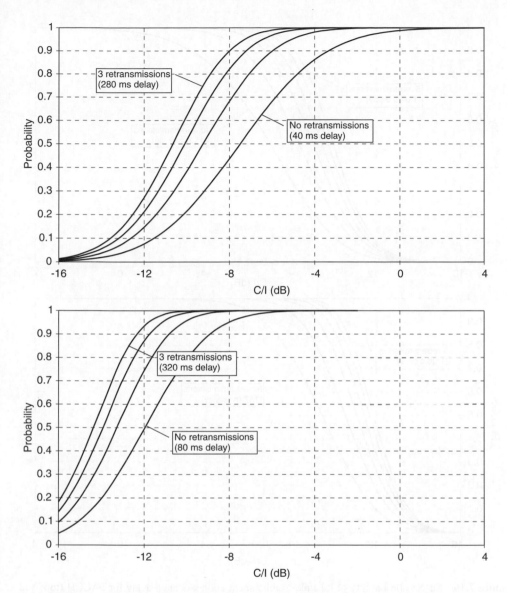

Figure 7.15 Success probability of PHYSICAL INFORMATION and associated delay for FACCH (top) and DL Repeated FACCH (bottom).

acknowledged mode, hence to calculate the probability of a successful transmission of the HANDOVER COMPLETE, we use Equation (7.3), where p_d and p_u are the FER of the FACCH on the downlink and uplink respectively (given in Figure 7.11 for normal bursts) and N is 1. . .35 (where the maximum number of retransmissions, i.e. the maximum number of transmissions minus one given in Table 7.2 for information transfer on the FACCH/Fullrate). The maximum transmission delay for an acknowledged LAPDm command frame on the

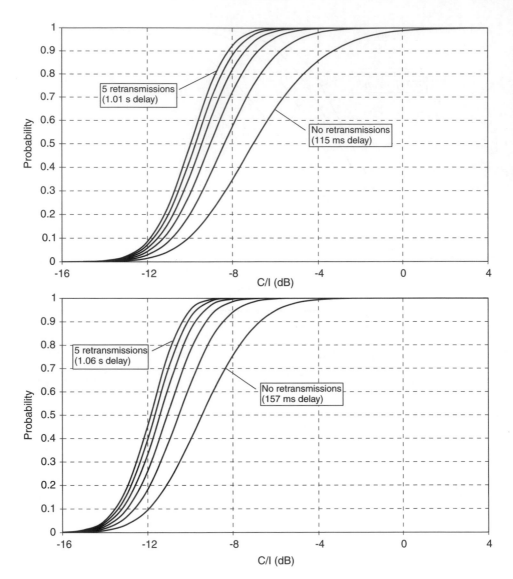

Figure 7.16 Success probability of L2 link establishment and associated delay for FACCH (top) and DL Repeated FACCH (bottom).

FACCH/Full-rate is $115 + n \times 180$ ms and this increases to $157 + n \times 180$ ms with Repeated Downlink FACCH, where n is the number of retransmissions. When taking into account the above assumptions, it is possible to calculate the success probability and transmission delay of the HANDOVER COMPLETE message on the FACCH/Full-rate. This is shown in Figure 7.17 for FACCH and Repeated Downlink FACCH.

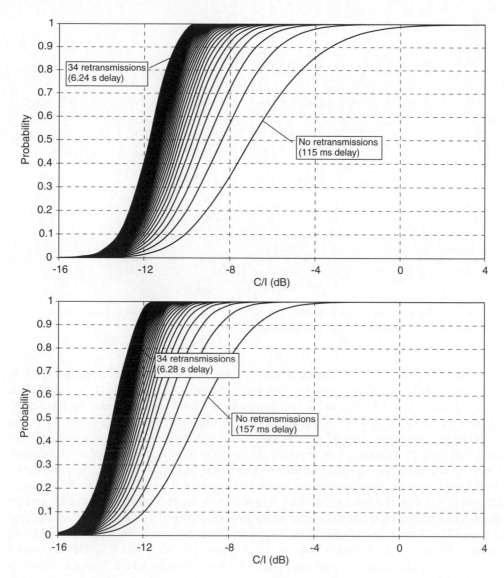

Figure 7.17 Success probability of HANDOVER COMPLETE and associated delay for FACCH (top) and DL Repeated FACCH (bottom).

7.3.3.11 Handover Completion

The probability of a successful handover is the combined probability of all the aforementioned messages. If we assume a message success probability of no less than 99% (which is equivalent to a 95% handover success probability or higher), we can calculate from the required number of transmissions the time needed to perform the handover. This is shown in Figure 7.18 both with and without Repeated Downlink FACCH.

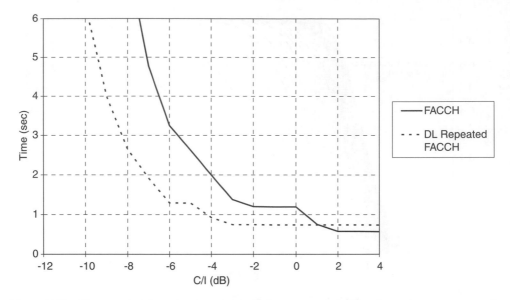

Figure 7.18 Time needed for a handover with 99% success probability for each message transfer (equivalent to a 95% handover success probability).

In relatively good channel conditions, the completion time with Repeated Downlink FACCH is shown to take longer. This is because of the worst-case assumptions on the delay calculation, where an additional 9 TDMA delay had been assumed for Repeated Downlink FACCH to account for the duplicate FACCH frame. In practice, this additional delay will not be seen in good channel conditions, because the first instance of the Repeated FACCH pair is quite likely to be succeed. Furthermore, it is expected that a network will enable Repeated Downlink FACCH only when the channel conditions deteriorate. In Figure 7.18, as the interference level increases, the handover completion time reduces exponentially thanks to the Repeated Downlink FACCH. Note that the discontinuities seen in the plot are the result of the integer number in retransmissions. If we assume a user will manually end the call after having endured a 5 second handover interruption time, then at this limit the user should be able to tolerate about 2 dB higher interference thanks to the Repeated Downlink FACCH. This additional robustness will particularly benefit interference limited networks, where handover failure is often the dominant cause for a dropped call.

The average number of transmissions needed to receive a message consisting of K frames could alternatively be calculated by using Equation (7.4)

$$K(1 - p_d)(1 - p_u) \tag{7.4}$$

where p_d and p_u denote the probability of the frame being received in error in downlink and uplink respectively. From the average number of transmissions it is possible to determine the average time needed for each message in handover and the combined delay for all the messages.

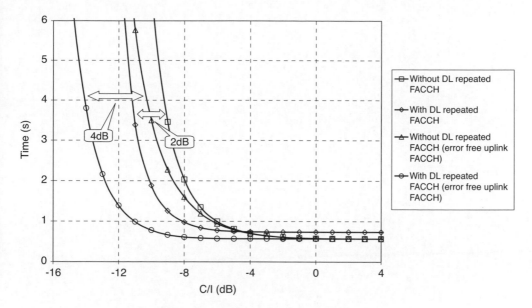

Figure 7.19   Average time needed for handover.

This is shown in Figure 7.19 as a function of C/I both with and without Repeated Downlink FACCH. Like the calculation based on the individual success probabilities, a 2 dB gain is clearly visible with Repeated Downlink FACCH.

The difference in the gain at this level with the gain at the physical layer level (where Repeated Downlink FACCH shows about 4 dB gain measured at the 66% FER level in Figure 7.12), can be attributed to the impact on the handover procedure from the uplink FACCH transmissions. For example, if we consider error-free conditions in uplink, the gain is increased to 4 dB with Repeated Downlink FACCH. Handover performance clearly relies and a balanced link between the uplink and downlink.

References

[1]   3GPP TS 45.003, v6.9.0, "Channel Coding", January 2006.
[2]   3GPP TS 26.975, v6.0.0, "Performance Characterization of Adaptive Multi-Rate (AMR) Speech Codec", December 2004.
[3]   ITU-T Recommendation P.800 Telephone Transmission Quality: Methods for Objective and Subjective Assessment of Quality, 1996.
[4]   3GPP TDoc GP-050943, "Repeated SACCH", Nokia, TSG-GERAN #24.
[5]   3GPP TS 44.006, v6.8.0, "Data Link (DL) Layer Specification", December 2008.
[6]   3GPP TS 44.018, v6.24.0, "Radio Resource Control (RRC) Protocol", February 2009.
[7]   3GPP TS 45.008, v6.21.0, "Radio Subsystem Link Control", February 2009.
[8]   3GPP TDoc GP-010246, "ACCH Enhancements for Full-rate AMR Speech Traffic Channels", Nokia, TSG-GERAN #3.
[9]   3GPP TS 44.005, v6.1.0, "Data Link (DL) Layer General Aspects", November 2005.
[10] 3GPP TS 24.008, v6.19.0, "Core Network Protocols", June 2008.

8

Orthogonal Sub-Channels with AMR/DARP

Mikko Säily, Jari Hulkkonen, Kent Pedersen, Carsten Juncker,
Rafael Paiva, Renato Iida, Olli Piirainen, Seelan Sundaralingam,
Alexandre Loureiro, Jon Helt-Hansen, Robson Domingos and
Fernando Tavares

8.1 Introduction

The number of GSM subscriptions worldwide reached 4 billion by the end of 2009 [1]. Voice is still one of the most important mobile services and its traffic continues to grow. In order to accommodate this growth, new voice enhancement concepts have continuously been developed for GSM. Both standardized and vendor-specific features have been developed to increase GSM networks' voice capacity. The Spectral Efficiency (SE) has improved many times since the introduction of GSM Phase 1 voice at the beginning of the 1990s. Performance has also improved, for example with more robust speech coding, advanced receivers and concepts allowing more users per channel. Moreover, Radio Resource Management (RRM) methods have been developed in order to optimize the available resources in changing traffic conditions and to fulfill the quality requirements of network operators. High capacity, spectral efficiency, cell coverage, low call drop rate, low blocking probability and high handover success rate are the basic requirements for a network to provide end users with a high quality voice service. In the past few years energy efficiency has also become an increasingly important aspect for network operators.

In this chapter, a performance evaluation of GSM voice evolution is presented. The main focus is on the latest voice performance features, namely Orthogonal Sub-Channel (OSC), Voice over Adaptive Multi-user channels on One Slot (VAMOS), Adaptive Multi-Rate (AMR), Single Antenna Interference Cancellation (SAIC) and Downlink Advanced Receiver Performance (DARP). First, an overview of GSM voice evolution is given in Section 8.2. Then, the performance of AMR with SAIC/DARP is presented in Section 8.3. A detailed investigation

GSM/EDGE: Evolution and Performance   Edited by Mikko Säily, Guillaume Sébire and Eddie Riddington
© 2011 John Wiley & Sons, Ltd

of the Orthogonal Sub-Channel then follows in Section 8.4 and finally a summary is given in Section 8.5.

8.2   Overview of GSM Voice Evolution

8.2.1   Voice Capacity

The GSM voice spectral efficiency (SE) is used to quantify the number of voice users a network (or site or cell) is able to carry in a given spectrum. Spectrum is a limited resource for all radio technologies and its use requires a license and significant investment. The lower frequency bands are especially attractive since they provide better coverage than those at higher frequencies. However, one challenge with the GSM 900 MHz band is that the allocations are typically quite small (between 5 MHz to 10 MHz) and scattered. "Re-farming" the 900 MHz spectrum from GSM to HSPA has started in a few countries, but HSPA requires a minimum of 4.2 MHz unscattered spectrum [2], which means that a HSPA 900 MHz deployment may require more than 50% of the original spectrum available for GSM. Given that today most of the voice traffic is still carried over GSM networks, GSM voice capacity has become an important issue for network operators. Re-farming of a spectrum must avoid any degradation of service.

Tighter frequency reuse factors can be used to improve a network's SE by allowing more TRXs per cell for the same available spectrum. GSM networks utilizing AMR and enhanced RRM algorithms can be operated with very tight frequency reuses. For example, AMR performance with tight frequency reuse factors has been studied in [3], where it is shown that AMR can improve the network SE up to 150%.

Figure 8.1 describes how spectral efficiency of GSM voice has developed from GSM Full-rate speech to the VAMOS concept introduced in 3GPP Release 9. Figure 8.1 includes potential features for a further voice evolution as well as how the voice performance KPIs have been evolving.

8.2.1.1   Adaptive Multi-Rate (AMR)

Next to the introduction of GSM Phase 1 voice service, AMR has been one of the most significant voice enhancement features deployed. It provides better voice quality by introducing a new speech codec which adapts the speech and channel coding to the changes in radio signal quality. This increased robustness to channel errors can be utilized for both system capacity and quality improvements. The improved voice quality has also enabled the use of half-rate speech via AMR. Then, a major step towards a high speech quality was achieved with the introduction of the AMR-WB speech codec. AMR-WB is evaluated in Chapter 9.

8.2.1.2   Downlink Advanced Receiver Performance (DARP)

DARP represents a significant step in downlink performance evolution. With DARP techniques the mobiles can tolerate more interference and thereby operate in more challenging interference conditions. This allows network operators to deploy tighter frequency reuse factors, which leads to higher spectral efficiency. Together with the robust AMR speech codecs, DARP is

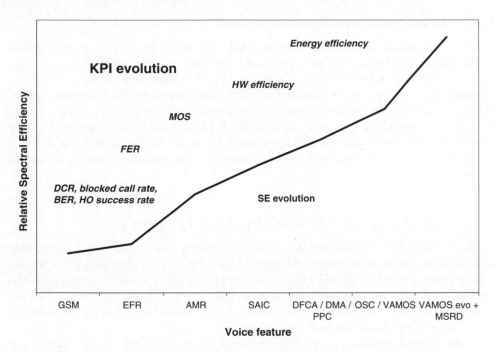

Figure 8.1 GSM network voice SE evolution.

able to provide good speech quality with very low C/I values. Detailed performance results are shown in Section 8.3.

8.2.1.3 Dynamic Frequency and Channel Allocation (DFCA), Progressive Power Control (PPC)

DFCA and PPC use AMR flexibility to further improve network quality and capacity. These are examples of vendor-specific RRM algorithms that the network operators can use to optimize their networks. DFCA is described in Chapter 10 and PPC and other non-standardized capacity enhancement techniques are described in Chapter 12.

8.2.1.4 Orthogonal Sub-Channels (OSC), VAMOS

The OSC and VAMOS concepts have recently been introduced. At the beginning of 2009, Nokia Siemens Networks successfully demonstrated the OSC concept in a network trial where four calls were carried in one GSM radio time slot [4]. The OSC concept does not require standardization but DARP Phase 1 capable mobile stations are needed [5] for successful operation. VAMOS is a GERAN Release 9 feature which benefits from a new set of Training Sequences (TSC) and two new mobile classes. VAMOS is presented in detail in Chapter 4. Both OSC and VAMOS double the number of voice channels per TRX which allows an increase in the network HW efficiency. Performance results are shown in Section 8.4.

8.2.1.5   Mobile Station Receiver Diversity (MSRD)

MSRD is a potential enhancement to push GSM voice capacity to the next level. MSRD is specified in 3GPP Release 7 and mobiles able to fulfill the MSRD performance requirements are denoted as DARP Phase 2 mobiles. MSRD is described further in Chapter 2. Currently this technique is not deployed in small mobile handsets. However, new systems like HSPA and LTE are pushing terminals towards receiver diversity. Dual antenna receiver diversity is commonly used in base stations, which has resulted in the downlink being the limiting link for voice capacity. The introduction of mobile station diversity in downlink would challenge the uplink performance which could be further enhanced with 4-way antenna diversity.

8.2.2   Voice Performance KPIs and Hardware Efficiency

The *Drop Call Rate (DCR)* and the *Handover (HO) success rate* have always been important KPIs since they directly affect the end user's experience. A lost connection has a negative impact on the end user's perceived quality of service and hence a well-performing network should have a DCR which is lower than a pre-defined limit (such as 2%). The *Blocked Call Rate (BCR)* is another important criterion which should also be lower than a predefined limit (again, such as 2%) during the busy hour traffic.

Following the introduction of AMR, more interest has been directed to voice quality. Low *Frame Erasure Rate (FER)* indicates good voice quality, but FER alone cannot account for the difference in quality between the different speech codecs. Therefore, in case of AMR an indicator reflecting a subjective speech quality is needed such as defined by the *Mean Opinion Score (MOS)*. While this is not available from the measurement reports, tools exist in the market to measure and quantify the subjective speech quality of the available link data.

The main benefit from OSC and VAMOS is an improvement in *hardware (HW) efficiency*. Figure 8.2 shows the evolution of the number of voice channels supported per GSM TRX. The GSM carrier is divided into 8 time slots allowing a maximum of 8 full-rate channels to be allocated to one TRX. A half-rate channel only uses every second time slot thereby doubling the capacity to 16. In OSC and VAMOS two users are allocated to the same channel. Therefore, these techniques can provide up to 32 half-rate connections per TRX.

In case of the half-rate codec modes, higher C/I values are required to achieve a performance which is comparable to the full-rate codec modes. Therefore, connections in poor radio conditions must use full rate. Similarly, OSC has higher C/I requirements compared to the normal half-rate channel. In networks where a high SE is required and interference conditions are challenging, only a proportion of the users can be allocated to normal half-rate and OSC half-rate channels. In this task, advanced RRM algorithms play an important role to allocate users to the most fitting channel and mode.

HW efficiency is also closely related to energy efficiency. Energy consumption is becoming increasingly important in GSM/EDGE networks. Also, cost efficiency is important especially in low ARPU regions. These are addressed in more detail in Chapter 13.

In multimode networks (i.e. networks that support two or more radio access technologies such as GERAN, UTRAN or E-UTRAN) frequency and potentially also power resources may be shared between systems. With higher GSM SE more resources can be allocated to the new systems, and/or for the GSM/EDGE data traffic.

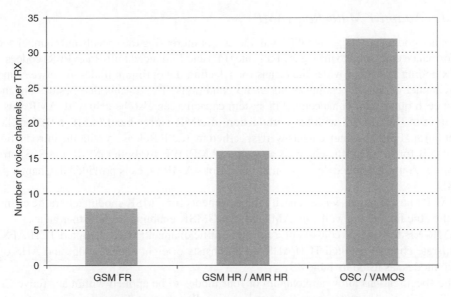

Figure 8.2   Evolution of GSM voice channels per TRX.

Overall, as the GSM voice is still one if not the most important mobile service, and as it continues to grow, high GSM SE is of key importance to network operators.

8.3   AMR and SAIC Performance

In November 2002, TSG GERAN initiated a feasibility study with the scope of studying the performance of Single Antenna Interference Cancellation (SAIC) in the mobile station. The basic idea behind interference cancellation in the downlink is to enhance the receiver performance by means of advanced signal processing techniques which increase the robustness towards interference in the network. Single antenna reception was considered in order to minimize the cost of the mobile station implementation. As mentioned earlier, the enhanced performance of the mobile station receiver in interference limited scenarios might be utilized to enable tighter frequency re-uses in the network provided that the penetration of DAR-capable mobile station is sufficiently high.

The feasibility study of SAIC was completed in 2004 [6], following which performance requirements were specified in 3GPP under the Downlink Advanced Receiver Performance (DARP) Phase 1 feature. Mobile station receiver diversity (MSRD) was later introduced as DARP Phase 2. It is further described in Chapter 2.

DARP Phase 1 only targets GMSK-modulated signals, as the SAIC signal processing techniques showed the most promising gains for this modulation type [6]. A natural consequence of this is that DARP Phase 1 primarily targets improvements in voice capacity.

DARP Phase 1 has been the de facto standard for most mobile station vendors, and therefore the market penetration of such mobiles is high. Sections 8.3.1 to 8.3.3 describe the principles of AMR for GMSK-modulated speech traffic and quantify the link level gains that can be achieved for these channels with the use of DARP Phase 1.

8.3.1   Adaptive Multi-Rate (AMR)

AMR is a codec introduced in GSM with the aim of increasing the speech and channel coding performance compared to the legacy Full-rate (FR) and Enhanced Full-Rate (EFR) codecs. The AMR coding system provides the means for selecting the optimum trade-off between speech and channel coding depending on the radio conditions. This means that besides enhancing the speech quality, an enhancement in system capacity can also be achieved. AMR was first specified in Release 98, with the introduction of AMR narrowband (AMR-NB) for GMSK modulation. Further enhancements were specified in 3GPP Release 5 with the introduction of AMR-NB for 8PSK traffic channels as well as AMR-WB for both GMSK and 8PSK traffic channels. A more detailed description of the various AMR types is provided in Chapter 4 and in [3].

DARP Phase 1 targets voice capacity improvements for GMSK-modulated traffic channels, and thus the feature fits well with AMR-NB for GMSK enabling even further gains.

AMR-NB for GMSK consists of a full-rate traffic channel (TCH), denoted TCH/AFS and a half-rate channel denoted TCH/AHS. AFS consists of eight codec modes and AHS of six codec modes, as listed in Table 8.1.

The specifications allow a maximum of four modes to be applied within an Active Codec Set. The selection between these four modes is handled through link adaptation in the mobile station and in the network [7]. An AMR-capable mobile station continuously monitors its perceived link quality and suggests which codec mode should be applied in the downlink, based on a set of thresholds signaled from the network. The signaling of the requested codec mode as well as the applied codec mode is performed through in-band signaling using bits of the traffic channel. The applied codec mode can be changed every second speech frame which corresponds to every 40 ms (full-rate).

The link adaptation of AMR described above selects the best suited speech codec mode taking the performance of the mobile station into account as well as the overall carrier to

Table 8.1   AMR codec modes for TCH/AFS and TCH/AHS

TCH	Speech codec rate
TCH/AFS 12.2	12.2 kbit/s
TCH/AFS 10.2	10.2 kbit/s
TCH/AFS 7.95	7.95 kbit/s
TCH/AFS 7.4	7.4 kbit/s
TCH/AFS 6.7	6.7 kbit/s
TCH/AFS 5.9	5.9 kbit/s
TCH/AFS 5.15	5.15 kbit/s
TCH/AFS 4.75	4.75 kbit/s
TCH/AHS 7.95	7.95 kbit/s
TCH/AHS 7.4	7.4 kbit/s
TCH/AHS 6.7	6.7 kbit/s
TCH/AHS 5.9	5.9 kbit/s
TCH/AHS 5.15	5.15 kbit/s
TCH/AHS 4.75	4.75 kbit/s

interference levels in the network (cell). This can be exploited to increase system capacity while still maintaining a good speech quality, as is explained below.

As mentioned, DARP is a general receiver enhancement which improves the link performance of a GMSK-modulated channel. Thus, DARP introduces gains for all AMR modes allowing further capacity improvements. As will be shown in Section 8.3.3, the gains from DARP are constant between the different AMR codec modes.

8.3.2 Downlink Advanced Receiver Performance (DARP)

DARP Phase 1 is specified in 3GPP Release 6 and the specifications primarily consist of a new set of performance requirements for the mobile station [8], a new set of test scenarios and DARP capability signaling. The capability signaling was introduced in such a way that DARP may be implemented as a release independent feature, thereby allowing DARP to be introduced in pre-3GPP Release 6 mobile station implementations.

As mentioned earlier, DARP enables the mobile station to cope with higher interference in the network which can be exploited to increase network capacity provided that the market penetration of DARP mobile stations is sufficiently high. For an end-user, the feature may result in an increase in voice quality since the mobile station would allow a higher speech code rate than a legacy mobile station. That is, a DARP mobile station may be able to use TCH/AFS12.2 at certain signal conditions where a legacy mobile station would apply TCH/AFS7.95. Examples of the link level gains of DARP with AMR are provided in Section 8.3.3.

In order to specify a new set of performance requirements for the DARP mobile station, a new set of test scenarios were defined by 3GPP TSG GERAN: the DARP Test Scenarios (DTS) [8]. Their scope is to provide a test environment that approximates to the interference environment of a real network while allowing the test and measure of a mobile station in a laboratory environment. During the feasibility study phase of SAIC, a number of link and system level network configurations were developed with the aim of evaluating the gains of SAIC by means of simulations [6]. The configurations were based on data derived from live network implementations for a number of different network deployments. The DARP test scenarios are inspired by these configurations, but have been simplified in order to be better suited for implementation in test equipment.

The DARP test scenarios seek to model the following conditions:

- single and multiple interferers;
- training sequence correlation between wanted signal and interferer;
- synchronous and asynchronous network configurations.

Tables 8.2 to 8.5 list the five different DARP test scenarios as defined in 3GPP TS 45.005 [8]. Each test scenario is characterized by a number of parameters. The "interfering signal" column defines the type of interference, such as co-channel, adjacent channel or white noise (AWGN). The "interferer relative power level" defines the power level of each signal relative to the dominant co-channel interferer (for a single interferer scenarios namely DTS-1 and DTS-4, this parameter is not applicable). For DTS-2 it can be observed that for example Co-channel 2 has 10 dB lower signal level than that of the dominant interferer. The "TSC" column describes whether the interferer includes a training sequence code or consists of random bits. It is well

Table 8.2 DARP test scenario for synchronous interference; single interferer

Reference test scenario	Interfering signal	Interferer relative power level	TSC	Interferer delay range
DTS-1	Co-channel 1	0 dB	none	no delay

Table 8.3 DARP test scenario for synchronous interference; multiple interferers

Reference test scenario	Interfering signal	Interferer relative power level	TSC	Interferer delay range
DTS-2	Co-channel 1	0 dB	none	no delay
	Co-channel 2	−10 dB	none	no delay
	Adjacent 1	3 dB	none	no delay
	AWGN	−17 dB	–	–
DTS-3	Co-channel 1	0 dB	random	−1 to +4 symbols*)
	Co-channel 2	−10 dB	none	no delay
	Adjacent 1	3 dB	none	no delay
	AWGN	−17 dB	–	–

Note: * The delay should be an integer number of symbols, arbitrarily chosen within the given interval and fixed throughout each test case.

Table 8.4 DARP test scenario for asynchronous interference; single interferer

Reference test scenario	Interfering Signal	Interferer relative power level	TSC	Interferer delay
DTS-4	Co-channel 1	0 dB*)	none	74 symbols

Note: * The power of the delayed interferer burst, averaged over the active part of the wanted signal burst. The power of the delayed interferer burst, averaged over the active part of the delayed interferer burst is 3 dB higher.

Table 8.5 DARP test scenario for asynchronous interference; multiple interferers

Reference test scenario	Interfering signal	Interferer relative power level	TSC	Interferer delay
DTS-5	Co-channel 1	0 dB*)	none	74 symbols
	Co-channel 2	−10 dB	none	no delay
	Adjacent 1	3 dB	none	no delay
	AWGN	−17 dB	–	–

Note: * The power of the delayed interferer burst, averaged over the active part of the wanted signal burst. The power of the delayed interferer burst, averaged over the active part of the delayed interferer burst is 3 dB higher.

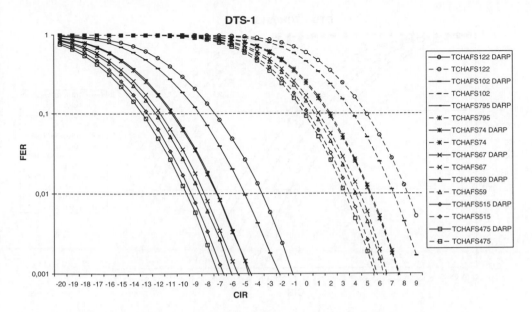

Figure 8.3 Link-level performance of legacy and DARP Phase 1 MS. AMR full-rate codecs (TCH/AFS). DTS-1 scenario.

known that the choice of training sequence affects the link level performance, because the different training sequences have different cross-correlation properties. If the network is not synchronized, random bits can provide a more accurate model of the average performance in a real network where the mobile station is prone to a mix of interferers at different delays. DTS-3 consists of the same interferer signal levels as DTS-2, but correlates the TSC to emulate the impact of a synchronous interferer. For asynchronous interferers, the dominant interferer is offset by 50% of a normal burst as shown in the column "Interferer delay".

The DARP test scenarios are to be seen as additional tests applicable for the DARP mobile station. That is, all legacy test cases still apply.

All DARP test cases are to be conducted using a typical urban channel profile with a mobile station speed of 50 km/h. Examples of the link performance of DTS-1 and DTS-2 are presented in the following section.

8.3.3   AMR DARP Link Performance

The link level performance of a DARP Phase 1 capable mobile station using AMR-NB is exemplified in Figures 8.3 to 8.6 and compared to that of a legacy mobile station implementation. Single and multiple interferer scenarios are considered (DTS-1 and DTS-2) as well as TCH/AFS and TCH/AHS.

It can be observed from Figures 8.3 and 8.4 that in case of single interference (DTS-1) the DARP Phase 1 mobile station has a significant performance gain compared to the legacy mobile station. The gain is in the range of 12–13 dB when looking at the 1% frame erasure rate.

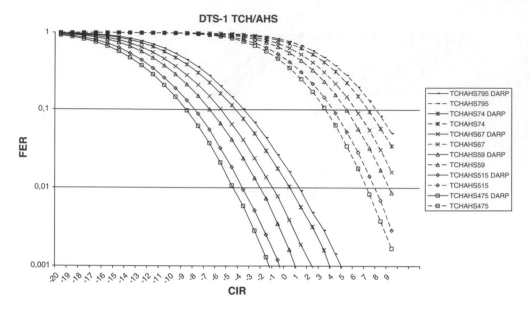

Figure 8.4 Link-level performance of legacy and DARP Phase 1 MS. AMR half-rate codecs (TCH/AHS). DTS-1 scenario.

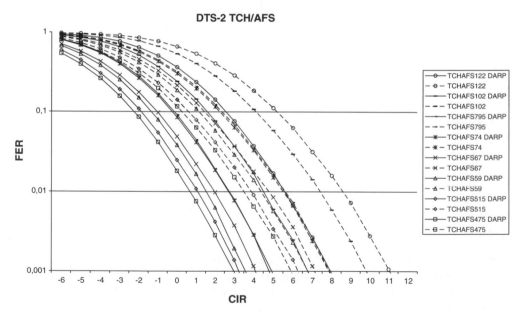

Figure 8.5 Link-level performance of legacy and DARP Phase 1 MS. AMR full-rate codecs (TCH/AFS). DTS-2 scenario.

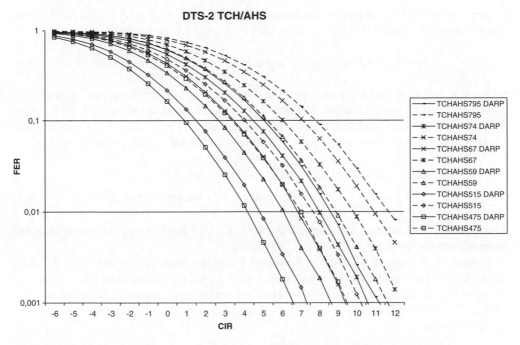

Figure 8.6 Link-level performance of legacy and DARP Phase 1 MS. AMR half-rate codecs (TCH/AHS). DTS-2 scenario.

The multi-interferer scenarios of Figure 8.5 and Figure 8.6 show that also here the DARP Phase 1 mobile station performs better than the legacy mobile station. The gains are more modest than those achieved in the single interferer scenarios, but are in the order of 3 dB which is quite remarkable considering the level and variations in the interference type.

Today, both AMR and DARP Phase 1 are standard features of most mobile station vendors. Although the results presented in this section have focused on AMR-NB, the gains are equally applicable to AMR-WB. DARP is a general receiver improvement for GMSK modulated channels and thus the feature will also provide significant gains for AMR-WB.

8.4   OSC and VAMOS Performance

The following sections describe OSC and VAMOS fundamentals and performance. First, the general principles are described. Then, link and system level performance results are presented. Finally, OSC performance gains are demonstrated by means of drive test measurement results.

8.4.1   The Principles of the DL Concept

In this section, the basic principles for OSC and VAMOS downlink are described. First, we consider a frequency flat channel, that is a channel without inter-symbol interference (ISI).

The GMSK modulation can be seen as a binary modulation $x_i = \{-1, 1\}$, and for this case the maximum likelihood (ML) decision for the i th symbol can be formed as

$$\text{argmax}_{x_i}(\ln(p(y_i|x_i, h))) = \text{argmax}_{x_i}(2 \cdot \text{real}(y_i^H h x_i) - x_i^H h^H h x_i) \qquad (8.1)$$

where y_i is the i th received sample, h is a scalar channel and x_i is the modulated symbol.

The decision metric is clearly real. Now, consider another binary modulated symbol with symbols, $z_i = \{-1, 1\}$, and define a new transmitted symbol as $\hat{x}_i = x_i + z_i j$ carrying two binary information symbols, where $j = \sqrt{-1}$. It follows the ML metric can be formed as

$$\text{argmax}_{(x_i, z_i)}(\ln(p(y_i|\hat{x}_i, h))) = \text{argmax}_{x_i}(2 \cdot \text{real}(y_i^H)x_i - x_i^H h^H h x_i)$$
$$+ \text{argmax}_{z_i}(-2 \cdot \text{imag}(y_i^H h)z_i + z_i^H h^H h z) \qquad (8.2)$$

Thus, the terms related to x_i and z_i are independent of each other and they can also be decoded independently of each other without any interference.

In practice, the situation is more complicated due to the frequency selectivity of the radio channels. The multi-path fading channels cause inter-symbol interference and also interference between the two information streams. This has to be taken into account in the actual receiver design. There are several ways to reduce the impact of the interference and some of the options are listed below:

1. *Filtering.* A digital filter, typically FIR, can be placed in the signal path before the detector. This filter converts the channel impulse response to minimum phase, that is close to a frequency flat channel. An example of such a solution is the pre-filter of an MMSE-DFE receiver. This type of filter may already exist in many of today's EGPRS receivers.
2. *Single Antenna Interference Cancellation (SAIC).* Typically new GSM mobiles are equipped with DARP capabilities. As the interfering information stream of an OSC or VAMOS sub channel looks like a co-channel interferer, it is possible to apply SAIC to reduce the impact of the interference. Thus, DARP Phase 1 mobiles do not require additional design to receive voice transmitted over OSC or VAMOS signals.
3. *Successive Interference Cancellation (SIC).* In a SIC receiver the sub-channels, or users, are received in an iterative fashion. The dominant sub-channel is received first and according to the detected information the impact of the detected stream is removed before the detection of the other sub-channel and so on.
4. *Joint Detection (JD).* In joint detection, both information streams are decoded at the same time. An example is a QPSK type receiver, where both information streams are demodulated as one signal.

8.4.1.1 Constellation Power Control

Depending on the channel quality, the base station transmitter may wish to adjust the power between the different sub-channels. The symbol amplitudes can be represented as $\alpha = [0, 1]$ where $\hat{x}_i = \alpha x_i + \sqrt{1 - \alpha^2}z_i j$ is a variable used to scale the power. The total transmitted power is still constant, but the power distribution between users can be adjusted. Even more specifically, if α is defined from the set $\alpha \in \{\sin(\pi/8), \sin(2\pi/8), \sin(3\pi/8)\}$, a particular

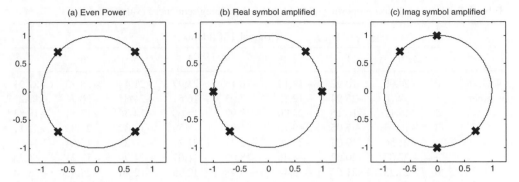

Figure 8.7   Constellation power control utilizing 8-PSK modulation constellation.

power control scheme is obtained in which the constellation points of 8-PSK modulation are utilized (see Figure 8.7). This allows the possibility to reuse the existing EGPRS transmitters.

8.4.2   The Principles of UL Concept

In the uplink, mobiles transmit a normal GMSK signal. Since two VAMOS or OSC users share the same physical radio resources, it is left for the BTS to separate the user sub-channels. One could say that the system resembles a multi-user multiple-input multiple-output (MIMO) system where the separation of the user signals is allowed by independent multi-path channel characteristics. The base station should be equipped with antenna diversity in order to perform the signal detection. The signal separation is further facilitated by the fact that different training sequences are applied in the transmitted sub-channels.

The training sequences should preferably be orthogonal to each other to guarantee the quality of the channel estimates. In order to improve the correlation properties a new improved set of training sequences was specified in [9]. In the following, the correlation properties of the training sequences when using a 5 tap least squares (LS) channel estimator are studied. The correlation properties of the original GSM training sequences (TSC) are presented in Table 8.6.

Table 8.6   The correlation properties of GSM training sequences (values in dB)

		TSC							
		0	1	2	3	4	5	6	7
TSC	0	0.00	−5.97	−26.42	−19.39	−15.11	−15.38	−7.47	−25.39
number	1	−5.98	0.00	−20.76	−28.88	−21.05	−22.31	−9.45	−18.19
	2	−26.42	−19.39	0.00	−5.97	−23.00	−9.21	−23.02	−20.39
	3	−20.76	−28.88	−5.98	0.00	−28.42	−4.37	−15.78	−17.04
	4	−14.65	−18.04	−20.03	−30.47	0.00	−12.29	−12.02	−6.59
	5	−17.71	−23.46	−10.11	−4.60	−11.14	0.00	−23.18	−10.97
	6	−6.86	−8.61	−17.21	−13.79	−9.80	−28.95	0.00	−10.46
	7	−17.46	−13.04	−19.67	−15.61	−6.31	−10.67	−10.89	0.00

Table 8.7 The correlation properties of GSM and new training sequences (values in dB)

		Original TSC number							
		0	1	2	3	4	5	6	7
New TSC	0	−24.90	−10.92	−18.12	−16.15	−23.97	−21.44	−28.73	−21.21
number	1	−9.16	−25.94	−12.13	−7.05	−16.61	−24.81	−20.71	−17.23
	2	−18.12	−16.15	−24.90	−10.92	−15.00	−14.36	−17.51	−18.83
	3	−12.13	−7.05	−9.16	−25.94	−8.59	−14.37	−25.82	−21.11
	4	−21.56	−19.03	−25.29	−12.22	−27.21	−24.23	−15.64	−14.80
	5	−22.07	−20.62	−8.46	−18.02	−11.07	−24.99	−19.21	−14.76
	6	−23.73	−21.67	−15.44	−8.48	−12.95	−12.24	−28.82	−28.74
	7	−22.44	−22.73	−22.40	−20.36	−16.72	−15.44	−18.45	−24.25

The correlation properties between original and the new set of sequences are given in Table 8.7. Clearly, the new set has the best correlation properties when paired with the original TSC with the same sequence number. It should be noted that the original set of training sequences already can provide suitable pairs for an UL multi-user MIMO setup, but some pre-selection is needed in order to find the best pairs.

Similarly to the case of downlink, there are different options to receive the multi-user signals. Those include Interference Rejection Combining (IRC), Successive Interference Cancellation (SIC) and Joint Detection (JD).

8.4.3 Downlink VAMOS Link Performance with AMR

As mentioned earlier, VAMOS enables two full-rate or four half-rate users to be allocated to the same TDMA slot which thereby yields an increase in network capacity. In order to evaluate the performance of VAMOS level 1 and level 2 mobile station receivers a new set of VAMOS test scenarios (MTS) were included in the 3GPP VAMOS working assumptions [10]. As with the DARP test scenarios, the scope of the MTS is to provide a test environment which matches that of a real interference-limited network, by modeling the conditions where there are single and multiple interferers, and synchronous and asynchronous network deployments.

The main difference between the VAMOS and DARP test scenarios is the applied interferer modulation type. Tables 8.8 to 8.11 list the four different VAMOS test scenarios assumed in [10], and characterized by a number of parameters. The "interfering signal" column defines

Table 8.8 VAMOS test scenario for synchronous interference; single interferer. Reproduced by permission of 3GPP TS 45.005 v6.15.0 Annex L. © 2009. 3GPP™ TSs and TRs are the property of ARIB, ATIS, CCSA, ETSI, TTA who jointly own the copyright on them. They are subject to further modifications and are therefore provided to you "as is" for information purposes only. Further use is strictly prohibited

Reference test scenario	Interfering signal	Interferer relative power level	TSC	Interferer delay range
MTS-1	Co-channel 1	0 dB	none	no delay

Table 8.9 VAMOS test scenario for synchronous interference; multiple interferers. Reproduced by permission of 3GPP TS 45.005 v6.15.0 Annex L. © 2009. 3GPP™ TSs and TRs are the property of ARIB, ATIS, CCSA, ETSI, TTA who jointly own the copyright on them. They are subject to further modifications and are therefore provided to you "as is" for information purposes only. Further use is strictly prohibited

Reference test scenario	Interfering signal	Interferer relative power level	TSC	Interferer delay range
MTS-2	Co-channel 1	0 dB	none	no delay
	Co-channel 2	−10 dB	none	no delay
	Adjacent 1	3 dB	none	no delay
	AWGN	−17 dB	–	–

Table 8.10 VAMOS test scenario for asynchronous interference; single interferer. Reproduced by permission of 3GPP TS 45.005 v6.15.0 Annex L. © 2009. 3GPP™ TSs and TRs are the property of ARIB, ATIS, CCSA, ETSI, TTA who jointly own the copyright on them. They are subject to further modifications and are therefore provided to you "as is" for information purposes only. Further use is strictly prohibited

Reference test scenario	Interfering signal	Interferer relative power level	TSC	Interferer delay
MTS-3	Co-channel 1	0 dB*)	none	74 symbols

Note: * The power of the delayed interferer burst, averaged over the active part of the wanted signal burst. The power of the delayed interferer burst, averaged over the active part of the delayed interferer burst is 3 dB higher.

Table 8.11 VAMOS test scenario for asynchronous interference; multiple interferers. Reproduced by permission of 3GPP TS 45.005 v6.15.0 Annex L. © 2009. 3GPP™ TSs and TRs are the property of ARIB, ATIS, CCSA, ETSI, TTA who jointly own the copyright on them. They are subject to further modifications and are therefore provided to you "as is" for information purposes only. Further use is strictly prohibited

Reference test scenario	Interfering signal	Interferer relative power level	TSC	Interferer delay
MTS-4	Co-channel 1	0 dB*)	none	74 symbols
	Co-channel 2	−10 dB	none	no delay
	Adjacent 1	3 dB	none	no delay
	AWGN	−17 dB	–	–

Note: * The power of the delayed interferer burst, averaged over the active part of the wanted signal burst. The power of the delayed interferer burst, averaged over the active part of the delayed interferer burst is 3 dB higher.

the type of interference, such as co-channel, adjacent channel or white noise (AWGN). The "interferer relative power level" defines the power level of each signal relative to the dominant interferer. For a single interferer scenario such as MTS-1 and MTS-3 this level is 0 dB as there is one interferer, and therefore with equal QPSK sub-channel powers the external interferer may actually be 3 dB stronger than the desired signal on each of the two VAMOS sub channels. The modulation for Co-channel 1 will be either: GMSK or VAMOS type. For MTS-2 and MTS-4 the modulation for Co-channel 1, Co-channel 2, and Adjacent 1 will be either: GMSK or VAMOS type. All interferers should use the same modulation type. The "TSC" column describes whether the interferer includes a training sequence code or it consists of random bits. The "Interferer delay" column denotes the delay in symbols.

The VAMOS working assumption is that the VAMOS test scenarios are to be seen as additional tests applicable for the VAMOS mobile station. That is, all legacy DARP test cases still apply.

Examples of the VAMOS link performance using MTS-1 and MTS-2 are presented in the following section, using a typical urban channel profile with a mobile station speed of 3 km/h.

8.4.3.1 Equal Sub-Channel Power Imbalance Ratios

Sub-Channel Power Imbalance Ratio (SCPIR) defines the user-specific power given by the modulation constellation and these details are described for VAMOS in Chapter 4. A VAMOS capable mobile station should be able to operate with equal SCPIR values and over a range of different SCPIR configurations. The VAMOS link level performance with AMR is divided into two sections. First, the performance is exemplified for a VAMOS level 1 and a VAMOS level 2 mobile station when operating with 0 dB SCPIR and then the performance is exemplified for a VAMOS level 1 and a VAMOS level 2 mobile station in a VAMOS configuration over a range of (non-zero) SCPIRs.

The link level performance of both a VAMOS level 1 and a VAMOS level 2 capable mobile stations for all AMR-NB codec modes are exemplified in Figures 8.8 to 8.15. Single and multiple interferer scenarios are considered (MTS-1 and MTS-2) using both GMSK and QPSK modulated interference as well as TCH/AFS and TCH/AHS channel modes.

It can be observed from Figures 8.8 to 8.11 that in the case of a single interferer (MTS-1) both the VAMOS level 1 and the VAMOS level 2 mobile station performance depend on the interferer modulation type. Furthermore it can be seen that the VAMOS level 2 mobile stations performs better than the VAMOS level 1 mobile station when the interferer is QPSK modulated. The gain is in the order of 1 dB (measured at the 1% FER point) when SCPIR = 0 dB. Assuming the VAMOS level 1 mobile station is developed from a DARP Phase 1 mobile station platform but updated to support the new VAMOS TSCs (as described in Chapter 4), then the presented performance results in Figures 8.8 to 8.11 will also reflect the performance of DARP Phase 1 mobile station.

The multi-interferer scenarios (MTS-2) of Figures 8.12 to 8.15 also show that both VAMOS level 1 and VAMOS level 2 mobile station performance depend on the interferer modulation type. Furthermore, it is noted that when SCPIR = 0 dB, the VAMOS level 2 mobile station performs better than the VAMOS level 1 mobile station both with GMSK and with QPSK modulated interference. The performance differences are in the range of 1 dB to 1.8 dB depending on the modulation type of the interferer.

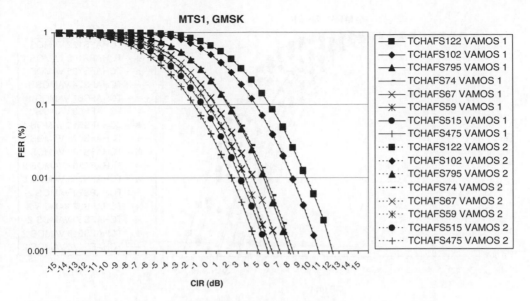

Figure 8.8 Link performance of a VAMOS level 1 and VAMOS level 2 MS. AMR full-rate codecs (TCH/AFS). MTS-1 scenario, GMSK modulated interference.

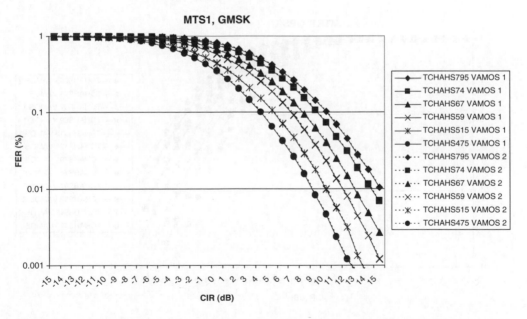

Figure 8.9 Link performance of a VAMOS level 1 and VAMOS level 2 MS. AMR half-rate codecs (TCH/AHS). MTS-1 scenario, GMSK modulated interference.

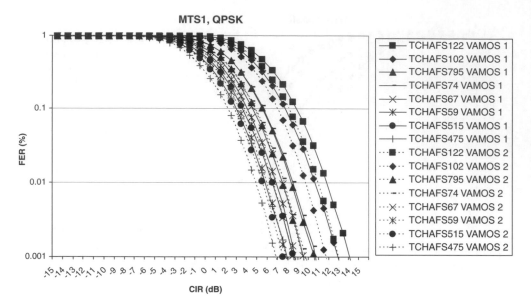

Figure 8.10   Link performance of a VAMOS level 1 and VAMOS level 2 MS. AMR full-rate codecs (TCH/AFS). MTS-1 scenario, QPSK modulated interference.

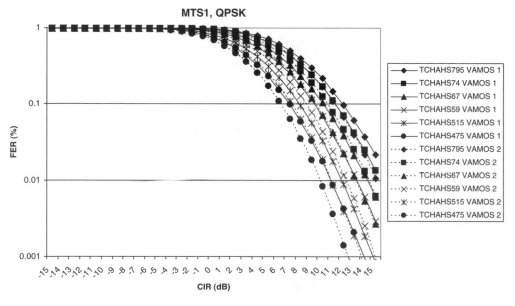

Figure 8.11   Link performance of a VAMOS level 1 and VAMOS level 2 MS. AMR half-rate codecs (TCH/AHS). MTS-1 scenario, QPSK modulated interference.

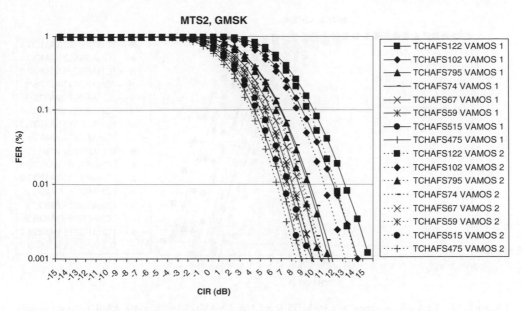

Figure 8.12 Link performance of a VAMOS level 1 and VAMOS level 2 MS. AMR full-rate codecs (TCH/AFS). MTS-2 scenario, GMSK modulated interference.

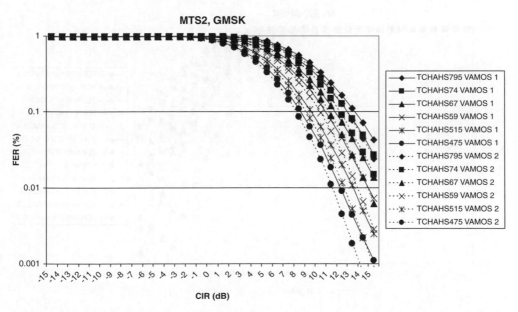

Figure 8.13 Link performance of a VAMOS level 1 and VAMOS level 2 MS. AMR half-rate codecs (TCH/AHS). MTS-2 scenario, GMSK modulated interference.

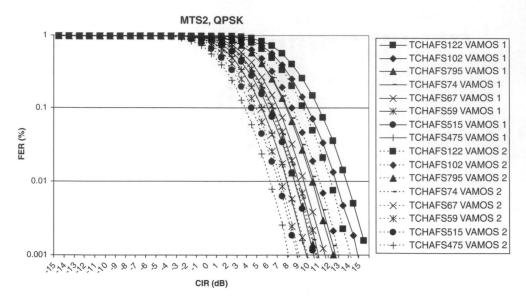

Figure 8.14 Link performance of a VAMOS level 1 and VAMOS level 2 MS. AMR full-rate codecs (TCH/AFS). MTS-2 scenario, QPSK modulated interference.

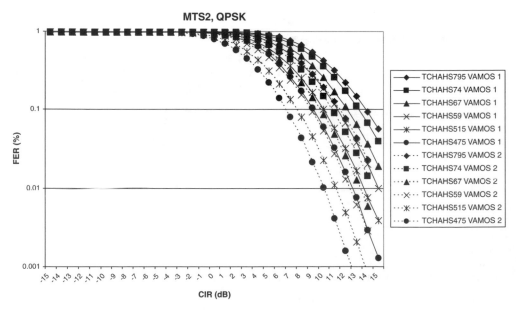

Figure 8.15 Link performance of a VAMOS level 1 and VAMOS level 2 MS. AMR half-rate codecs (TCH/AHS). MTS-2 scenario, QPSK modulated interference.

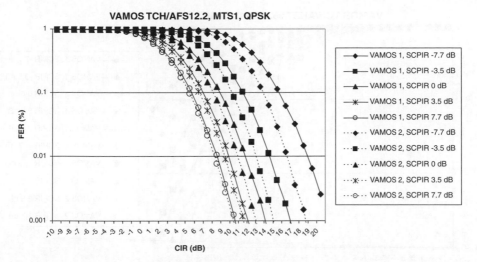

Figure 8.16 Link performance of a VAMOS level 1 MS and a VAMOS level 2 MS for SCPIRs [−7.7, −3.5, 0, 3.5, 7.7] dB. AMR full-rate codec 12.2 (TCH/AFS12.2). MTS-1 scenario with QPSK modulated interference.

8.4.3.2 Different Sub-Channel Power Imbalance Ratios

The link level performance of both VAMOS level 1 and VAMOS level 2 mobile stations using a subset of the AMR-NB codec modes are exemplified in Figures 8.16 to 8.23. Single and multiple interferer scenarios are considered (MTS-1 and MTS-2) using QPSK modulated interference.

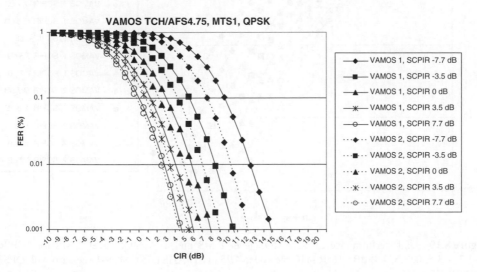

Figure 8.17 Link performance of a VAMOS level 1 MS and a VAMOS level 2 MS for SCPIRs [−7.7, −3.5, 0, 3.5, 7.7] dB. AMR full-rate codec 4.75 (TCH/AFS4.75). MTS-1 scenario with QPSK modulated interference.

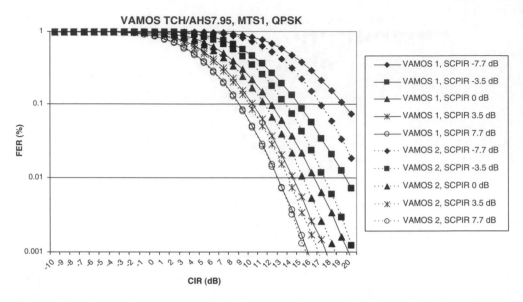

Figure 8.18 Link performance of a VAMOS level 1 MS and a VAMOS level 2 MS for SCPIRs [−7.7, −3.5, 0, 3.5, 7.7] dB. AMR half-rate codec 7.95 (TCH/AHS7.95). MTS-1 scenario with QPSK modulated interference.

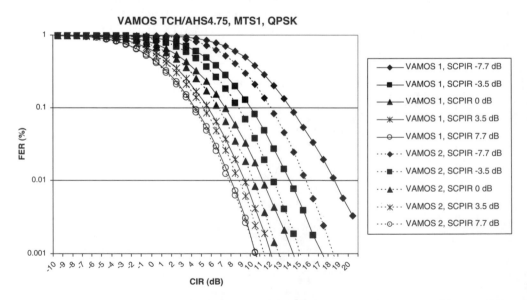

Figure 8.19 Link performance of a VAMOS level 1 MS and a VAMOS level 2 MS for SCPIRs [−7.7, −3.5, 0, 3.5, 7.7] dB. AMR half-rate codec 4.75 (TCH/AHS4.75). MTS-1 scenario with QPSK modulated interference.

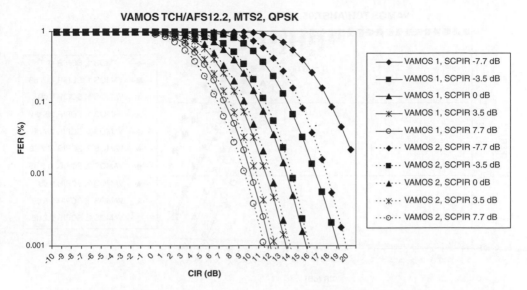

Figure 8.20 Link performance of a VAMOS level 1 MS and a VAMOS level 2 MS for SCPIRs [−7.7, −3.5, 0, 3.5, 7.7] dB. AMR full-rate codec 12.2 (TCH/AFS12.2). MTS-2 scenario with QPSK modulated interference.

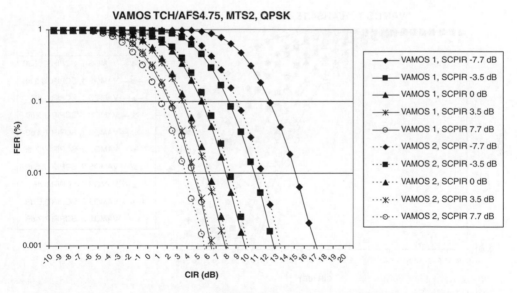

Figure 8.21 Link performance of a VAMOS level 1 MS and a VAMOS level 2 MS for SCPIRs [−7.7, −3.5, 0, 3.5, 7.7] dB. AMR full-rate codec 4.75 (TCH/AFS4.75). MTS-2 scenario with QPSK modulated interference.

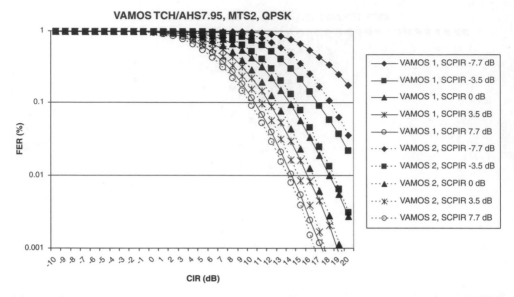

Figure 8.22 Link performance of a VAMOS level 1 MS and a VAMOS level 2 MS for SCPIRs [−7.7, −3.5, 0, 3.5, 7.7] dB. AMR half-rate codec 7.95 (TCH/AHS7.95). MTS-2 scenario with QPSK modulated interference.

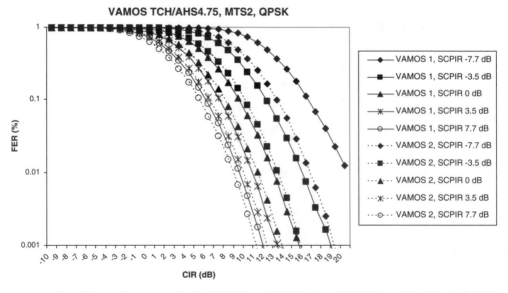

Figure 8.23 Link performance of a VAMOS level 1 MS and a VAMOS level 2 MS for SCPIRs [−7.7, −3.5, 0, 3.5, 7.7] dB. AMR half-rate codec 4.75 (TCH/AHS4.75). MTS-2 scenario with QPSK modulated interference.

From Figures 8.16 to 8.19 it can be observed that in the case of MTS-1 with QPSK modulated interference, the VAMOS level 2 mobile station performs better than the VAMOS level 1 mobile station for all SCPIR values and for both AMR-NB codecs. The actual performance gain depends both on the SCPIR level and on the AMR-NB codec. From the performance curves it can also be seen that the gain is more than 2 dB (measured at the 1% FER point) when SCPIR = −7.7 dB.

The MTS-2 scenarios from Figures 8.20 to 8.23 also show that the VAMOS level 2 mobile station performs significantly better than the VAMOS level 1 mobile station for all SCPIR values and for all AMR-NB codecs when the multi-interference scenario is used. Again, the largest gains were observed for the negative SCPIR levels, where the gain is above 3 dB (measured at the 1% FER point) when SCPIR = −7.7 dB.

In general, the conclusion can be made that the VAMOS level 1 mobile station is more sensitive to the applied SCPIR level than the VAMOS level 2 mobile station.

8.4.4   Uplink VAMOS Link Performance with AMR

The link-level performance of UL OSC is exemplified in Figures 8.24 to 8.26. Figure 8.24 illustrates how the UL OSC receiver is performing for AMR half-rate channels. Figures 8.25 to 8.26 exemplify the impact of Sub-Channel Power Imbalance Ratios, training sequence cross-correlation and the modulation of interference to the uplink performance.

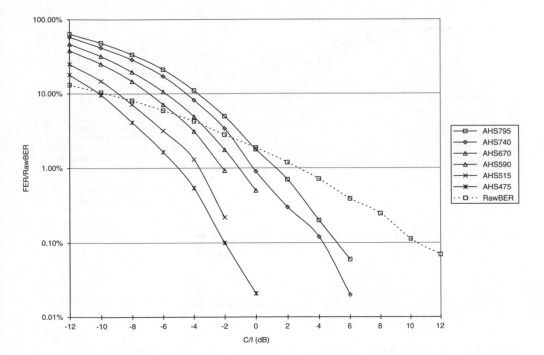

Figure 8.24   Link performance of OSC AMR Half rate simulated in TU3iFH 900 MHz Single Co-channel interference scenario.

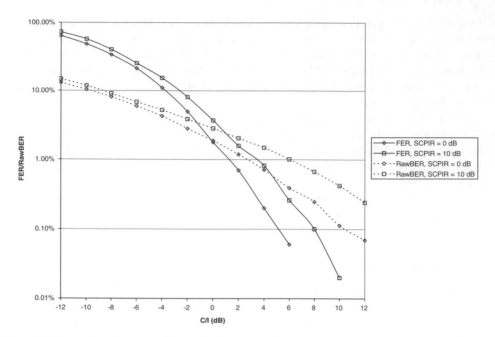

Figure 8.25 Performance Impacts of Sub-Channel Power Imbalance Ratio simulated for AHS 7.95 TU3iFH 900 MHz Single Co-channel interference scenario.

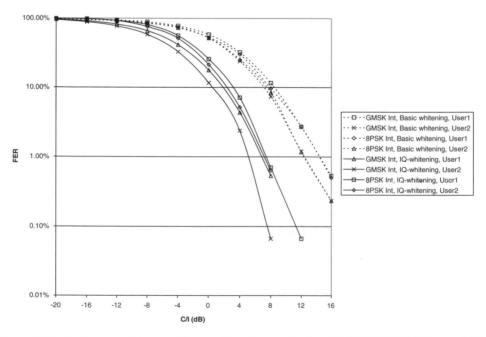

Figure 8.26 Performance Impacts with type of interference simulated for TU3iFH 900 MHz sensitivity scenario.

8.4.4.1 Performance Impacts of Sub-Channel Power Imbalance Ratio in Uplink

In Figure 8.25 the uplink receiver performance is evaluated when using average SCPIR values up to 10 dB with legacy training sequence pairs. In this scenario the UL receiver is able to decode the weaker channel while achieving the 3GPP performance requirements.

8.4.4.2 Performance Impacts of Interference Type

Noise whitening and interference rejection have been proved to be effective means, when separating the OSC users. One technique for performing noise whitening is based on the de-correlation of colored noise for a binary signal by separating the received signal into its real and imaginary parts (IQ). The IQ-whitening can cancel GMSK modulated interference more efficiently due to its binary nature. With OSC there can be two categories of interference:

1. User interference where OSC sub-channel 0 (User1) is an interference source for the second sub-channel and vice versa, where both OSC users are GMSK modulated signals.
2. Residual interference coming from other users sharing the same radio resources (e.g. GMSK or 8PSK, or a mix of two or more GMSK and/or 8PSK modulated signals).

For OSC, the user interference is usually more dominant than the residual interference. With an OSC receiver based on IQ-noise whitening, the other user can be more or less removed completely given that it is equivalent to a GMSK modulated interferer. When both users have the same power level, an efficient whitening mechanism is required to detect the desired user from the other sub-channel.

Figure 8.26 shows the IQ-whitening and non IQ-whitening receiver performance when GMSK and 8PSK residual interferences are simulated with the legacy training sequence pair $\{0, 6\}$ for TU3iFH radio propagation scenario. The IQ-whitening receiver performs significantly better in both GMSK and 8PSK residual interference cases when compared to the non-IQ-whitening receiver. The gain of the IQ-whitening receiver is much higher for the GMSK interference whereas the non-IQ whitening receiver performs almost the same for both GMSK and 8PSK interference.

8.4.5 System Level Performance

This section presents system level performance evaluations for OSC and VAMOS. The results are based on dynamic network level simulations. First, simulation scenarios and assumptions are described. Then the results are presented, starting from the basic OSC performance and, after that, evaluating some enhancements for the VAMOS concept.

8.4.5.1 Simulation Configurations and Assumptions

The evaluated scenarios are based on VAMOS working assumptions described in [10]. Results are shown for two scenarios. Both network configurations are based on a regular grid with

Table 8.12 Simulation scenarios

Parameter	Scenario 1	Scenario 2
Frequency band	900 MHz	1800 MHz
Cell radius	500 m	500 m
Bandwidth	4.4 MHz	11.6 MHz
Guard band	0.2 MHz	0.2 MHz
Number of channels excluding guard band	21	57
Number of TRX	4	6
BCCH frequency reuse	4/12	4/12
TCH frequency reuse	1/1	3/9
Frequency Hopping	Synthesized	Baseband
Number of hopping frequencies per cell	9	5
Channel mode	Typical Urban (TU)	Typical Urban (TU)

75 sectors. Table 8.12 and Figure 8.27 summarize the characteristics of each network configuration. Scenario 1 is an interference- limited scenario with hopping layer frequency reuse factor of 1/1. Scenario 2 is a loose frequency reuse scenario and is more limited by hardware capacity than by interference. BCCH layer reuse factor was 4/12 in both scenarios and BCCH TRX was included in the simulations.

Mobile stations were assumed to be DARP Phase 1 compliant with the exception that the simulations of constellation power control were done with 50% DARP Phase 1 and 50% non-DARP mobile stations.

The simulated channel modes are summarized in Table 8.13. *REF* represents the reference case with AMR 5.9 Half-Rate (HR) codec, while *OSC* is the case with the same codec, which also includes the Double Half-Rate (DHR) OSC channel mode. The simulations were done using only half-rate channel modes, because those channel modes better indicate the achievable gains of OSC on top of current AMR codecs. The channel mode adaptation algorithm to adapt between HR and DHR modes takes into account RxQual and RxLevel measurements and cell load conditions.

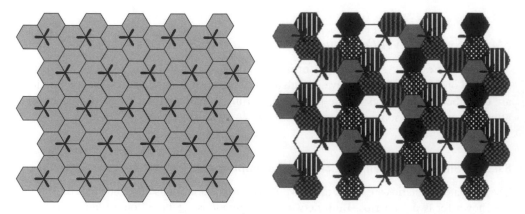

Figure 8.27 Hopping schemes used in Scenario 1 (1/1) and Scenario 2 (3/9).

Table 8.13   Channel mode and codecs used in simulations

Label	Channel modes	Codec
REF (Reference)	HR	AMR HR 5.9
OSC	HR ↔ DHR	AMR HR 5.9

Two quality indicators were used: the first was Blocked Call Rate (BCR) with 2% QoS criterion; the second was the Bad Quality Calls (BQC) with 5% QoS criterion. Connection quality was measured with Frame Erasure Rate (FER). It was assumed that the quality of a call was bad if connection average FER was higher than 3%.

A summary of test cases used for the simulations in this section is presented in Table 8.14, and details for each one are given below:

1. The first simulation is evaluating the baseline performance. Results are shown for the reference AMR with DARP Phase 1 and OSC with regular RRM. The advanced RRM are introduced by the other cases.
2. In the second case, performance improvement with an optimized pulse shape is evaluated. Linearized Gaussian Minimum Shift Keying (LGMSK) is the basic pulse shape for VA-MOS. The specification of an optimized pulse shape for VAMOS is under study as part of the 3GPP MUROS Feasibility Study [10]. Typically, the choice of pulse shape is expected to represent a compromise between increased adjacent channel interference and improvement in reception robustness. Results are shown for two types of pulse shapes. Linearized GMSK pulse shape and a Root-raised-cosine (RRC) pulse shape filter with 3 dB-bandwidth of 240 kHz and roll-off value of 0.3. These pulse shapes were introduced in Chapter 4. In these simulations, adjacent channel protection values for the pulse shape filters are 18.2 dB for LGMSK, 13.8 dB for RRC.
3. In this simulation, the optimized channel allocation for OSC is studied. The baseline simulations used the allocation algorithm of a normal GSM network using interference estimation to choose the best target channel. This algorithm sorts the available channels by their interference levels. An OSC-based algorithm sorts the available channels by the path loss difference between the incoming user and the other half-rate user which is already using the channel, thus it minimizes the power requirement differences of each OSC mobile station in downlink. The hybrid mode uses the priority provided from both allocation algorithms.

Table 8.14   Simulation cases

Simulation cases and evaluated features	Used pulse shape	Used channel allocation algorithm
1) Reference and baseline	LGMSK	Normal
2) Optimized pulse shape	LGMSK/RRC	Normal
3) Optimized channel allocation	RRC	Normal, Hybrid and OSC based
4) User diversity	RRC	Hybrid
5) Sub-channel specific power control	RRC	Hybrid
6) FACCH strategies	RRC	Hybrid

Frame number	Subchannel 0	Subchannel 1	Traffic type	Frame number	Subchannel 0	Subchannel 1	Traffic type
0	USER 1	USER 2	TCH	0	USER 1	USER 2	TCH
1	USER 3	USER 4	TCH	1	USER 3	USER 2	TCH
2	USER 1	USER 2	TCH	2	USER 1	USER 4	TCH
3	USER 3	USER 4	TCH	3	USER 3	USER 4	TCH
4	USER 1	USER 2	TCH	4	USER 1	USER 2	TCH
5	USER 3	USER 4	TCH	5	USER 3	USER 2	TCH
6	USER 1	USER 2	TCH	6	USER 1	USER 4	TCH
7	USER 3	USER 4	TCH	7	USER 3	USER 4	TCH
8	USER 1	USER 2	TCH	8	USER 1	USER 2	TCH
9	USER 3	USER 4	TCH	9	USER 3	USER 2	TCH
10	USER 1	USER 2	TCH	10	USER 1	USER 4	TCH
11	USER 3	USER 4	TCH	11	USER 3	USER 4	TCH
12	**USER 1**	**USER 2**	**SACCH**	12	**USER 1**	**USER 2**	**SACCH**
13	USER 3	USER 2	TCH	13	USER 1	USER 2	TCH
14	USER 1	USER 4	TCH	14	USER 3	USER 2	TCH
15	USER 3	USER 2	TCH	15	USER 1	USER 4	TCH
16	USER 1	USER 4	TCH	16	USER 3	USER 4	TCH
17	USER 3	USER 2	TCH	17	USER 1	USER 2	TCH
18	USER 1	USER 4	TCH	18	USER 3	USER 2	TCH
19	USER 3	USER 2	TCH	19	USER 1	USER 4	TCH
20	USER 1	USER 4	TCH	20	USER 3	USER 4	TCH
21	USER 3	USER 2	TCH	21	USER 1	USER 2	TCH
22	USER 1	USER 4	TCH	22	USER 3	USER 2	TCH
23	USER 3	USER 2	TCH	23	USER 1	USER 4	TCH
24	USER 1	USER 4	TCH	24	USER 3	USER 4	TCH
25	**USER 3**	**USER 4**	**SACCH**	25	**USER 3**	**USER 4**	**SACCH**

Figure 8.28 Traffic multi-frame structure for normal OSC traffic and User diversity pattern 2 [11].

4. This set of simulations evaluates User diversity for OSC, which is a feature suggested in 3GPP for increasing interference diversity on OSC calls. It changes the frame structure of OSC calls in order to mix different mobiles on the same OSC channel. There are several mixing patterns proposed for this feature [11]. In these simulations, the pattern 2 described in Figure 8.28 was used. In this pattern, the frame structure of users 1 and 3 is kept unchanged, and therefore those sub-channels are able to hold legacy users which are not aware of frame structure modification. For users 2 and 4 the frame structure is modified in a way that those either interfere with user 1 or with user 3. The important point about this feature is the DTX silence period. When DTX is enabled, the BTS changes the constellation from QPSK to GMSK if one of the users in the OSC channel is silent. In this case, the active user gains from the inactivity of its pair and therefore better quality is obtained. When user diversity is enabled, DTX gains are spread over all users which are paired with the inactive user. As an example, consider that user 1 in Figure 8.28 is in DTX mode. During normal OSC transmission, user 2 will have no interferer; therefore, it will behave as a normal half-rate call. When user diversity is enabled, half of the bursts received by user 2 and user 4 will be using GMSK, therefore both users benefit from DTX state of user 1. Additionally, user diversity tends to spread DTX gains over the call period since it is not expected that users 1 and 3 will always be inactive simultaneously.

Table 8.15 Downlink constellation patterns for sub-channel power unbalancing

Constellation patterns	Double Half-Rate (DHR)	
	Rectangle bursts	Square bursts
EQUAL	0	4
RE1SQ3	1	3
RE2SQ2	2	2
RE3SQ1	3	1
RE4SQ0	4	0

5. Sub-channel specific power control is evaluated in this simulation set. In addition to normal power control and channel allocation methods, DL Sub-Channel Power Imbalance Ratios (SCPIRs) provides a means for implementing different downlink powers for each sub-channel, as described in Section 8.4.1.1. One important application of SCPIRs is to improve the radio conditions for legacy receivers [12]. Receivers which are not DARP Phase 1-compliant suffer from high quality degradation as they are multiplexed on an OSC channel. Therefore, by using this technique it is possible to apply an equivalent power unbalancing in downlink to improve the legacy receiver's link quality while leading to small increase in FER of advanced receivers. This approach presents a good compromise in order to make it feasible to multiplex non-DARP terminals in OSC channels. In these simulations, rectangular constellation points were used. In addition to the different SCPIR constellations, it is possible to take advantage of the burst interleaving of GSM traffic in order to achieve different sub-channel power steps. This is done by transmitting part of the half-bursts that compose the interleaving period of each block using a rectangular constellation, for example switching between modulation constellation points of 8PSK and QPSK modulation on each received time slot [13], for example the modulation method does not change, only the points selected from the constellation. Table 8.15 illustrates the constellation patterns used in simulations. Additionally, this set of simulations was created using 50% of DARP Phase 1 and 50% of non-DARP receivers.

6. This set of simulations evaluates the performance of different FACCH transmission strategies for OSC. One concern with OSC is how to manage signaling as calls are paired in the same channel. Degradation in the expected bit error rate of the FACCH caused by OSC multiplexing might be an important factor that could lead to an increase in the dropped call rate. There are some proposed alternatives for this issue and this text will present the Double FACCH Stealing and Soft FACCH Stealing techniques. Both of them improve the transmission of FACCH signaling at the cost of some degradation in voice quality for the second user on an OSC channel. When using Double FACCH Stealing, the user transmitting FACCH has full priority over the channel, and therefore the channel is not shared at this time with its OSC pair. As a consequence, the paired user loses the TCH block under transmission. The Soft FACCH Stealing feature presents a compromise between the full priority for transmitting FACCH and the voice quality of the second user. This soft-stealing technique uses an unbalanced OSC constellation to apply equivalent higher power for the user that is transmitting the FACCH block. Simulations were performed to test the different

Table 8.16   Simulation results for the cases 1 to 3 for Scenario 1

Channel mode/test case	Spectral efficiency [Erl/MHz/site]	Hardware efficiency [Erl/TRX]	EFL (%)	OSC gain (%)	Limiting KPI
1) REF AMR + DARP phase 1	36.2	12.7	30.2	–	BCR
1) OSC baseline	41.3	14.4	34.4	14	BQC
2) OSC + Optimized pulse shape	45.6	16.0	38.0	26	BQC
3) OSC + Optimized channel allocation	47.1	16.5	39.2	30	BQC

FACCH signaling alternatives for OSC using Scenario 2. These simulations were performed using the RE1SQ3, RE2SQ2, RE3SQ1 and RE4SQ0 constellation patterns, described in Table 8.15.

8.4.5.2   Simulation Results

A summary of the simulation results for cases 1 to 3 are presented in Tables 8.16 and 8.17. In both scenarios the network was blocking limited. In the tight frequency reuse scenario (Scenario 1), the OSC baseline showed 14% capacity gain compared to the reference case. When the optimized pulse shape was used, the gain increased to 26%, and when the optimized channel allocation was employed, this increased further to 30%.

In the loose frequency reuse scenario (Scenario 2), the OSC gains were significantly higher. Already the baseline results showed a gain above 50% for OSC. And, with the optimized pulse shape the gain increased to 70%.

In terms of hardware efficiency, it can be seen that the reference case is limited to about 13 Erls/TRX. In Scenario 1 up to 16.5 Erls/TRX was achieved and in Scenario 2 up to 23 Erls/TRX with OSC when using the optimized pulse shape and channel allocation. Given that the 2% hard blocking limit for OSC is about 28 Erls/TRX, then it is reasonable to assume that additional methods and optimization algorithms could be considered to improve OSC performance in this scenario.

It can be seen from the results that network capacity has increased significantly when using the RRC pulse shape. Although the RRC pulse shape has a lower adjacent channel protection than the LGMSK pulse shape, there is BER performance improvement when this pulse shape is used, leading to an improvement for both the tight and loose reuse scenarios. This can be

Table 8.17   Simulation results for the cases 1 to 3 for Scenario 2

Channel mode/test case	Spectral efficiency [Erl/MHz/site]	Hardware efficiency [Erl/TRX]	EFL (%)	OSC gain (%)	Limiting KPI
1) REF AMR + DARP phase 1	21.4	13.3	17.8	–	BCR
1) OSC baseline	32.6	20.3	27.2	52	BQC
2) OSC + Optimized pulse shape	36.4	22.6	30.3	70	BCR
3) OSC + Optimized channel allocation	36.4	22.6	30.3	70	BCR

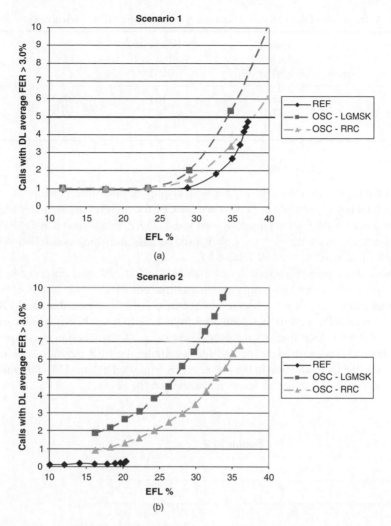

Figure 8.29 Bad quality calls for different pulse shapes in Scenario 1 and Scenario 2.

observed in the Bad Quality Calls statistics shown in Figure 8.29, where a significant FER improvement is seen when the RRC pulse shape is applied.

A set of simulations were created to compare different channel allocation algorithms. Table 8.18 shows the results for the normal, purely OSC-based algorithm and for the hybrid algorithm. Evaluation was made in scenario 1. It is noted that the hybrid algorithm achieved the best performance. This algorithm included both interference estimation and path loss measurement.

User diversity simulation results are demonstrated in Figure 8.30, where a significant improvement in bad quality calls rate is shown when user diversity is used. In Figure 8.30 an average BQC improvement of 35% is observed, since the mean FER for OSC was reduced due to the improved interference diversity received by each OSC user.

Table 8.18 Comparison of the Channel Allocation Algorithm using RRC in Scenario 1

Channel allocation	Spectral efficiency [Erl/MHz/site]	Hardware efficiency [Erl/TRX]	EFL (%)	Limiting KPI
Normal	45.6	16.0	38.0	BQC
Hybrid	47.1	16.5	39.2	BQC
OSC Based	44.5	15.6	37.1	BQC

The simulation results shown in Figure 8.31 display the possible benefits of using sub-channel specific power control for OSC when 50% of the mobiles are non-DARP. In all the simulations a fixed constellation pattern is used when DARP Phase 1 and non-DARP receivers are multiplexed on the same channel. Labels used to indicate which constellation pattern is used in each curve are described in Table 8.15.

From these simulations, a capacity gain of about 2% and 12% was observed in Scenarios 1 and 2, respectively, when using the constellation pattern RE1SQ3. In addition to this gain in the overall call quality, it was observed in Scenario 2, for instance, that the FER on OSC frames for non-DARP receivers has decreased from 22% to 6% while it has increased from 0.3% to 7% for the DARP Phase 1 receiver. Therefore, sub-channel power control for OSC has been shown to be an efficient method for equalizing the FER among users in different conditions, leading to a reduction in the power levels requested by non-DARP mobile stations and, thus, a decrease in the overall interference level on the network.

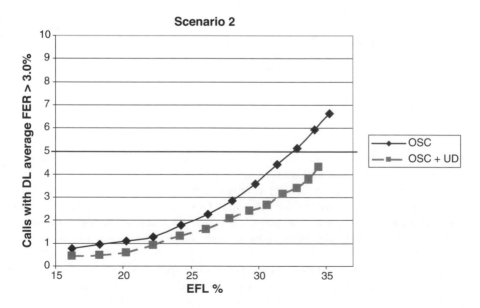

Figure 8.30 Bad quality calls statistics for User Diversity using Scenario 2.

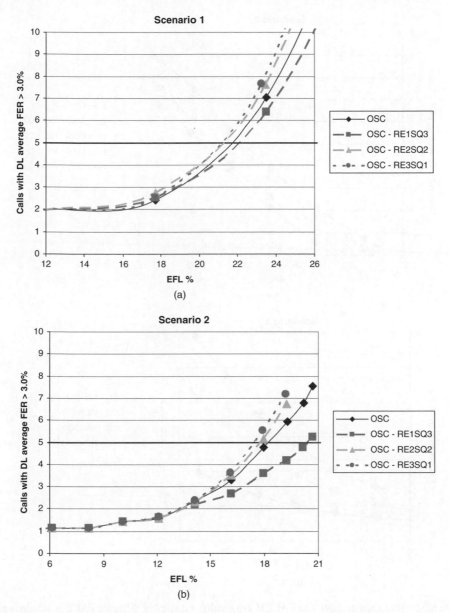

Figure 8.31 Bad Quality Calls for different constellation patterns in Scenarios 1 and 2 with 50% of non-SAIC terminals.

Simulations were done to test the different FACCH signaling alternatives for OSC using Scenario 2. Soft-stealing simulations were done using the RE1SQ3, RE2SQ2, RE3SQ1 and RE4SQ0 constellation patterns, described in Table 8.15.

Figures 8.32 and 8.33 show the Dropped Calls, FACCH error and Bad Quality Calls statistics of those simulations. In those figures, it is possible to observe that FACCH stealing

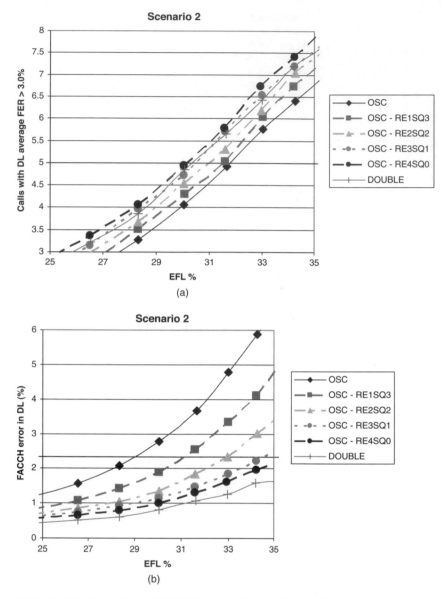

Figure 8.32 Bad Quality calls and FACCH error calls statistics of different FACCH schemes using Scenario 2.

schemes lead to lower FACCH error, thus decreasing the Dropped Calls percentage, while increasing the percentage of Bad Quality calls. Although it might be expected that the Double FACCH Stealing would provide the worst BQC results, the worst result observed considering Bad Quality Calls is provided by the Soft FACCH stealing using RE4SQ0 constellation pattern. The higher FACCH error probability with Soft Stealing using RE4SQ0 compared to

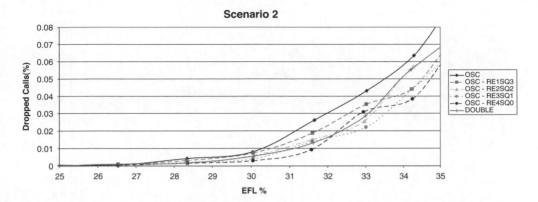

Figure 8.33   Dropped calls statistics of different FACCH schemes using Scenario 2.

Double Stealing could cause FACCH retransmissions, thereby decreasing the call quality of the second user.

Significant gains in the Dropped Calls can already be achieved by using only one rectangular burst with Soft SACCH stealing (RE1SQ3). This scheme has presented 20% improvement in the proportion of dropped calls near the 30% EFL network load point, while increasing the amount of BQCs by only 6%. This could be a good compromise between speech degradation and dropped calls. With the RE2SQ2 soft-stealing scheme, Dropped calls were improved by 30% to around 30% EFL point, while increasing BQC by 10%. These results demonstrate the trade-off between DCR and call quality. Selection of the optimal FACCH stealing scheme for a given network depends on the network characteristics and on the desired target values and balance between Dropped and BQC rates.

8.4.6   OSC Measurements Results

8.4.6.1   OSC Drive Test Motivation and Set-Up

Orthogonal-Sub Channels is a concept where the different implementations of legacy terminals could have a significant impact on the performance – especially given that such terminals have not been optimized to receive and demodulate the composite signal of two users. Therefore it is important to verify the simulations by performing field trials and drive tests, where the legacy terminals, base stations and multipath radio propagation environment are used to verify the expected performance and functionality.

To demonstrate OSC in a real-time environment, the OSC feature [5] was implemented in a GSM base station. A set of drive tests were performed where the OSC feature was enabled with four users sharing the same physical radio resource. An Abis protocol analyzer collected the DHR call traces and further post-processing was performed to obtain the RxLevel and FER results from OSC traces and to estimate the speech MOS.

Power control was enabled in the UL and disabled in the DL. Codec mode adaptation was enabled and the calls were directly initiated in DHR mode. Intra-cell handovers were not allowed and all the mobile phones were located in the same drive test car.

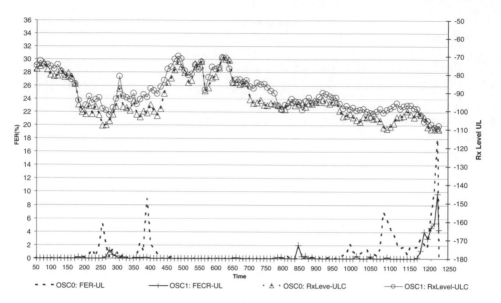

Figure 8.34   UL Double Half-Rate FER measurement traces samples.

8.4.6.2   OSC Drive Test Results

OSC Double Half-Rate (DHR) measurement trace samples are shown for the downlink and uplink in Figures 8.34 to 8.37. The results taken are from two DHR users who share the same physical radio resource in half-rate time slot.

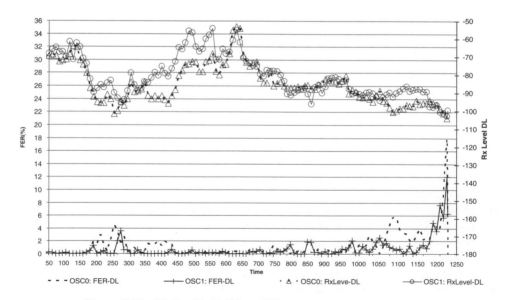

Figure 8.35   DL Double Half-Rate FER measurement traces samples.

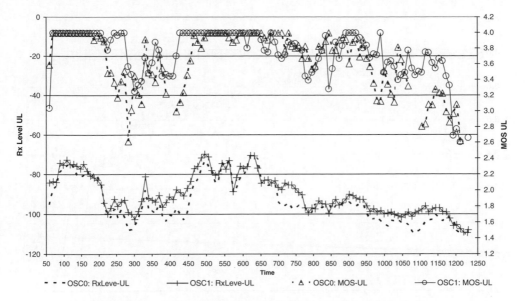

Figure 8.36   UL Double Half-Rate MOS measurement traces samples.

FER and RxLevel performance are presented for uplink and downlink in Figure 8.34 and Figure 8.35 respectively. FER was seen to be maintained at a good speech quality level and in average below 2% when RxLevel was higher than −95 dBm. This indicates that a low FER can be guaranteed for the paired OSC users when both radio links have a good signal level.

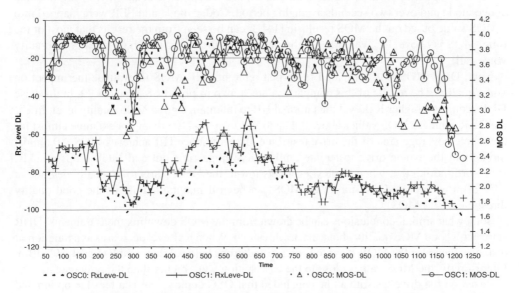

Figure 8.37   DL Double Half-Rate MOS measurement traces samples.

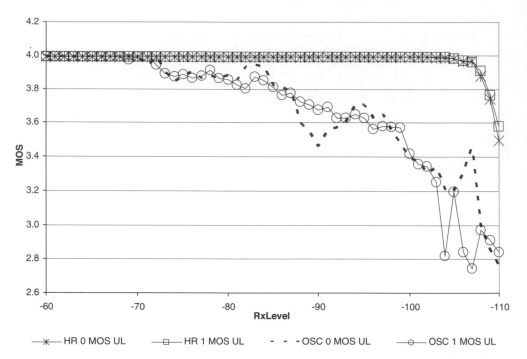

Figure 8.38   UL MOS, Double Half-Rate and Half-Rate, versus RxLevel measurement traces samples.

MOS estimation, seen in Figure 8.36 and Figure 8.37, was calculated from the median codec mode and over two seconds averaging period. Codec mode and FER were mapped into a MOS value based on the MOS results in [14]. It can be seen from the results that most of the time, MOS is above 3.2, which is usually determines the target threshold for a good quality voice call.

Next, DHR MOS was compared to normal reference HR MOS to evaluate the impact on voice quality. The results of this comparison are shown in Figures 8.38 and 8.39. In this case DHR requires a higher RxLevel than normal HR to maintain a good MOS quality level. In UL the DHR achieves MOS values above 3.2 up to RxLevel of -105 dBm, for example almost for the whole coverage area of the drive test route. The reference HR achieves MOS 4.0 almost throughout the whole drive test route until the coverage runs out and MOS drops quickly. It is worth noting that when the RxLevel gets lower, the DHR starts to gradually degrade at -70 dBm as expected, but the DHR MOS can be still maintained above the good quality threshold of 3.2 MOS.

Somewhat similar conclusions can be drawn from the MOS downlink measurements. DHR starts with good MOS quality above the 3.2 threshold. When RxLevel decreases as a function of distance, then also the MOS starts to decrease and drops below 3.2 around RxLevel -95 dBm. Interestingly, the MOS values close to the base station were higher than the reference.

Based on the drive tests,it can be concluded that OSC deployment is a feasible option and can provide good MOS and coverage in legacy networks with doubled hardware capacity.

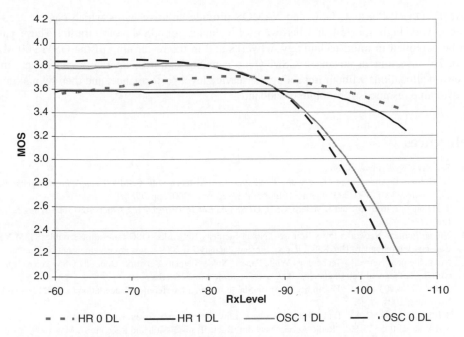

Figure 8.39   DL MOS, Double Half-Rate and Half-Rate, versus RxLevel measurement traces samples.

8.5   Conclusion

Voice capacity of the GSM networks has increased multifold since the introduction of GSM Phase 1 voice service. In this chapter, the focus was on the latest voice capacity features, namely AMR, DARP Phase 1, Orthogonal Sub-Channels and VAMOS, which is standardized in 3GPP Release 9.

The AMR and DARP Phase 1 features provide improvements in link performance which enable the introduction of OSC and VAMOS. In the downlink, the interfering sub-channel is equivalent to another GMSK modulated signal and hence it is possible for a DARP Phase 1 mobile to reduce the impact of this interference. In the uplink, the concept can be seen as a multi-user MIMO system where the separation of the signals is taken care of by independent multipath channel characteristics. Robust AMR codec modes make sure that OSC can be used in the lower C/I conditions.

OSC enables a maximum of 32 voice calls per one TRX and thus significantly improves GSM voice HW efficiency. As shown with the system level simulations, OSC provides high capacity gains in blocking limited scenarios. In the interference-limited networks, gains are more limited. In 3GPP Release 9 a new set of training sequences and downlink constellation power control have already been specified and a wider transmit pulse shape is under study. With these new enhancements spectral efficiency can be significantly improved. For example, simulation results showed that 30% capacity gains in the interference-limited network are achievable. OSC has also been tested in the field where measurements demonstrated that four calls can be carried in one GSM radio time slot. Furthermore, there are several enhancement methods under study, which can further improve the capacity of voice evolution concepts.

AMR, DARP Phase 1, OSC and VAMOS provide additive gains which can be used to achieve very high spectral and hardware efficiencies for GSM voice traffic. These gains can be exploited in multi-technology networks and in the re-farming of the GSM 900 MHz band. By investing in advanced GSM voice capacity features, operators can target small spectrum allocations without impacting the service level performance, and thereby maximize the available resources for mobile broadband.

References

[1] See gay www.gsmworld.com.
[2] H. Holma, T. Ahonpaa and E. Prieur, "UMTS900 Co-Existence with GSM900", in *Vehicular Technology Conference, 2007*. VTC2007-Spring. IEEE 65th, 22–25 April 2007, pp. 778–782.
[3] T. Halonen, J. Romero, and J. Melero, *GSM, GPRS and EDGE Performance*, Chichester: John Wiley & Sons, Ltd., 2002.
[4] Nokia Siemens Networks Press Release, "Nokia Siemens Networks' OSC innovation Doubles GSM Voice Capacity with Standard Handsets", Espoo, Finland, January 21, 2009, www.nokiasiemensnetworks.com.
[5] Nokia Siemens Networks Technology White Paper, "Nokia Siemens Networks Doubling GSM Voice Capacity with the Orthogonal Sub Channel", 2009, www.nokiasiemensnetworks.com.
[6] 3GPP TR 45.903 v6.0.1, "Feasibility Study on Single Antenna Interference Cancellation for GSM Networks", November 2004.
[7] 3GPP TS 45.009 v7.7.0, "Radio Access Network; Link Adaptation", November 2001.
[8] 3GPP TS 45.005 v7.18.0, "Radio Access Network; Radio Transmission and Reception", May 2009.
[9] 3GPP TS 45.002 v9.2.0, "Multiplexing and Multiple Access on the Radio Path", November 2009.
[10] 3GPP TR 45.914 v2.0.2, "Circuit Switched Voice Capacity Evolution for GSM/EDGE Radio Access Network (GERAN)", May 2009.
[11] 3GPP GP-081162, "Optimized User Diversity Patterns for OSC", Nokia Siemens Networks, TSG GERAN #39, August 2008.
[12] 3GPP GP-080171, "Sub Channel Specific Power Control for Orthogonal Sub Channels", Nokia Siemens Networks, TSG GERAN #37, February 2008.
[13] 3GPP GP-081179, "Downlink Power Control with Orthogonal Sub Channels," Nokia Siemens Networks, TSG GERAN #39, August 2008.
[14] Ming-Ju Ho and A. Mostafa, "AMR Call Quality Measurement Based on ITU-T P.862.1 PESQ-LQO", in *Vehicular Technology Conference, 2006*. VTC-2006 Fall. 2006 IEEE 64th; 25–28 Sept. 2006, pp. 1–5.

9

Wideband AMR Performance

Robert Müllner, Carsten Ball, Kolio Ivanov, Markus Mummert,
Hubert Winkler and Kurt Kremnitzer

9.1 Overview

The Adaptive Multi-Rate (AMR) Wideband (WB) codec, a new development in speech codec performance which was standardized in 2001 [1–3], has found applications in mobile GSM and UMTS networks, wire-line services, VoIP and multimedia applications. Its attractiveness lies in its superior speech quality compared to traditional narrowband (NB) telephony thanks to an extended audio bandwidth.

In this chapter on Wideband AMR Performance, these speech quality improvements are evaluated in detail together with their effects on radio network performance and capacity.

The chapter considers the performance of AMR-WB in GERAN networks from several perspectives. First, listening tests are described which assess the perceived quality difference between wideband and narrowband coded speech. A network planning study is then described which investigated the impact of AMR-WB on coverage and quality in the network. Different frequency reuse patterns were applied ranging from 4/3 down to 1/1. Finally, the network performance of AMR-WB is analyzed by network simulation, taking into account results from the network planning study and listening tests as well as dynamic effects such as fading, power control, and AMR codec mode adaptation. These simulations complement the results of the planning study and provide a detailed evaluation of quality and capacity in AMR-WB networks when compared to conventional AMR-NB.

9.2 Introduction

The AMR-NB speech codec, which is already well established in mobile networks [4–9], outperforms older codec standards such as "Enhanced Full-Rate" (GSM-EFR), "Full-Rate" (GSM-FR), and "Half-Rate" (GSM-HR) both in respect to speech quality and error robustness [10]. The next major step towards excellent speech quality was achieved with the introduction

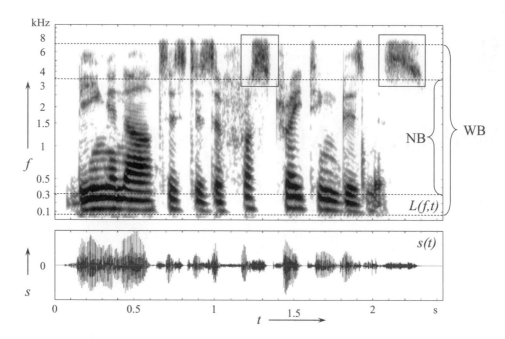

Figure 9.1 AMR-WB time signal s(t) and its spectrogram L(f,t).

of AMR-WB, which was standardized for use both in GSM and UMTS mobile networks [1, 2, 11] and which has been adopted by the ITU-T as Recommendation G.722.2, thereby facilitating its introduction across other networks [3].

3GPP Release 8 defines five codec modes ranging from 6.60 kbps to 23.85 kbps [8]. In GSM/EDGE, all five of these codecs are supported on the 8-PSK full-rate speech channel (O-TCH/WFS), three are supported on the GMSK full-rate speech channel (TCH/WFS) and three are supported on the 8-PSK half-rate channel (O-TCH/WHS). The source rates compare with the eight codec modes of AMR-NB which range from 4.75 kbps to 12.2 kbps and to the source rate of ISDN narrowband speech which employs G.711 A-law coding [12] which is delivered at a rate of 64 kbps.

9.3 Audio Bandwidth Extension for More Natural Sounding Speech

The benefit of AMR-WB over traditional narrowband telephony lies in the audio bandwidth extension. Extending the upper frequency coding range from 3400 Hz to 7000 Hz improves intelligibility, while a lower frequency extension from 300 Hz down to 50 Hz contributes to a more natural sound impression. Figure 9.1 shows a time signal s(t) and its spectrogram L(f,t) as a function of time and frequency for an AMR-WB audio signal under error-free radio conditions [13]. The combination of P.341[1] and MSIN[2] filtering was applied to reflect the

---

[1] WB send filter, ITU-T P.341/G.191, eliminates frequencies higher than 7 kHz and attenuates frequencies lower than 50 Hz; affects disturbing noise but hardly voice.

[2] Attenuation of low frequencies, ITU-T G.191, −3 dB at 190 Hz, 40 dB/decade.

bandwidth limitations caused by technical restrictions in mobile phones. These include both the miniaturization of loudspeakers and acoustic design considerations such as suppression of low-frequency noise and optimization of intelligibility. In the spectrogram, the blackening signifies the amount of signal energy in the time/frequency domain. A frequency scale has been chosen which is proportional to the Bark scale (to obtain an aurally-adequate spectrogram) [14]. While a significant amount of energy can be seen in the low frequency extension range, its contribution in terms of information is only low. The main benefit of AMR-WB coding lies in the extension in the upper frequency range and at these frequencies, information can be seen which is not conveyed in the narrowband area.

9.4 End-User's Quality Perception by Listening Tests

9.4.1 Test Method

The difference in the perceived quality between wideband and narrowband coded speech can be quantified by performing listening tests [15]. One such listening test method that can be used to compare different bandwidths within the same experiment is the "MUlti Stimulus test with Hidden Reference and Anchors" (MUSHRA) [16–18]. MUSHRA has been designed to provide a reliable and repeatable measure of the quality of intermediate-quality signals [16]. In this chapter, a modified MUSHRA test methodology is adopted whereby the listeners have been taken from a selection of standard mobile phone users instead of professional listeners. While a quality assessment by people recruited on the street represents a deviation from the strict MUSHRA concept, it was chosen in order to evaluate better the end-user's perception.

9.4.2 Experimental Design

In MUSHRA, "multiple stimulus" means differently processed versions of a sound source are presented simultaneously on a display to be played at random by pressing a button. This allows the test subject to listen to one version and to switch quickly to another version. The original unprocessed version of the source is also presented and is identified as the reference version. This guarantees that the listener knows how each version should sound. That would not be the case in Absolute Category Rating (ACR) tests [19]. Among versions to be assessed, the MUSHRA test also includes a hidden reference and so-called anchors. Thus the original unprocessed version also appears as one of the versions to be graded by the subject. Anchors are low-pass filtered originals with a band limit at 3.5 kHz (NB anchor) and 7 kHz (WB anchor), respectively [17]. Both the hidden reference and anchors provide a reference grid that covers the grading scale.

The subjects were instructed to rate the speech quality of each signal on a scale of 0 to 100 with the understanding that 100 corresponded to the reference [15]. The term "speech quality" is not defined so the listener's understanding could be a combination of intelligibility and general quality impression. The fact that a hidden reference appears among the versions to be graded is not revealed to the subjects. Subjects were asked to proceed in two steps. In the first, the quality is classified according to a coarse scale of five equidistant ranges representing "excellent", "good", "fair", "poor", and "bad" quality. In the second step, a refinement of the setting was requested, this time using the continuous scale between 0 and 100. It was left to the subject to decide whether his/her assessments were satisfactory and when to terminate the experiment.

The experiments were controlled by a computer which was connected via an external sound card and amplifier for individual volume control to a monaural headphone. The test subjects were trained in a previous experiment to make them familiar with the user interface and to expose them to the range of the sound quality. All tests were carried out in a silent environment by an independent research institute.

9.4.3   Test Conditions and Stimuli

A total number of 150 test subjects (78 male, 72 female) were recruited at three different locations in Germany. They represented a uniform distribution of business users and private users in four age categories: 16–25, 26–40, 41–55, and over 55 years. People with hearing impairments were not included.

In the main experiment, the type rating of AMR-WB against AMR-NB quality in error-free conditions was conducted with a total number of 80 subjects. Four additional experiment types in error-prone channel conditions within specific GSM channels were also conducted, with a total of 32 subjects each. For each experiment, subjects were randomly selected to reflect proportions of the whole sample. Disregarding the training experiments, specific combinations of experiment type and speaker were presented not more than once per session. Audio source material (recorded at 44.1 kHz) comprised eight different utterances in German language, spoken by male or female speakers in clean or typical background noise, as shown in Table 9.1. Each sample lasted about 8 s.

The stimuli used in the test were generated from source material using rate conversion, filtering, transcoding, and level adjustment [15]. Characteristics of pre-filtering and possible transcodings to determine stimulus are summarized in Table 9.2.

For error-prone channel transcoding, the encoded signal was passed through a radio channel simulator, which was configured for the following conditions: single interferer, Typical Urban 3 km/h, ideal frequency hopping, GSM900 with carrier to interference ratios (C/I) of 1, 4, 7, 10, and 13 dB. Performance of the simulator was similar to the one used within ETSI/3GPP AMR performance characterization tests [10,11]. In each condition only the best-suited AMR codec mode was presented, as determined in advance by experienced listeners.

Table 9.1   Speaker characterization

Speaker	Description	
1	male	clean speech
2	male	clean speech
3	female	clean speech
4	female	clean speech
5	male	speech with train station noise
6	male	speech with crowded street noise
7	female	speech with suburban street noise
8	female	speech inside car/vehicle noise

Table 9.2 Stimulus types used in test

Stimulus type	Pre-filter	Pre-filter characteristics	Transcoding
Reference	(none)	Original sound recorded at 44.1 kHz	(none)
7 kHz WB-anchor	C7K	Low pass eliminating f > 7 kHz, steep roll-off similar to P.341	(none)
3.5 kHz NB-anchor	C3K5	Low pass eliminating f > 3.5 kHz, rather steep roll-off [1]	(none)
AMR-WB	P.341 + MSIN	P.341: WB send filter, ITU-T P.341/G.191, eliminates f > 7 kHz and attenuates f < 50 Hz; affects disturbing noise but hardly voice. MSIN: attenuation of low frequencies, ITU-T G.191, −3 dB at 190 Hz, 40 dB/decade	TCH/WFS, O-TCH/WFS & O-TCH/WHS
AMR-NB	MIRS	Modified IRS send filter, ITU-T P.830/G.191, attenuates low frequencies and past 3.4 kHz while accentuating mid-range with a peak near 3 kHz	TCH/AFS
G.711 (ISDN)	MIRS	Modified IRS send filter, ITU-T P.830/G.191, attenuates low frequencies and past 3.4 kHz while accentuating mid-range with a peak near 3 kHz	G.711 A-law

Note: [1]Gaussian, dB:kHz −1:3.45, −3:3.50, −7:3.55, −12:3.60, −33:3.70, −64:3.80.

9.4.4   Listening Test Results

9.4.4.1   Error-Free Radio Channel Condition

Figure 9.2 shows the mean and 95% confidence intervals of the results from the main experiment corresponding to clean speech using male and female voices.[3] Sample size is almost 40, representing half of the 80 subjects who assessed clean speech but with some subjects rejected due to an excessive variance in their hidden reference [15].

The hidden reference received the highest score but fell short of the expected value (100). AMR-WB speech quality assessments stood out considerably from narrowband results, even though they did not match up to the original sound alias reference. The quality improvement of around 20 corresponds to a full step on the five-grade scale that the experiments maintained in parallel with the continuous scale. The 7 kHz anchor ranked only slightly higher indicating that losses due to wideband transcoding had only minor effects. These AMR-WB assessments did not expose a difference between the 23.85 and 12.65 kbps transcoding.

The 3.5 kHz anchor adhered to the narrowband group, where no significant differences were observed as well. Thus any AMR-NB transcoding losses in relation to G.711 (ISDN) were of no importance.

---

[3] Although channel codec types are mentioned, results for error-free channel conditions are a feature of the source coding only and independent of the employed channel coding.

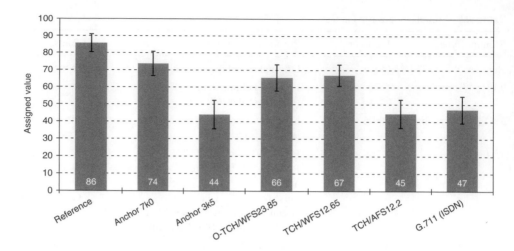

Figure 9.2   Speech quality assessment for clean speech. Error-free channel, male, and female speakers.

Note: Mean and 95% confidence interval of the assigned values is based on N=38 samples. Because of error-free condition, results are in fact independent of the channel type.

In the results corresponding to speech in background noise conditions shown in Figure 9.3, the shape when compared to the clean speech results is virtually preserved. This confirms the validity of prior observations. However, the level of the scores is reduced by a common offset (20 on average). Thus, the score of the hidden reference misses the expected projection considerably, a fact which will be discussed later.

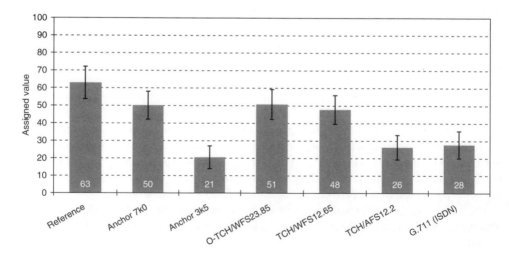

Figure 9.3   Speech quality assessment for noisy speech. Error-free channel, male, and female speakers.

Note: Mean and 95% confidence interval of the assigned values is based on N=39 samples. Because of error-free condition, results are in fact independent of the channel type.

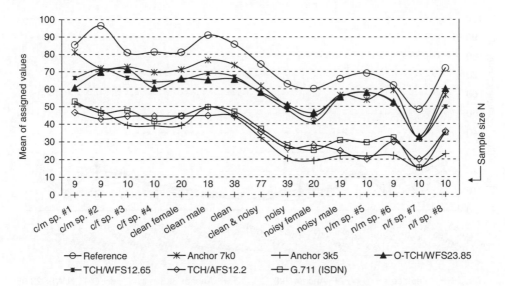

Figure 9.4 Dependency of speech quality on speaker group for error-free channel: mean of assigned values.

To further analyze the outcome of the main experiment, the dependency of the mean of the assigned values has been depicted in terms of speaker group in Figure 9.4. Assessments for single speakers or combinations of speakers with clean speech are shown on the left and assessments for noisy speech are shown on the right. For example, the mean values in Figure 9.2 can be found directly to the left of "clean & noisy", each plotted one above the other. Similarly, the values in Figure 9.3 can be found to the right of center. Across speaker groups the assessments of the same stimulus type are linked by lines to serve visual orientation. This does not necessarily imply dependence between connected points. Added on the abscissa is the sample size corresponding to the number of subjects contributing.

In a similar manner, Figure 9.5 shows the half-width of the 95% confidence interval to illustrate reliability of the mean values given above. Naturally, highest reliability is achieved if assessments are averaged for all speakers, thus including the assessments of 77 subjects.

The size of the confidence interval and its variance among different codec modes and filter combinations increase with decreasing sample size, as statistics become less reliable. Average standard deviation was determined to be round about 23, almost independent of the speaker group.

When shifting the focus from clean speech to noisy speech in Figure 9.4, it becomes apparent that the assignment levels vary more across speakers, revealing dependency on the nature of the background noise (spectral distribution and level). Also it appears that the 7 kHz anchor loses its small advantage over AMR-WB, hinting that wideband codec losses are even less important in the presence of background noise. However, this tendency is not confirmed for error-prone channels, presented below.

It is important to note that while disregarding assessment offset variations, Figure 9.4 demonstrates that observations made previously for clean and noisy speech are essentially reflected in any subgroup.

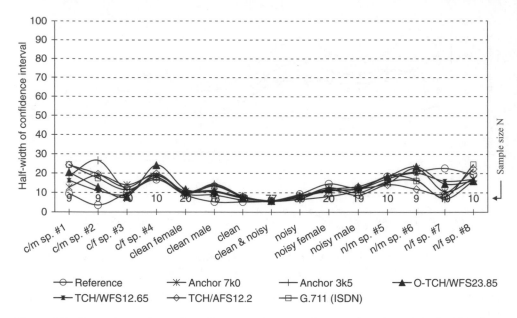

Figure 9.5 Dependency of speech quality on speaker group for error-free channel: half-width of confidence interval.

When dividing results into individual speakers, two effects should be kept in mind: (a) low statistical reliability due to small sample size and (b) effects peculiar to individual speakers and background conditions. In any case, AMR-WB was always found to be of higher quality than narrowband.

9.4.4.2   Error-Prone Radio Channel Conditions

Four additional experiments reflecting error-prone radio conditions for three AMR-WB channels and one AMR-NB channel are shown in Figures 9.6 to 9.9. Each figure depicts mean values for clean speech represented by squares, noisy speech represented by asterisks, and all speakers are represented by circles. The three stimulus types on the left correspond to the hidden reference and anchors and the remaining types correspond to deteriorating radio conditions for the given channel. Marks for 95% confidence intervals are attached to the data points of the "all speakers" group[4] and data points were linked by lines for better visual distinction only. Due to the fact that sample size is much smaller than in the main experiment, intervals are somewhat larger.

Although confidence intervals are of a considerable size, the curves of deteriorating speech quality are clearly localized for almost all channels and speaker groups. In other words, the mean values hardly deviate from a visually smooth curve for the deteriorating side of the

---

[4] For better visual representation 95% confidence intervals for clean and noisy speech are omitted in Figures 9.6 to 9.9.

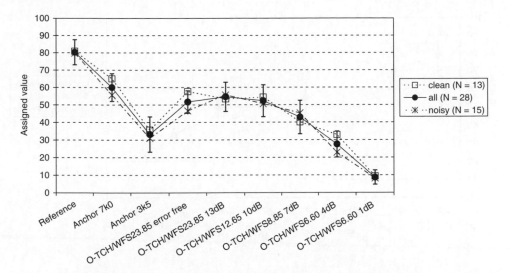

Figure 9.6 Speech quality assessments for error-prone radio channel and O-TCH/WFS channel type.

Note: Mean of the assigned values for speaker groups clean, noisy, all and 95% confidence interval for all speakers.

diagram. This indicates that the variance of assessments only partly accounts for independent random error. Obviously, the subject's assessment in relation to each other are likely more reliable than the assessments for a particular stimulus between subjects. The subjects' individual interpretations of scale create a spread that contributes to variance and, in consequence, to standard deviation and confidence width.

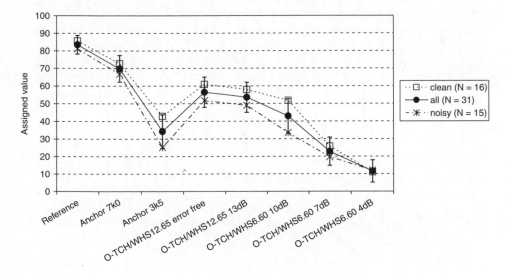

Figure 9.7 Speech quality assessments for error-prone radio channel and O-TCH/WHS channel type.

Note: Mean of the assigned values for speaker groups clean, noisy, all and 95% confidence interval for all speakers.

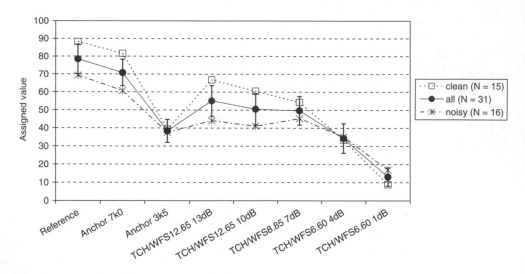

Figure 9.8 Speech quality assessments for error-prone radio channel and TCH/WFS channel type.

Note: Mean of assigned values for speaker groups clean, noisy, all and 95% confidence interval for all speakers.

Speech quality of 3.5 kHz anchor plus/minus a small margin may serve as an orientation for the level of AMR-NB in an error-free channel in Figures 9.6 to 9.8. This quality level also corresponds to G.711, as evidenced by error-free channel results. The procedure delivers the following results: except for poor radio conditions, wideband speech quality always outperforms error-free channel narrowband. The quality turnover points fall between 7 dB and 4 dB C/I for O-TCH/WFS and TCH/WFS and between 10 dB and 7 dB for O-TCH/WHS.

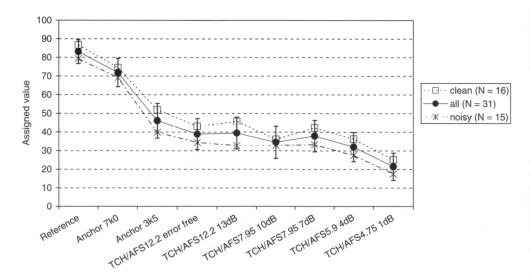

Figure 9.9 Speech quality assessments for error-prone radio channel and TCH/AFS channel type.

Note: Mean of the assigned values for speaker groups clean, noisy, all and 95% confidence interval for all speakers.

For fair comparison with AMR-NB, its deterioration must also be taken into account. The true turnover points are therefore even lower.

A common finding among the results was that the hidden reference was rated significantly lower than the expected value of 100, especially in the case of noisy speech. Obviously some subjects graded on an absolute scale rather than in relation to the reference, which naturally discriminates speech in a distracting background. A likely explanation is that the meaning of the reference was not always understood, or perhaps was in doubt, when the background noise made a bad overall impression. Subjects recruited on the street cannot be expected to be as attentive to the test procedure as experienced listeners would be in a genuine MUSHRA test.

However, this does not appear to compromise the basic validity of the results: The ordering of the stimuli as well as the relative differences in their assessments is not affected when only a smaller fraction of the full scale is used.

To summarize, the end-user study attests to a substantial gain in speech quality if wideband codecs like AMR-WB replace conventional narrowband codecs such as AMR-NB. The challenge for network engineers is to exploit this benefit to the full in real networks.

9.5 Impact of AMR-WB on Network Planning[5]

For network analysis it is essential to identify to what extent the inherent quality gain by speech coding enhancements is transferred to network performance improvement. Channel coding of AMR-WB plays an important role in this investigation. The dependency of the perceived speech quality on C/I was analyzed by listening tests and is shown in Figures 9.6 to 9.8.

The scope of this section is to identify the prevailing C/I conditions in various geographical parts of the network and to map these conditions on to the operating ranges of the different AMR-WB codec modes. The average C/I was calculated in a real network planning study for dense urban and sub-urban deployment areas in Munich, Germany [20]. In addition, the uplink (UL)/downlink (DL) power budget balance was analyzed for its impact on the coverage.

9.5.1 Network Planning Model

The designed network was originally planned and optimized for indoor speech service using plain GSM-FR speech. Sites were distributed to achieve an indoor received signal level RxLev≥-104 dBm (corresponding to the receiver sensitivity of GSM Mobile Stations (MS) according to 3GPP standard [21]) for 95% of the cell area. The link budget is presented in Table 9.3. For 8-PSK modulated AMR-WB channel types O-TCH/WFS and O-TCH/WHS a back-off of 2 dB in BTS output power, that is 46.3 dBm, was selected to account for peak-to-average ratio (PAR) power excess. In UL a maximum MS transmission power of 27 dBm for 8-PSK modulation was used [21]. Specific antenna characteristics are listed in Table 9.4. The antenna down-tilt in the designed network area was optimized for improving the homogeneity

---

[5] This section is based on "AMR-Wideband: Enjoying Superior Voice Quality at Full Coverage and Competitive Capacity in GERAN Networks", by R. Müllner, C.F. Ball, K. Ivanov, D. Hartmann and H. Winkler which appeared in IEEE VTC, Spring 2005, Stockholm. © 2005 IEEE.

Table 9.3 Link budget for GSM900 in an urban area deployment and static indoor subscribers

Uplink	Value	Downlink	Value
MS output power	33 dBm	Base Transceiver Station	48.3 dBm
Body loss	3 dB	(BTS) output power	
Indoor penetration loss	20 dB	BTS feeder cable loss	3 dB
Fading margin	12.5 dB	Combiner loss	5.7 dB
Maximum UL path loss	125.2 dB	BTS antenna gain	18 dBi
BTS antenna gain	18 dBi	Fading margin	12.5 dB
Polarization diversity gain	4 dB	Maximum DL path loss	126.1 dB
BTS feeder cable loss	3 dB	Indoor penetration loss	20 dB
Duplexer loss	25 dB	Body loss	3 dB
BTS receiver sensitivity	−111.2 dBm	MS receiver sensitivity	−104 dBm

in the near area and maintaining the overall coverage. Typical antenna heights for dense urban European cities were assumed.

Based on these values, the BTS deployment distancewas calculated using a tuned COST-231 two-slope propagation model. To achieve the desired indoor coverage (which corresponded to an outdoor RxLev of −68.5 dBm or higher), the maximum site-to-site distance was 900 m. Figure 9.10 shows the considered network area and deployed base stations.

9.5.2 Network Coverage Analysis for AMR-WB

The weaker path in the link budget is the UL, especially for 8-PSK modulation. Hence only UL plots are shown in the following three figures, where utilization of the different AMR-WB codec modes based on coverage analysis is shown.

Figure 9.11 shows the indoor coverage for O-TCH/WFS. The utilization of various codec modes within the network area depends on receiver sensitivity, which is specific for each codec mode. The legend indicates the required outdoor receive level for operating the corresponding codec mode in indoor environments. Areas in white indicate insufficient indoor coverage.

However, full outdoor coverage is still provided for the whole area. Figure 9.12 demonstrates the situation for O-TCH/WHS. Due to the higher signal level requirements of the O-TCH/WHS codec modes, the regions with insufficient coverage (white) are larger than those in Figure 9.11. In these areas the utilization of the 8-PSK modulated codec modes is not possible and a switch to GMSK modulation has to be performed. This provides higher output power. Finally,

Table 9.4 Antenna characteristics

Polarization type	Cross-polarized	Front to back ratio	39.2 dB
Horizontal beam width	64.5°	Down-tilt	4°
Vertical beam width	7.0°	Antenna height	20–30 m

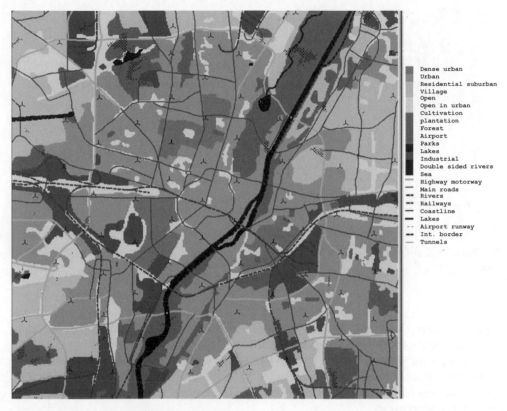

Figure 9.10 Network area of the AMR-WB planning study.

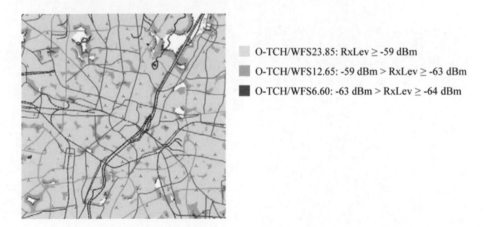

Figure 9.11 Indoor coverage of O-TCH/WFS.

Note: RxLev indicated in the legend is the minimum outdoor receive level for indoor utilization of the different codec modes.

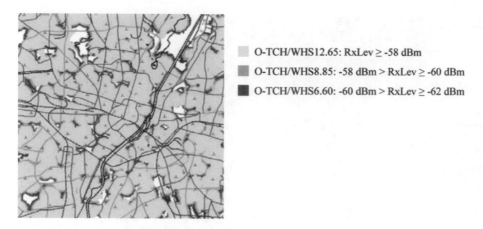

O-TCH/WHS12.65: RxLev ≥ -58 dBm

O-TCH/WHS8.85: -58 dBm > RxLev ≥ -60 dBm

O-TCH/WHS6.60: -60 dBm > RxLev ≥ -62 dBm

Figure 9.12   Indoor coverage of O-TCH/WHS.

Note: RxLev indicated in the legend is the minimum outdoor receive level for indoor utilization of the different codec modes.

Figure 9.13 shows that throughout the whole network area, the signal strength is sufficiently high to utilize the highest GMSKmodulated full-rate codec mode TCH/WFS12.65.

The conclusion is that O-TCH/WFS and O-TCH/WHSprovide indoor coverage in almost the entire network area planned for conventional GSM-FR speech codec. Fallback to TCH/WFS allows full area indoor coverage. Deployment of further base stations or re-planning of the network is not necessary.

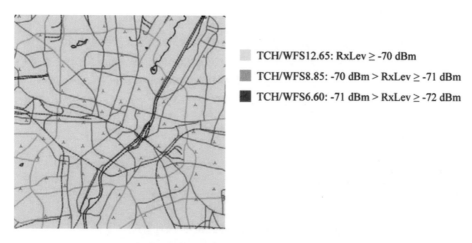

TCH/WFS12.65: RxLev ≥ -70 dBm

TCH/WFS8.85: -70 dBm > RxLev ≥ -71 dBm

TCH/WFS6.60: -71 dBm > RxLev ≥ -72 dBm

Figure 9.13   Indoor coverage of TCH/WFS.

Note: RxLev indicated in legend is the minimum outdoor receive level for indoor utilization of the different codec modes.

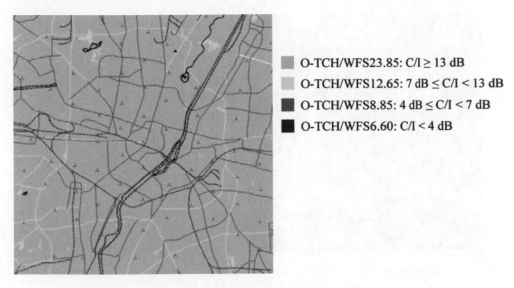

O-TCH/WFS23.85: C/I ≥ 13 dB

O-TCH/WFS12.65: 7 dB ≤ C/I < 13 dB

O-TCH/WFS8.85: 4 dB ≤ C/I < 7 dB

O-TCH/WFS6.60: C/I < 4 dB

Figure 9.14   Network quality plot for O-TCH/WFS in 4/3 frequency reuse.

9.5.3   Network Quality Analysis

In the following investigation, the average C/I for the geographical areas of the network considered above was calculated using a radio network planning and frequency assignment tool. Figures 9.14 to 9.16 show the worst case local mean DL C/I for relaxed 4/3 and 3/3 versus tight 1/1 frequency reuses, which have been mapped on to the operating ranges according to

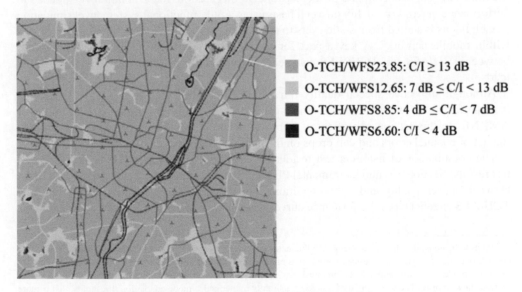

O-TCH/WFS23.85: C/I ≥ 13 dB

O-TCH/WFS12.65: 7 dB ≤ C/I < 13 dB

O-TCH/WFS8.85: 4 dB ≤ C/I < 7 dB

O-TCH/WFS6.60: C/I < 4 dB

Figure 9.15   Network quality plot for O-TCH/WFS in 3/3 frequency reuse.

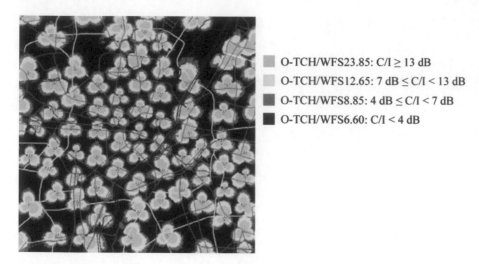

Figure 9.16   Network quality plot for O-TCH/WFS in 1/1 frequency reuse.

[11, 22] for the different O-TCH/WFS codec modes of 23.85, 12.65, 8.85 and 6.60 kbps. As expected, the higher codec modes provide full coverage in the relaxed frequency reuse.[6]

9.6   Network Quality and Capacity Advantage of AMR-WB over AMR-NB

In the previous section, the performance of the network in terms of C/I was evaluated for the different geographical parts of the network. In this section, the evaluation is extended to performance in terms of quality on a cell basis using a fully functional system level simulation model. Focus is shifted from static users to moving subscribers including the dynamic effects within mobile networks. A key aspect for capacity analysis is the performance difference between TCH/AFS and TCH/WFS channel types in interfering situations at various traffic loads.

Although subjective speech quality improvements are provided by AMR-WB, the link level performance for codec modes of comparable source bit rates remains similar to that of AMRNB. In this section, the effect of improved audio perception in combination with the impact of channel errors and call drops on the network performance is analyzed [13].

The combination of listening test results for AMR-NB [10], AMR-WB [11], MUSHRA test results (Section 9.4) and instrumental PESQ [23] measurements provide a general relation between speech quality and radio conditions. For the comparison between TCH/WFS and TCH/AFS channel types, performance curves were generated using C/I as the common basis.

---

[6] It has to be noted that the instantaneous C/I fluctuates around the mean C/I value due to fading effects. Hence true codec mode adaptation is also necessary within areas of the same shades of gray and not only at the border regions. In this study moderate codec mode adaptation thresholds were applied. Dynamic simulations described in the following section show improved network quality for selecting slightly higher codec mode adaptation thresholds, that is more frequent utilization of more robust codec modes.

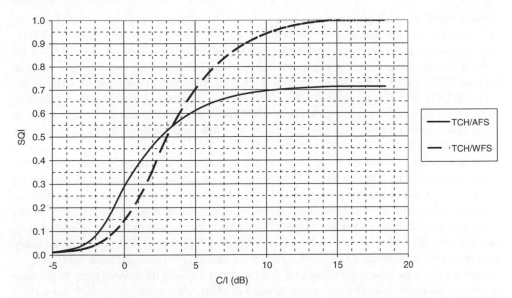

Figure 9.17 Speech quality indicator (SQI) for the envelope of the best performing codec mode of TCH/AFS and TCH/WFS versus C/I applying TU3 channel model at GSM900, ideal frequency hopping.

MUSHRA test results span the grid points for the two performance curves and define the quality difference between AMR-NB and AMR-WB at a given C/I. Because MUSHRA data spacing is too coarse to interpolate continuous slopes over C/I with sufficient accuracy, the shapes of the curves were derived using additional data sources. These comprised results of subjective listening tests [10, 11] as well as our own experiments evaluating instrumental "objective" speech quality based on PESQ. After approximating continuous shapes from the sources' data they were fitted to the MUSHRA data by adjusting offset and scaling on the quality dimension and offset-only on the C/I dimension. Figure 9.17 shows the envelope of the best performing codec modes on GMSK modulated full-rate traffic channels using TCH/WFS and TCH/AFS as a function of C/I. The quality measure is Speech Quality Indicator (SQI), ranging from 0 to 1, with 1 representing the maximum achievable quality for TCH/WFS under error-free radio conditions.[7] TCH/WFS outperforms TCH/AFS for C/I higher than 3.5 dB. The reason for the better performance of TCH/AFS below this turnover point is the availability of more robust codec modes with source bit rates down to 4.75 kbps. The fundamental relationship between SQI and C/I for TCH/WFS and TCH/AFS shown in Figure 9.17 was applied in the performance analysis which follows.

9.6.1 Simulation Model

The quality analysis and capacity analysis for TCH/AFS and TCH/WFS were performed by system level simulations for GSM900. A hexagonal cell layout in an urban deployment

---

[7] Each SQI value represents the quality of the best performing codec mode for this channel type at given radio conditions, that is movement on the envelope of the performance characteristics of the codec modes is assumed.

scenario for slow moving subscribers was assumed. Simulations were performed for three different frequency reuse scenarios. For reuses 4/3 and 3/3, a frequency spectrum of 12 MHz and 10MHzrespectivelywas assumed which included theBroadcastControlCHannel (BCCH) frequencies. Speech service capacity was evaluated on TCH hopping carriers only since in this case GPRS/EDGE and signaling channels were mainly configured on the BCCH carrier.

The BCCH carrier was planned in 4/3 reuse (consuming a bandwidth of 2.4 MHz), resulting in up to four TCH carriers per cell, that is 32 GSM full-rate traffic channels. The speech traffic on the four hopping transceivers (TRX) were fixed at the hard blocking limit (22 Erlang for 32 traffic channels @ 1% blocking probability).

The tight reuse scenario 1/1 focuses on network operators with limited frequency spectrum of for example 5MHz. A bandwidth of 12 hopping frequencies was assumed for TCH planning in reuse 1/1.[8] Latencies in the power control algorithm were modeled accurately. Real codec mode adaptation with hysteresis and latency was applied in the simulation model.[9] The system load is given as Effective Frequency Load (EFL) which is the average traffic per cell divided by the product of 8 GSM time slots and number of hopping frequencies. The focus of this study was set on DL simulations. The essential parameter settings are summarized in Table 9.5.

Speech calls were generated according to a Poisson process with arrival rate λ depending on the offered traffic load and mean call duration of 90 s. After positioning a new subscriber in the 183 cell area, an independent slow fading value between each BTS to MS connection is used. The MS chooses the cell with the lowest path loss and connects to this BTS. During the simulation run aMS can change its serving cell by handover based on a power budget or quality decision criterion using real handover modeling. Lognormal fading is modeled by a Gaussian distribution with a standard deviation of $\sigma = 6$ dB and a correlation distance of 50 m. The fast fading values are modeled by Rayleigh distribution derived from link level simulations.

The value is updated for each speech burst (≈ 5 ms) assuming constant C/I during burst length and resulting in a time dependent fast fading, uncorrelated between each frequency and BTS-MS path. The C/I per speech burst is calculated as the ratio of the carrier signal between serving BTS and MS and the sum of all interfering signals considering co-channel and adjacent channel interference.

9.6.2   Quality Criteria

9.6.2.1   Speech Quality Indicator (SQI)

GSM radio network planning typically aims at a mean C/I of 9 dB at the cell border as a minimum requirement for networks planned in relaxed frequency reuse. According to Figure 9.17, the corresponding SQI for AMR-NB is 0.69. In tight reuse networks lower quality is tolerated for the benefit of increased capacity. Hence in the tight reuse scenario a slightly lower quality requirement of SQI = 0.66 was assumed corresponding to 7 dB C/I. In the simulation

---

[8] Four TRX can be configured per cell if using static Mobile Allocation Index Offset (MAIO). Applying dynamic MAIO allocation in reuse 1x1, up to 12 TRX can be deployed depending on the operator's capacity requirements. Thus installing a sufficient number of TRX per cell, hard blocking due to lack of resources can be minimized.

[9] It is assumed that always one link limits connection quality, which in this case is the GSM side. Hence the impact of the remote link on codec mode adaptation was ignored. Impacts of in-band signaling decoding errors as well as advanced features like TX diversity, switched beams (adaptive antennae) and interference cancellation algorithms such as SAIC [24] were not taken into account.

Table 9.5 Parameters of the system level simulation model

Parameter	Value
GSM band	900 MHz
Cell layout	hexagonal, regular cell layout
Number of sites	61 (183 cells)
Site-to-site distance	900 m
Reuse pattern	BCCH: 4/3 and TCH: 4/3, 3/3, 1/1
TCH frequencies per cell	4 for 4/3 and 3/3 reuse; 12 for 1/1 reuse
TCH TRX per cell	4 for 4/3 and 3/3 reuse; up to 6 for 1/1 reuse
Path loss slope	37.6 dB per decade of distance
Handover margin	4/3 and 3/3 reuse: 3 dB (0 dB intra-site); 1/1 reuse: 1 dB (0 dB intra-site)
Adjacent channel suppression	18 dB
Fast fading profile	TU3 (slow moving subscribers at a speed of 3 km/h)
Slow fading standard deviation	6 dB
Correlation distance	50 m
Power control	level and quality based
Channel types	TCH/AFS and TCH/WFS
AMR-NB codec modes	TCH/AFS12.2, TCH/AFS7.95, TCH/AFS5.90, TCH/AFS4.75
AMR-WB codec modes	TCH/WFS12.65, TCH/WFS8.85, TCH/WFS6.60
Mean call duration	90 s
Voice activity factor (DTX)	0.6

model the C/I is averaged in a rectangular filter of 1.92 s length (4 SACCH periods) and transformed into a SQI value according to the relation shown in Figure 9.17.

The quality criterion to be fulfilled is that 95% of the calls need to show a mean SQI per call higher than 0.69 for relaxed frequency reuse and higher than 0.66 for tight 1/1 frequency reuse.

9.6.2.2 Frame Erasure Rate (FER)

SQI is based on assessments related to general speech quality impression rather than intelligibility. Increasing the traffic load increases the interference in a network resulting in an increased amount of frame errors and consequently a loss of information. To minimize the loss of information, a certain level of FER should not be exceeded. In this study a maximum value of 2% per call was assumed. From a link level point of view, the support of robust codec modes with source bit rates down to 4.75 kbps provide a C/I gain of 2 dB at FER = 2% for TCH/AFS compared to TCH/WFS. A further advantage of TCH/AFS is related to faster codec mode adaptation [25] in tandem mode. The higher latency of codec mode adaptation using Tandem Free Operation (TFO) [26, 27], which is the prerequisite for AMR-WB, leads furthermore to an increased probability of sequential frame errors. Sequentially lost frames create a subjective impression of speech degradation more severe than reflected by SQI, the latter being more trustable for sporadic frame errors. The most severe impact is given if complete words are missing in a conversation (e.g. the word "not" in a negotiation!). These frame errors in a row are likely to occur if the most robust codec mode is already applied. This is more likely with TCH/WFS which has less robust codec modes than TCH/AFS (see Table 9.5).

Assuming a certain FER impairing intelligibility, the well-established criterion of mean FER per call lower than 2% to be fulfilled for 95% of the subscribers was considered the quality limit.

9.6.2.3   Bad Quality Probability (BQP)

Alternatively, the speech quality degradation caused by frame errors can be quantified by the percentage of speech samples taken within certain sampling intervals (e.g. over 4 SACCH periods, i.e. 1.92 s, with FER higher than a specific threshold, e.g. FERmax=2.5%). Since such speech samples indicate bad speech quality, the probability (percentage) of their occurrence is termed BQP. The target speech quality performance is deemed satisfactory if the BQP in the network is lower than 5%.

In this study, the established satisfied user criterion in tight reuse networks: 1-BQP(FER > 2.5%) > 95% was applied.

9.6.2.4   Call Drop Rate (CDR)

A further capacity-related criterion is CDR, which can be due to excessive failures on the Associated Control CHannels (ACCH). Temporary ACCH overpower [28] was applied to reduce the C/I gap in the link level performance between AMR voice on TCH and ACCH in order to improve the probability of successful decoding of the xACCH (FACCH/SACCH) frames[10] (refer to Chapter 12, Section 12.3). The criterion reflecting user satisfaction is CDR lower than 2%.

9.6.3   Performance Results for Networks in Relaxed Frequency Reuse Planning

Figure 9.18 shows the cumulative distribution function (CDF) of mean SQI per call for TCH/AFS and TCH/WFS in 4/3 reuse. The low interference level in 4/3 reuse networks and the large averaging window size result in steep SQI distributions for both channel types. The quality criterion mean SQI > 0.69 is fulfilled for 100% of the TCH/AFS and TCH/WFS calls.

For TCH/AFS the 5th percentile of mean SQI is 0.71, that is more than 95% of the calls provide the maximum achievable quality for this channel type. For TCH/WFS the 5th percentile of the mean SQI distribution is 1.00, that is for both channel types the highest network homogeneity was observed at a significant quality difference of $_-$SQI $= 0.29$ between AMR-WB and AMR-NB.[11]

Figure 9.19 presents the CDF of the mean FER per call in 4/3 reuse. The quality criterion means FER per call lower than 2% is fulfilled for 99.8% of the calls for TCH/AFS and 99.7%

---

[10] Repeated xACCH including soft combining was not applied in the simulations, because speech frame stealing by signaling channels is not considered in the fundamental transformation of C/I on SQI in Figure 9.17.

[11] In the 3/3 frequency reuse scenario, the 5th percentile of the distribution of mean SQI is again 0.71 for TCH/AFS and 1.00 for TCH/WFS.

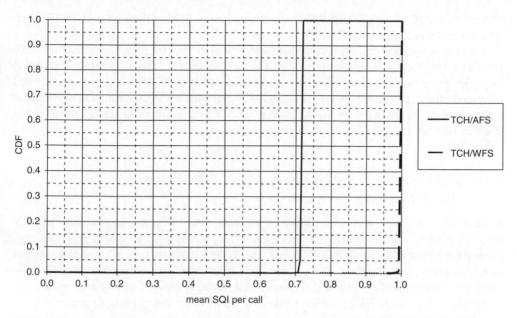

Figure 9.18 CDF of mean SQI per call for TCH/AFS and TCH/WFS. Four hopping TCH TRX per cell in 4/3 frequency reuse.

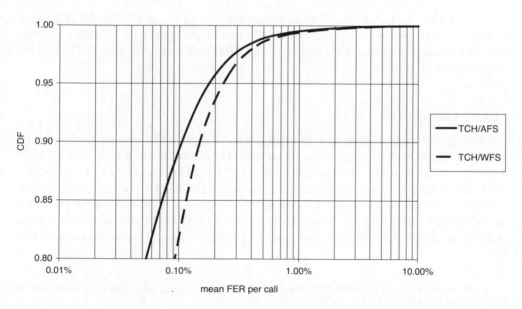

Figure 9.19 CDF of mean FER per call for TCH/AFS and TCH/WFS. Four hopping TCH TRX per cell in 4/3 frequency reuse.

of the calls for TCH/WFS. At the 95th percentile the mean FER of 0.18% for TCH/AFS is in the same order as mean FER of 0.23% for TCH/WFS.[12]

In 4/3 reuse, the BQP criterion is fulfilled for 99.7% of the TCH/AFS samples and for 99.5% of the TCH/WFS samples, respectively. The CDR is below 0.01%.[13] As a conclusion, in relaxed frequency reuse the FER of both channel types (TCH/AFS and TCH/WFS) is comparable and sufficiently low. The major performance difference indicated by SQI is related to the inherent speech quality improvement of the AMR-WB speech codec. The capacity of relaxed reuse schemes is limited by hard blocking, that is lack of resources.

9.6.4   Performance Results for Networks in Tight Frequency Reuse Planning

As the traffic load increases in high capacity tight reuse networks with a large number of TRX installed in each cell, the interference level caused especially by the cells of neighbor sites increases dramatically. In tight frequency reuse networks a different type of capacity limitation comes into play: soft-blocking. The soft-blocking limit defines the maximum traffic load at which the allocation of additional calls would violate the operator's quality expectations. In the following the soft-blocking limit is defined by the combination of multiple quality criteria SQI, FER, BQP, and CDR.

9.6.4.1   Quality Criterion SQI

Figures 9.20 and 9.21 show the CDFs of mean SQI per call for TCH/AFS and TCH/WFS, respectively. The quality criterion of mean SQI per call higher than 0.66 is fulfilled for more than 95% of the calls for an EFL up to 21% for TCH/AFS and up to 32% for TCH/WFS, respectively.

9.6.4.2   Quality Criterion FER

Figure 9.22 shows the CDF of the mean FER per call for TCH/AFS at varying offered load. The results for TCH/WFS are shown in Figure 9.23. The quality criterion for mean FER lower than 2% for more than 95% of the calls is fulfilled by TCH/AFS for an offered EFL up to 20%, nearly achieving the maximum EFL for the SQI criterion (21% EFL). For TCH/WFS the FER criterion is fulfilled for an offered EFL up to 17%, yielding a significant gap between FER and SQI based capacity limits. Selecting the more restrictive quality criterion (out of mean SQI and mean FER per call) limits the capacity of TCH/WFS in tight reuse networks to 17% EFL.[14]

---

[12] In the 3/3 frequency reuse scenario, 95% of the calls are characterized by a mean FER lower than 0.31% for TCH/AFS and lower than 0.38% for TCH/WFS, respectively.

[13] In 3/3 reuse, the BQP criterion is fulfilled for TCH/AFS for 99.4% of the samples. For TCH/WFS the BQP criterion is met for 99.2% of the samples. The CDR is below 0.01%.

[14] Higher capacity with respect to the FER criterion could be achieved by selecting higher codec mode adaptation thresholds. The drawback is a reduction in the potentially achievable quality, since selecting a more robust codec mode at good C/I levels results in subjective speech quality degradation.

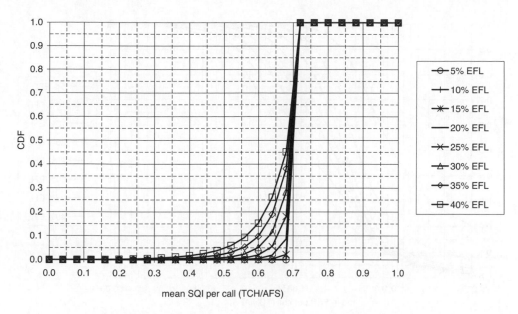

Figure 9.20   CDF of mean SQI per call for TCH/AFS in tight 1/1 frequency reuse depending on system load.

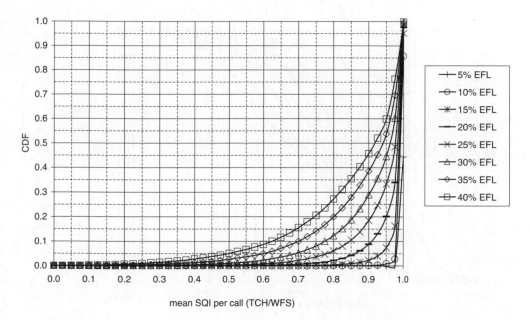

Figure 9.21   CDF of mean SQI per call for TCH/WFS in tight 1/1 frequency reuse depending on system load.

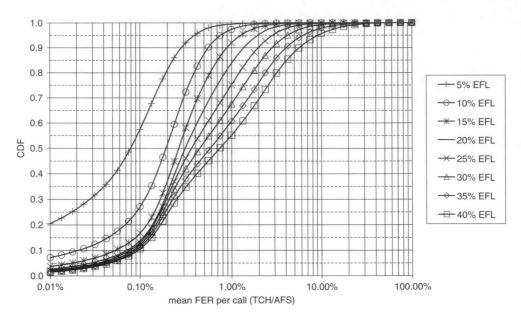

Figure 9.22   CDF of mean FER per call for TCH/AFS in tight 1/1 frequency reuse depending on system load.

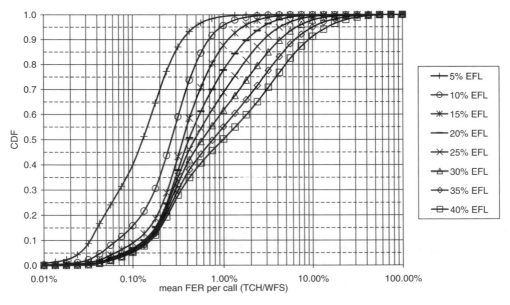

Figure 9.23   CDF of mean FER per call for TCH/WFS in tight 1/1 frequency reuse depending on system load.

Table 9.6 Bad Quality Probability (BQP) and Call Drop Rate (CDR) for tight 1/1 reuse networks using 12 hopping TCH frequencies per cell (%)

EFL	BQP TCH/AFS	BQP TCH/WFS	CDR
5	0.76	1.06	0.00
10	1.49	2.23	0.00
15	2.10	3.42	0.00
20	2.72	4.77	0.04
25	3.60	6.52	0.28
30	4.70	8.55	1.04
35	6.07	10.82	2.78
40	7.32	12.75	5.50

9.6.4.3 Quality Criteria BQP and CDR

The BQP criterion is less restrictive than the mean FER $< 2\%$ criterion. The BQP criterion is met for TCH/AFS at EFL of 31% and for TCH/WFS at EFL of 21%, respectively. BQP and CDR results for 1/1 reuse at varying offered load are listed in Table 9.6. Considering the BQP criterion along with the SQI criterion results in a soft-capacity limit of 21% EFL for both channel types TCH/AFS and TCH/WFS. The requirement for CDR $< 2\%$ allows for 33% EFL and is therefore less restrictive than the FER $< 2\%$ and BQP criteria for both channel types.

However, without temporary ACCH overpower (results not listed in Table 9.6) the capacity of the network would be limited by the dropping criterion to 25% EFL.

9.6.5 Combination of Quality Criteria

Figure 9.24 summarizes the results for TCH/AFS obtained with different satisfied user criteria as a function of the EFL. Aiming at 95% satisfied users, the mean SQI criterion well matches the FER criterion. The BQP and CDR are less restrictive criteria hence allow a higher capacity.

The results for TCH/WFS are shown in Figure 9.25. The graphs reveal a large gap between the SQI and FER/BQP based quality criteria. Applying the less restrictive BQP criterion in combination with mean SQI > 0.66 yields a maximum EFL of 21%, identical to that achieved by TCH/AFS. Replacing the BQP criterion by the FER criterion results in amaximum capacity of 17% EFL for TCH/WFS. The capacity limits for the different quality criteria are summarized in Table 9.7.

This analysis shows that applying a single criterion for network dimensioning is not sufficient. Different quality criteria should be combined covering subjective speech quality and intelligibility. Considering SQI and FER/BQP for network dimensioning the audio bandwidth advantage of AMR-WB is exploited best if it serves to improve the subscriber's quality experience rather than boosting capacity.

9.7 Conclusion

Listening test results prove a substantial advantage in speech quality for wideband coded speech over narrowband. For both clean and noisy speech, AMR-WB is perceived to be of

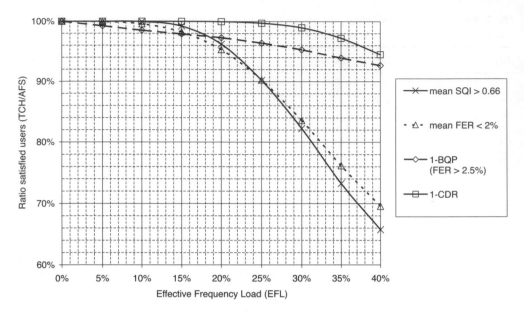

Figure 9.24 Trade-off between quality and capacity for TCH/AFS in tight 1/1 frequency reuse applying different quality criteria.

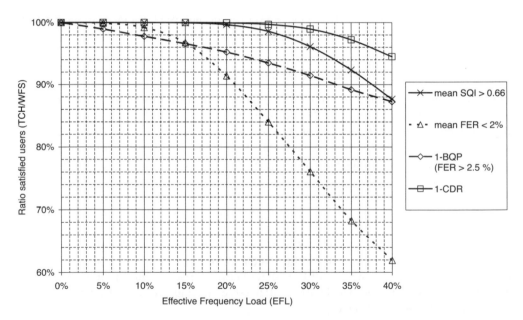

Figure 9.25 Trade-off between quality and capacity for TCH/WFS in tight 1/1 frequency reuse applying different quality criteria.

Table 9.7   Soft-blocking limits

Quality criterion	EFL for TCH/AFS (%)	EFL for TCH/WFS (%)
mean SQI > 0.66 for 95% of the calls	21	32
mean FER < 2% for 95% of the calls	20	17
BQP(FER > 2.5%) < 5%	31	21
CDR < 2% (temp. ACCH overpower)	33	33
CDR < 2% (w/o temp. ACCH overpower)	25	25

considerably higher quality thanAMR-NBand G.711, which were rated to be of approximately equal quality. For error-free channels, the quality difference corresponds to a full step on the five grade MOS scale. In error-prone radio conditions, if C/I is not too low, the quality of wideband speech for full-rate GMSK and 8-PSK channels was assessed to be higher than narrowband speech in error-free conditions. The turnover point when quality becomes worse ranges between 7 dB and 4 dB for full-rate GMSK and 8-PSK channels and between 10 dB and 7 dB for half-rate 8-PSK channel, respectively. For radio conditions above this point, speech bandwidth is more important than coded-speech bit rate. In the presence of background noise, this effect persists confirming that the benefits of AMR-WB are not reduced.

This study analyzed to what extent this inherent quality gain is transferred into network performance improvements. Coverage and quality impacts were investigated by planning a real AMR-WB network. Based on the knowledge of AMR-WB performance in a geographical cell layout, the quality differences within the network were studied in detail by network simulations using an idealized cell layout. Both approaches in combination provided a comprehensive analysis of quality and capacity in AMR-WB networks.

Deployment of AMR-WB in GERAN networks does not require additional sites, as shown by coverage analysis. In the critical case of dense urban indoor scenarios some stray coverage holes were only identified for 8-PSK modulation due to limited MS output power. In these regions a switch from 8-PSK to GMSK modulation will be triggered to ensure full coverage.

To analyze network quality and capacity in detail, four measures were used. A quality measure called SQI was introduced founded on listening tests from above. Since SQI relates to speech quality rather than to intelligibility, two additional criteria for determining themaximum network capacity were applied. FER and BQP criteria reflect frame errors that lead to loss of information during conversation. Finally, the Call Drop Rate was included in the analysis.

Results of the simulations show that AMR-WB gains a significant quality advantage over AMR-NB in relaxed frequency reuse networks. Capacity is limited by the number of available resources only. In high capacity tight reuse networks at varying traffic load, the SQI and FER-based criteria are well aligned for TCH/AFS, whereas a large gap between these criteria has been observed for TCH/WFS, which precludes full exploitation of subjective quality advantages for capacity boost.

For providing both high speech quality and intelligibility in tight reuse networks, the capacity of the network should be limited by admission control, taking into consideration a combination of criteria covering both aspects in order not to lose sight of intelligibility.

Otherwise, if the entire subjective quality gain of AMR-WB is traded in for capacity increase, user dissatisfaction is likely. This is because the quality span between busy and non-busy hours would be considerable, exceeding that experienced with TCH/AFS by far. It is thus recommended that the improved audio quality by AMR-WB should be used for quality improvement rather than for capacity increase.

References

[1]   I. Varga, R. Drogo de Iacovo, and P. Usai, "Standardization of the AMRWideband Speech Codec in 3GPP and ITU-T", *IEEE Communications Magazine*, May 2006.

[2]   3GPP TS 26.171, v8.0.0, "AdaptiveMulti-Rate –Wideband (AMR-WB) Speech Codec; General Description", December 2008.

[3]   ITU-T G.722.2, "Wideband Coding of Speech at around 16 kbit/s using Adaptive Multi-Rate Wideband (AMR-WB)", 2003.

[4]   W. C. Chu, *Speech Coding Algorithms*, New York: John Wiley & Sons, Ltd, 2003, p. 451.

[5]   H. Holma and A. Toskala, *WCDMA for UMTS*, New York: JohnWiley & Sons, Ltd, 2004, pp. 12–14, 369–376.

[6]   T. Halonen, J. Romero, and J. Melero, *GSM, GPRS and EDGE Performance*, New York: John Wiley & Sons, LTD, 2002, pp. 209–227.

[7]   J. Eberspächer, H.-J. Vögel, and C. Bettstetter, *GSM Switching, Services and Protocols*, John Wiley & Sons, Ltd, 2001, pp. 273–276.

[8]   3GPP TS 26.103, v8.1.0, "Speech Codec List for GSM and UMTS", December 2008.

[9]   3GPP TS 45.003, v8.2.0, "Radio Access Network; Channel Coding", May 2009.

[10] 3GPP TR 26.975, v8.0.0, "Performance Characterization of the Adaptive Multi-Rate (AMR) Speech Codec", December 2008.

[11] 3GPP TR 26.976, v8.0.0, "Performance Characterization of the Adaptive Multi-Rate Wideband (AMR-WB) Speech Codec", December 2008.

[12] ITU-T G.711, "Pulse Code Modulation (PCM) of Voice Frequencies", 1988.

[13] R. Müllner, C. F. Ball, K. Ivanov, H. Winkler, R. Perl, and K. Kremnitzer, "Exploiting AMR-WB Audio Bandwidth Extension for Quality and Capacity Increase", in *Proceedings of 16th IST Mobile & Wireless Communications Summit*, 2007, Budapest.

[14] T. Horn, "Image Processing of Speech with Auditory Magnitude Spectrograms", *Acustica* 84 (1998): 175–177.

[15] R. Müllner, and M. Mummert, "Speech Quality of AMR Wideband Coding vs. Narrowband as Perceived by the End-User", in *Proceedings of 15th IST Mobile & Wireless Communications Summit*, 2006, Mykonos.

[16] A. J. Mason, "The MUSHRA Audio Subjective Test Method", BBC Research & Development, White Paper WHP 038, 2002.

[17] ITU-R BS. 1534-1, "Method for the Subjective Assessment of Intermediate Quality Level of Coding Systems", 2001–2003.

[18] 3GPP TSG-SA4, Tdoc AHAUC-033, "Verification of Fixed-Point Implementation of AMR-WB+", 2005.

[19] ITU-T P.800, "Methods for Subjective Determination Of Transmission Quality", 1996.

[20] R. Müllner, C.F. Ball, K. Ivanov, D. Hartmann, and H. Winkler, "AMR-Wideband: Enjoying Superior Voice Quality at Full Coverage and Competitive Capacity in GERAN Networks", in *Proceedings of IEEE VTC*, Spring 2005, Stockholm.

[21] 3GPP TS 45.005 v8.5.0, "Radio Access Network; Radio Transmission and Reception", May 2009.

[22] 3GPP TSG GERAN #15, TDoc GP-031432, "Listening Test Results for AMR-WB", 2003.

[23] ITU-T P.862, "Perceptual Evaluation of Speech Quality (PESQ): An Objective Method for End-To-End Speech Quality Assessment of Narrowband Telephone Networks and Speech Codecs", 2001.

[24] C. F. Ball, K. Ivanov, H. Winkler, M. Westall, and E. Craney, "Performance Analysis of a GERAN Switched Beam System by Simulations and Measurements", in *Proceedings of IEEE VTC*, Spring 2004, Milan.

[25] 3GPP TS 45.009 v8.0.0, "Radio Access Network; Link Adaptation", December 2008.

[26] 3GPP TS 28.062 v8.1.0, "Inband Tandem Free Operation (TFO) of Speech Codecs; Service Description; Stage 3", March 2009.

[27] 3GPP TS 48.060 v8.0.0, "In-Band Control of Remote Transcoders and Rate Adaptors for Full Rate Traffic Channels", December 2008.

[28] K. Ivanov, C. F. Ball, R. Müllner, H. Winkler, R. Perl, and K. Kremnitzer, "Breaking through AMR Voice Capacity Limits Due To Dropped Calls By Control Channel Improvements In GERAN Networks", in *Proceedings of IEEE PIMRC*, 2005, Berlin.

10

DFCA and Other Advanced Interference Management Techniques

Sebastian Lasek, Krystian Majchrowicz and Krystian Krysmalski

10.1 Introduction

The radio frequency spectrum is a scarce resource and the spectrum itself constitutes a major investment the operators need to bear. Therefore spectral efficiency that quantifies the amount of traffic a network can serve per MHz is a critical metric and solutions that would ensure reuse of the existing spectrum in the most effective way have always been a top priority for mobile network operators. Advanced spectral efficiency techniques are essential especially in narrow-band scenarios, that is in networks having an allocated spectrum no higher than for example 5 or 7.2 MHz per link. However, they are gaining more importance in all networks facing rapid traffic increase where more capacity is required. There is also a growing interest in deploying other systems such as, for example WCDMA/HSPA into frequency bands already used by GSM technology (e.g. 900 MHz) – a policy known as spectrum re-farming. To preserve GSM capacity and service quality with reduced available spectrum, the implementation of features to boost GSM spectral efficiency need to be considered. Features such as frequency hopping, dynamic power control, discontinuous transmission (DTX) or more advanced solutions like AdaptiveMulti-Rate (AMR), Orthogonal Sub-Channel (OSC) and finally Dynamic Frequency and Channel Allocation (DFCA) are potential ways to enhance GSM spectral efficiency.

The main focus of this chapter is DFCA but for the sake of completeness, other solutions such as MAIO management in asynchronous and synchronous networks are discussed here as well since they are the intermediate steps on the way towards highly spectral efficient networks. Section 10.2 outlines the basics of the frequency hopping feature. In Section 10.3 the principles of intra-site interference management are described. Both static MAIO management that provides the control over frequency hopping and allows intra-site interference to be minimized

GSM/EDGE: Evolution and Performance   Edited by Mikko Säily, Guillaume Sébire and Eddie Riddington
© 2011 John Wiley & Sons, Ltd

and a dynamic MAIO allocation strategy that overcomes the limitation of the static MAIO assignment are covered. With the introduction of network synchronization the static MAIO allocation strategy can be extended to beyond the sectors of the same site which enables the control of inter-site interference as well. The detailed description of this technique is provided in Section 10.4. Finally, Section 10.5 presents a dynamic channel selection scheme called Dynamic Frequency and Channel Allocation (DFCA). DFCA is a proprietary radio resource management algorithm that aims to optimize the distribution of interference when taking into account the unique channel quality requirements of different users and services. In addition to the detailed description of the DFCA mechanism, this section also provides performance results extracted from comprehensive simulation campaigns and field trials performed in real networks around the world.

10.2 Frequency Hopping Basics

Frequency Hopping (FH) is a well-known and commonly used feature that provides a significant enhancement in the quality and capacity of GSM networks. When frequency hopping is activated, the radio link performance is improved as a result of two main effects: frequency diversity, which increases the fading de-correlation among the bursts that comprise a single speech frame and thereby the efficiency of channel coding; and interference diversity, which helps to mitigate the harmful impact of interference. In the GSM system, the so-called slow frequency hopping is utilized, where the frequency may change after each TDMA frame (4.615 ms).

The frequency hopping rule for determining the frequency used in each TDMA frame is based on two parameters: the mobile allocation index offset (MAIO) and the hopping sequence number (HSN). TheMAIO parameter is used to specify an offset within the hopping sequence and can take as many values as there are frequencies in the hopping list (i.e. the mobile allocation or MA list). The HSN parameter defines the hopping sequence and can take on 64 different values. When HSN is zero, the hopping mode is cyclic and when HSN is non-zero, the hopping mode is pseudo-random.

The FH algorithm is described in [1]. For a given set of parameters, the index to the absolute radio frequency channel number (ARFCN) in the MA list (the mobile allocation index or MAI) is obtained from the following algorithm, where FN is the Frame Number and N is the number of hopping frequencies (the number of elements in the MA list):

If HSN = 0 (cyclic hopping) then:
MAI, integer $(0 \ldots N-1)$:   MAI = (FN + MAIO) modulo N
else (HSN $\neq$ 0 random hopping):
MAI, integer $(0 \ldots N-1)$:   MAI = (S + MAIO) modulo N[1]

where S is calculated from the HSN, the time parameters, the number of frequencies and a predefined table which can be found in [1].

[1] The MAI equations are copied from 3GPP TS 45.002 v8.0.0 section 6.2.3 MAI equations. ©2008. 3GPP™ TSs and TRs are the property of ARIB, ATIS, CCSA, ETSI, TTA who jointly own the copyright on them. They are subject to further modifications and are therefore provided to you "as is" for information purposes only. Further use is strictly prohibited.

Two sequences that are determined by the same FN and HSN but different MAIOs will never overlap, that is the hopping sequences are orthogonal. Hopping sequences within a cell can be kept orthogonal provided appropriate MAIO planning is applied and the number of TRXs is lower than the length of the MA list. To keep the hopping sequences orthogonal between several cells (e.g. between all sectors of a site), those cells have to be frame number synchronized with a common FN. Two channels defined by the same time slot number, the same MA list but different HSN will interfere with each other only in 1/N bursts (in 1 out of N), where N is the number of frequencies in the MA list. In this way the interference can be averaged out over the network.

10.3   Intra-Site Interference Management

When radio frequency hopping and MAIO management are in use, there is a certain degree of interference control that can be used to avoid or minimize undesired co-channel and adjacent channel interference. In this section, we assume synchronization is between sectors of the same site only and therefore limit the discussion to intra-site MAIO management. We also consider a random frequency hopping scenario with tight frequency reuse pattern equal to 1/1.

10.3.1   Static MAIO Management

In a 1/1 frequency reuse network with random frequency hopping enabled, all the cells share the same set of hopping frequencies, therefore inter-cell interference diversity can be achieved by allocating different HSNs to all of them. When cells are synchronized with each other, there is also the potential to remove the co-channel and adjacent channel inter-cell interference through careful allocation of hopping parameters. In fact, if different MAIOs are assigned to cells that have the same HSN and frame numbering, all co-channel interference can be eliminated. Furthermore, cells can be prevented from operating on adjacent frequencies at the same time provided that the assignment of adjacent MAIOs is avoided.

Radio bursts within the same cell are always synchronized in a GERAN network. This means the co-channel and adjacent channel interference within a cell can be eliminated by proper MAIO planning in combination with fractional loading (where fractional loading is when the hopping frequencies exceed the hopping transceivers in a cell).

Typically, base stations that are located within the same BTS cabinet will share the same frame clock source and therefore will always be synchronized relative to each other. It is also quite straightforward to chain a common clock signal between BTS cabinets within the same site, and that way achieve local synchronization. Thus, if synchronization between sectors of the same site is supported, the frequencies used by each TRX within the site can be controlled by assigning identical HSNs to all co-sited sectors and by appropriate MAIO planning (see Figure 10.1).

A static MAIO allocation (SMA) is the strategy that consists of fixed MAIOs assignments to the hopping TRXs deployed in the same site. The principle is that available MAIOs are split into subsets comprising different MAIOs for each sector. As the number of MAIOs is limited to the number of frequencies in the MA list, the degree of control of both co-channel and adjacent channel interference within the site will be determined by the applied fractional load.

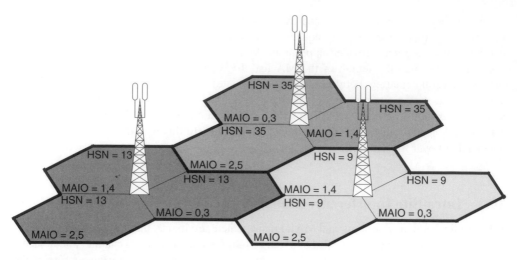

Figure 10.1 Example of static MAIO planning; 1/1 frequency reuse, 6 hopping frequencies, 2 hopping TRXs per cell.

Assuming consecutive frequencies in the MA list, the degree of interference control would be as shown in Table 10.1.

The Fractional Load (FL) is calculated as the ratio of the average number of hopping TRXs deployed per cell to the average number of frequencies allocated to a cell (the length of the MA list):

$$FL = \frac{N_{TRX/cell}}{N_{freq/cell}} \cdot 100\%.$$

Table 10.2 illustrates examples of static MAIO allocations for a site with 12 hopping frequencies and with different numbers of installed TRXs per cell (symmetrical configuration assumed).

Table 10.1 Interference control with MAIO management

Fractional Load	Intra-site interference	MAIO selection
FL ≤ 16.67%	No co-channel or adjacent channel interference	MAIO distance ≥ 2
16.67% < FL ≤ 33.33%	Adjacent channel interference between co-sited sectors; no co-channel interference	MAIO distance ≥ 1
33.3% < FL ≤ 100%	Co-channel interference between co-sited sectors; adjacent interference between co-sited sectors; adjacent channel interference within the sectors	MAIO distance < 1; reuse of MAIOs necessary

Table 10.2 Static MAIO Allocation for 3-sector site with 12 allocated frequencies and varying number of installed TRXs per cell

MAIO	0	1	2	3	4	5	6	7	8	9	10	11
1 TRX per cell/FL = 8,33% - no intra site co-channel or adjacent channel interference												
Sector 1	TRX1											
Sector 2						TRX1						
Sector 3											TRX1	
2 TRX per cell/FL = 16,67% - no intra site co-channel or adjacent channel interference												
Sector 1	TRX1						TRX2					
Sector 2			TRX1						TRX2			
Sector 3					TRX1						TRX2	
3 TRX per cell/FL = 25% - adjacent channel interference between co-sited sectors, no co-channel interference												
Sector 1	TRX1			TRX3			TRX2					
Sector 2			TRX1			TRX3			TRX2			
Sector 3					TRX1			TRX3			TRX2	
4 TRX per cell/FL = 33,33% - adjacent channel interference between co-sited sectors, no co-channel interference												
Sector 1	TRX1			TRX2			TRX3			TRX4		
Sector 2			TRX1			TRX2			TRX3			TRX4
Sector 3		TRX1			TRX2			TRX3			TRX4	

When assuming a single MAIO usage with 12 hopping frequencies, the maximum number of hopping TRXs per site is limited to 12. Should the fractional load exceed 33.33%, then a number of deployment strategies are possible. One option is to maintain a static MAIO management in all the sectors within the site, which will cause permanent co-channel interference between certain TRXs. Another option is to allocate different HSNs per sector, which will increase the overall interference but will randomize its occurrence. Another way to deal with the intra-site interference would be to assign the same HSN only to two sectors within the site and to use MAIO management to control the interference between them, while a different HSN is assigned in the third sector of the site.

The performance of the different strategies has been compared by means of system level simulations performed in a 1/1 frequency reuse scenario with $5 \times 5 \times 5$ site configuration (5 hopping TRXs per cell, 3 sector site) and 12 hopping frequencies in the MA list (the fractional load is therefore above 33.33%). The simulations also assumed an urban network deployment with slowmoving subscribers (3 km/h), a hexagonal deployment with a site-to-site distance of 1200 m, a propagation index of 3.76, a slow fading standard deviation of 8 dB and a handover margin of 3 dB. AMR link adaptation was activated and only four AMR FR codec modes (12.2, 7.4, 5.9, and 4.75 kbps) were included in active codec set. Also, the performance of the non-BCCH layer was studied only. In all of the scenarios, speech quality was evaluated in terms of Bad Quality Samples (BQS), where a Bad Quality Sample is defined as a sample of speech 4 SACCH periods in duration (1.92 s) with 2.1 % FER or higher. The system load is given as Effective Frequency Load (EFL), which is the average voice

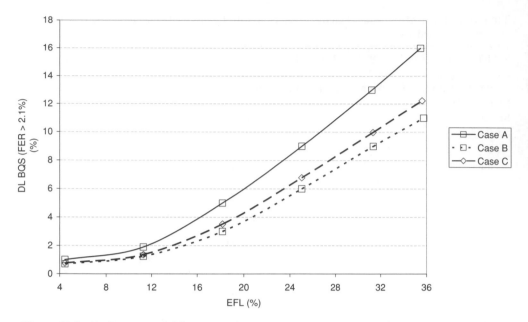

Figure 10.2   Performance of different interference management strategies; fractional load >33.33%.

traffic per cell divided by the product of 8 time slots and the number of hopping frequencies. For example in the considered scenario, voice traffic of 25 Erlang per cell results in EFL of 26%.

Figure 10.2 shows the simulation results for three different interference management strategies. Case A corresponds to the scenario with different HSNs in each sector and in this case, interference between all cells, even co-located ones, can only be randomized. Case B corresponds to the scenario where MAIO management is applied. As the number of hopping TRXs per site exceeds the number of hopping frequencies, the MAIOs have to be reused within the site. Finally, Case C is a combination of MAIO management that was done between two sectors of the site having the same HSN and the third sector being allocated a different HSN (to achieve a randomization of the remaining intra-site interference). It can be seen that Case A has the worst performance in terms of BQS and that Case B outperforms Case C. This is because in Case A the gain from MAIO management is not utilized at all, that is control over intra-site interference is lost as all sectors have different HSN assigned. In Case B, intra-site interference can be significantly reduced. Only when MSs are handled on the same time slot number and on a TRXs with the same MAIO assigned, might continuous co-channel interference occur. How strong the interference will be depends on where the MSs are located. In the simulations, the radio resources were allocated in the randomized manner. This means an intra-site co-channel assignment may happen also in low load conditions, that is even when resources on non co-MAIO TRXs are available. By considering the results of uplink idle channel measurements, such suboptimum resource assignments could be partially minimized. In which case, the performance gap between Case C and Case B will increase.

10.3.2   Dynamic MAIO Management

As mentioned in the previous section, the degree of control of both co-channel and adjacent channel interference within the site provided by a static MAIO allocation is conditioned by the fractional load. If the number of deployed on-site hopping transceivers exceeds half of the number of hopping frequencies, then adjacent intra-site interference will occur. If the number of hopping TRXs per site is higher than the number of hopping frequencies, then the MAIOs have to be reused within the site. This means continuous intra-site co-channel interference will be introduced between connections allocated on TRXs with the same MAIO. This is especially relevant for MSs located at the sector border where they may experience high co-channel and adjacent channel interference from the neighboring sector of the same site due to the overlapping antenna diagrams.

Furthermore, a static MAIO allocation strategy does not offer an efficient use of the MAIOs because of their fixed assignment for each TRX. For this reason a dynamic MAIO allocation (DMA) scheme has been introduced to allow a fully flexible and optimum selection of the time slot and MAIO pair for a new channel request.

In case of dynamic MAIO allocation, the list of MAIO values is not divided into subsets for each sector but it is considered as common resource at the site level. So, for each incoming call, DMA selects the best MAIO and time slot (TS) pair in a way that intra-site interference is minimized. In this way, instantaneous non-uniformity of traffic distribution within the site is exploited and non-optimum allocations may be avoided.

The DMA algorithm estimates the expected interference for all idle TS-MAIO combinations by taking into account the TS-MAIO pairs being currently in use at the site, and allocates to an upcoming call the TS-MAIO pair with the minimum expected interference level. This strategy is based on an optimization algorithm operating on the entire TS $\times$ MAIO domain, which means that all MAIOs can be used by any sector of the site. Figure 10.3 depicts a simplified scheme of the DMA algorithm.

For each incoming voice call (e.g. after call set-up or handover), DMA selects a channel on the basis of the current MAIO utilization state within the site. Initially, all TS-MAIO combinations are considered in the channel assignment problem and then the algorithm progressively restricts the solution space, until a single pair (TS-MAIO) is selected. In a first step for each free TS the best associated MAIO value is chosen (the MAIO selection step). Having determined the list of candidate channels (TS-MAIO pairs), the best channel is selected (TS selection step).

The benefits of the DMA feature are remarkable especially for an instantaneous inhomogeneous traffic distribution within the site, as this leads to an efficient use of the common MAIO pool. To study the effect of the Dynamic MAIO Allocation strategy on GSM voice performance in terms of capacity and quality, system level simulations have been performed in a 1/1 frequency reuse scenario with $5 \times 5 \times 5$ site configuration (hopping TRXs) and 12 hopping frequencies in the MA list. The performance of SMA and DMA were evaluated for varying load conditions expressed in EFL ranging from 10–35% (see Figure 10.4). Since the DMA algorithm tries to minimize the number of MAIO repetitions within the site and controls the number of adjacent MAIO allocations within the site, the best possible time slot and MAIO pair were always selected. DMA outperformed SMA by reducing the intra-site adjacent and also co-channel interference (once the load increased). It is worth noting that the difference in DMA and SMA performance starts to decrease at very high EFL ranges. The reason is that

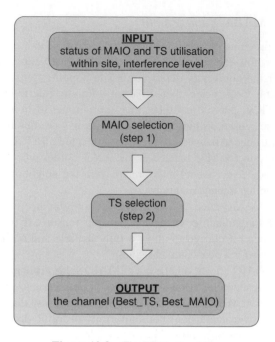

Figure 10.3   The DMA algorithm.

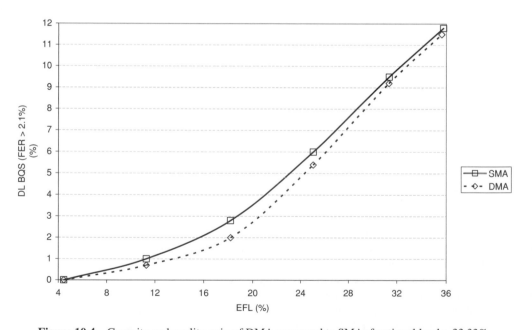

Figure 10.4   Capacity and quality gain of DMA compared to SMA; fractional load >33.33%.

the TS-MAIO solution space becomes limited at the very high network loads and it is more difficult to find a clean TS-MAIO pair. While a uniform traffic distribution over the sectors of the same site was assumed, an uneven traffic distribution in a real network environment will lead to even higher gain with DMA.

10.4   Inter-Site and Intra-Site Interference Management

The solutions that have been described in the previous sections only help to combat intra-site interference, while the dominant interference in a network is inter-site interference, as shown in Figure 10.5. Therefore, the key to achieving highly spectral efficient networks is to implement advanced ways of controlling inter- and intra-site interference.

Figure 10.6 shows a histogram derived from real downlink measurement reports provided by mobile stations in dedicated mode. Based on the measurement reports, information on a cell's interdependencies can be collected and cells having the strongest interference relationships are identified. It can be seen that co-sited sectors (cells B and C) are not the strongest interferers for the victim cell A which implies that the maximum possible interference suppression will not be achieved if concentrated on intra-site interference only.

Without network synchronization, interference control is only generally feasible between sectors of the same base station, while in a synchronized network the interference avoidance is feasible between any sectors, irrespective of the base station. One way to achieve networkwide synchronization is to provide the common master clock signal to all BTS sites in a cluster

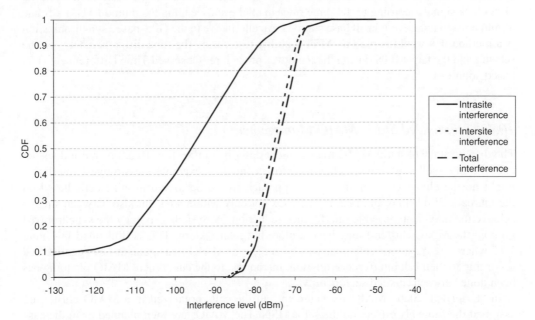

Figure 10.5   CDF of intra-site, inter-site and total network interference at 40% of EFL. Reproduced by permission of K. Ivanov, C.F. Ball, R. Müllner, and H. Winkler, "Smart Interference Reduction Dynamic MAIO Allocation Strategy for GERAN Networks" *VTC2004*, Fall. 2004 IEEE 60th. © 2004 IEEE.

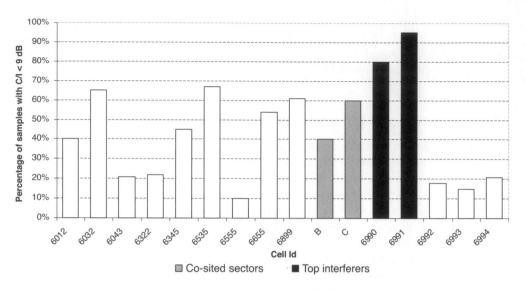

Figure 10.6   Example of C/I histogram derived from downlink measurements collected in cell A.

or network. To this end, a stable and globally available clock signal from the GPS satellites can be utilized. A GPS receiver in every BTS site provides a frame clock signal that can then be distributed to all BTS cabinets in the site. The frame clock can be calculated from the GPS clock signal according to the same rules in all the sites leading to common TDMA frame alignment and numbering in all base stations. An alternative to the GPS-based synchronization is a method that would exploit an MS capability to measure the time difference between the serving and the target BTS during the handover procedure (Observed Time Difference (OTD) based solution).

10.4.1   Extended Static MAIO Management

The disadvantage of static MAIO management described in Section 10.3.2 is that it does not take the inter-cell interference relations into account, for example the strongest interferers might not be eliminated. An alternative approach would be to synchronize cells based on the mutual cell relations and plan the MAIOs not only within sites but also between sites to achieve the maximum possible interference reduction. With such a strategy the synchronized groups (the MAIO clusters) may have a more irregular pattern. This is illustrated in Figure 10.7 where a grouping of cells that will be subject to MAIO management has been created according to their cell interference relation, in contrast to the case where MAIO planning has been done between co-sited sectors only.

In "Extended" Static MAIO management, all the cells organized in a MAIO cluster are assigned the same FN offset and the same HSN. The MAIOs are then planned as in the case of MAIO planning between sectors of the same site, that is cells within the MAIO cluster are treated as if they were of the same site. Given that the control of both co-channel and adjacent channel interference between cells will be conditioned by the applied fractional load,

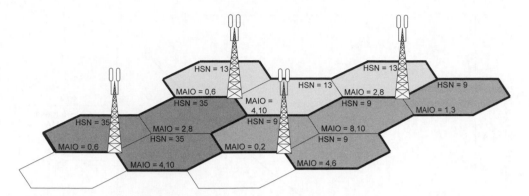

Figure 10.7   Example of extended static MAIO management in synchronized network; 1/1 frequency reuse, 12 hopping frequencies, 2 hopping TRXs per cell.

the grouping of the cells in MAIO clusters cannot be done according to the interference matrix[2] only. The maximum size of the MAIO cluster should be determined by the number of the hopping frequencies. This is needed to avoid strong continuous co-channel interference between certain TRXs in the cluster. Since the strongest interferers within a cluster are subject to MAIO management, more care is needed to avoid a repeated MAIO than when planning between co-sited sectors.

The drawback of this interference management method is the grouping of the cells in MAIO clusters is done statically which ignores the time-varying nature of the interference distribution. If the distribution of interference changes significantly, for example over a day, then the grouping of the cells in MAIO clusters should also follow these changes. Moreover any extensions to the cell configuration may require the whole grouping process to be repeated if MAIO clusters were optimized for maximum size. Indeed, in order to maximize the benefit from extended MAIO planning, the clusters should include as many hopping TRXs as there are hopping frequencies defined in the MA list.

To investigate the benefits of extended static MAIO planning, system level simulations were performed in a homogenous network with $4 \times 4 \times 4$ site configuration (hopping TRXs) and 12 hopping frequencies in the 1/1 reuse. Cells were grouped in MAIO clusters according to the interference matrix. The percentage of satisfied users was measured for changing load conditions in terms of EFL. Speech quality has been evaluated by means of Bad Quality Samples (the percentage of samples with FER higher than 2.1%). Simulation-based studies showed that capacity gains of 10–20% can be achieved in a regular network layout scenario with uniform traffic distribution when compared to intra-site static MAIO management (see Figure 10.8). The actual gain will depend on the network topology and how well MAIO clusterization is done. In real irregular networks, even higher gains may be expected.

10.4.1.1   Frame Number Offsets Considerations

A common TDMA frame numbering, when inter-site synchronization is implemented, is not desirable at the wide area level since it leads to a coincident broadcast of the broadcast

---

[2] Interference Matrix describes the level of interference between any two cells in a cellular network

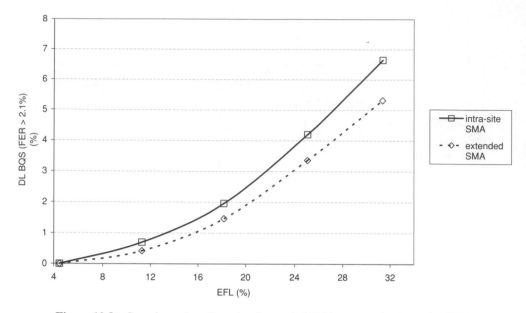

Figure 10.8   Capacity and quality gain of extended SMA compared to intra-site SMA.

common control channel (BCCH) and traffic (TCH) channel multi-frames in all the base stations [2]. In the current method of performing neighbor cell measurements in GSM, the MS measures the received signal strength (RXLEV) of neighbor cells on the cell BCCH frequencies in the time slots between uplink transmission and downlink reception. A neighbor cell is identified by decoding BSIC on the Synchronization Channel (SCH) and it is required that the MS decodes the BSIC at least every 10 seconds for that cell to be considered as a handover candidate. RXLEV measurements from the six strongest base stations in the BCCH Allocation (BA) list that satisfy this criterion are then included in each measurement report together with their broadcast frequencies and BSICs. When a common TDMA frame number is used for the BCCH, co-incident transmission of the synchronization bursts will occur in every 10 TDMA frames (see Figure 10.9). An MS in dedicated mode will attempt to search for neighbors during the idle frames and these idle frames occur once within a TCH multi-frame of 26 TDMA frames (see Figure 10.10). Thus, if the network is synchronized with a common frame numbering, BSIC decoding cannot be performed as often in a synchronous network since the number of instances where a SCH will be co-incident with an idle TCH frame will be reduced. This will affect BSIC decoding performance leading to a reduction in handover performance.

Another issue with common frame numbering is the Slow Associated Control Channel (SACCH) transmissions always occur at the same time over a synchronized area (see Figure 10.10). Since the SACCH is always transmitted regardless of voice activity, it would not benefit from DTX at all [2]. This is undesirable, particularly with Adaptive Multi Rate (AMR) where the benefits have been found to be limited by the signaling channel (SACCH, FACCH) performance

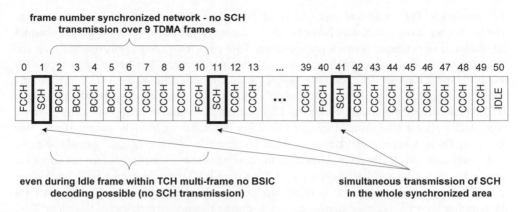

Figure 10.9 BCCH multi-frame structure.

The problems described above can be avoided by planning different frame number offsets throughout the synchronized area. In order to improve identification of the neighboring cells in a synchronized network, the frame numbering between cells should be shifted so that all the synchronization bursts between neighboring cells are transmitted at different times. This is achieved when the difference in FN offsets between two cells is neither zero nor a multiple of 51. With this rule, simultaneous SACCH transmissions are also avoided.

10.4.1.2 Training Sequence Code Considerations

Another important aspect is the influence of the Training Sequence Codes (TSC). The TSCs used in GSM were originally optimized to ensure good autocorrelation properties, while the cross-correlation properties were not so considered. For unsynchronized networks, this is not an issue since the TSCs from different base stations coincide only randomly. However, with base station synchronization, the training sequences will continuously coincide and this will have an impact on receiver performance. Channel estimation is performed to characterize the multipath fading profile and subsequently to mitigate the multipath fading by equalization. However, with base station synchronization, cross-correlation energy will be received from

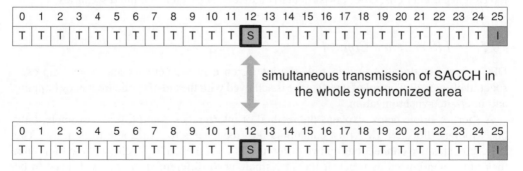

Figure 10.10 TCH multi-frame structure.

the interfering TSC code and this will result in a degradation to the channel estimation. The higher the cross-correlation between TSCs, the more detrimental the impact on channel estimation. For optimum network performance, TSC pairs assigned to cells operating on the same frequency shall have vanishing cross-correlation properties. The best way to cope with this problem is through TSC planning, where the likelihood of the worst TSC combinations occurring is reduced. Since all neighboring cells are potential interferers in networks applying a 1/1 frequency reuse pattern, the following TSC planning rules might be applied: cells grouped in a MAIO cluster (the same HSN and FN offset but different MAIOs assigned) may use the same TSCs, whereas cells between which co-channel collisions appear should use pairs with minimum cross-correlation. Moreover, the best pairs should be reserved for the strongest interferers that may be identified based on for example the interference matrix and/or the traffic load on the hopping layer. As mentioned previously, the distribution of interference on the hopping layer may change significantly (e.g. during the day), which means that static TSC planning will not be able to provide the optimum network performance. Therefore, by applying a dynamic TSC allocation scheme that follows the interference variations, the performance of the receivers may be improved.

10.5   Dynamic Frequency and Channel Allocation

The main characteristic of random frequency hopping is that two channels having different HSNs, but the same frequency list and the same time slot, will interfere in 1/N of the bursts, where N is the number of frequencies in the hopping sequence. This means that random frequency hopping randomizes or averages the interference throughout the network.

Dynamic Frequency and Channel Allocation (DFCA) on the other hand tries to control rather than randomize the interference and this can lead to a significant increase in the network spectral efficiency. DFCA is a proprietary radio resource management functionality that dynamically assigns the radio channel for each incoming call. Specifically, DFCA selects a radio channel based on information about the interference in the network combined with the time slot and frequency usage information. The idea behind DFCA is to assign a channel that provides sufficient quality (in terms of C/I), so that the Quality of Service requirement of each connection is met. Moreover, the radio channel is selected in such a way that outgoing interference caused by a new call towards the existing connections is minimized. The main advantage of DFCA compared to other channel allocation methods is that DFCA dynamically adapts to changing interference conditions. DFCA is particularly suited to interference limited scenarios.

10.5.1   Interference Control

DFCA is a functionality that controls the interference in the network and hence improves spectral efficiency. Control of interference is achieved with the aid of cyclic frequency hopping and inter-site synchronization.

A short example below explains the method of interference control. In the example, only one Mobile Allocation (MA) list containing the frequencies f0, f1, f2 and f3 is in use. The BSC allocates two connections on two different TRXs. If these TRXs belong to the same site, they are synchronized as usual. If the TRXs belong to different sites, they will need to be synchronized in order to control the interference between them.

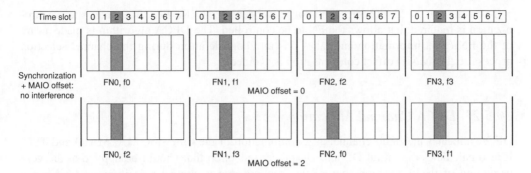

Figure 10.11 Interference control – no interference due to synchronization and correct MAIO plan.

In the example, the frequency of each TRX is taken from the same MA list and is changed in every TDMA frame to another frequency in the MA list. Synchronization between TRXs ensures that two calls will always use different frequencies: they use the same time slot number and the same MA list but they do not interfere because they have different MAIOs assigned (Figure 10.11). The interference is under control because the frequencies occur in certain time slot numbers systematically (i.e. periodically) – the frequency separation (resulting from MAIO assignment) between two interfering radio channels is the same and does not change during the whole call duration. Indeed, focusing on one particular time slot (e.g. time slot number 2 of TRX1 and TRX2, with TRX1 and TRX2 still using the same MA list), the condition of co-channel interference is met only when the MAIO offset is compensated by the FN offset:

$$(FN_A + MAIO_A) \bmod MA \; length = (FN_B + MAIO_B) \bmod MA \; length$$

This is an undesirable situation that may happen when TRX1 and TRX2 belong to different cells and are not synchronized. In such a case, even with the correct MAIO plan, the above equation can be satisfied due to differences in frame numbering (Figure 10.12).

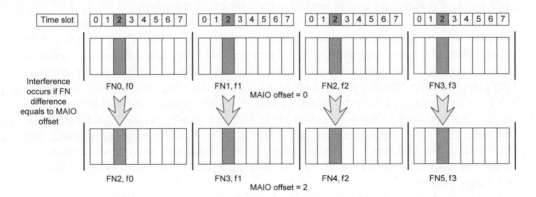

Figure 10.12 Interference control – co-channel interference due to lack of synchronization.

To ensure that the MAIO offset will not be compensated by the FN offset, inter-site synchronization is required that allows the FN to be controlled. The DFCA algorithm is made aware of the FN offset information and takes this into consideration during the channel selection process so that MAIO offset compensation is avoided.

10.5.2   DFCA Channel Allocation

DFCA allocates the radio channel as a combination of the time slot, MA, MAIO and TSC. It also calculates the initial DL and UL power reduction that should be applied at the very beginning of the connection (before the standard power control algorithm start). When a channel needs to be assigned as a result of a newly initiated connection or after handover, DFCA evaluates all the possible candidates and chooses the most suitable one. For this reason, the C/I is estimated for each available radio channel. The aim is to choose the best channel, in terms of both incoming interference (the disturbance that the new connection would be suffering) and outgoing interference (the disturbance that the new call would cause to other calls in other cells). The C/I of an idle radio resource can be estimated based on the mobile station measurement reports received by the network.

With pre-Release 99 mobiles the number of reported neighboring cells is limited to six. However, with Release 99 compliant mobiles, more neighboring cells can be reported thanks to the enhanced measurement reporting (EMR). In most cases, reporting of only six neighbors is sufficient for handover evaluation purposes, but for interference estimation the potential C/I needs to be determined from all the surrounding cells. While the C/I can be directly calculated for the neighboring cells that are included in the latest measurement report (the C/I is estimated based on the mobile Rx Level value of the serving and neighboring cells as well as the applied DL power reduction), for the remaining surrounding cells, a statistical C/I estimation can be performed. This is done by collecting long-term neighboring cell Rx Level measurement statistics from mobile measurement reports sent within a cell. From these long-term statistics, a C/I with a given outage probability can be determined for all the surrounding cells. This information is stored in the BSC as an Interference Matrix (IM). That way, the DFCA can estimate the C/I for all the potentially interfering cells even if some of them are not included in the latest measurement report. Since the downlink path loss is generally equal to the uplink pathloss, the C/I statistics can also be used to determined for the uplink.

There are four C/I estimations calculated for each available radio channel. The incoming C/I describes the interference coming from existing connections that would affect the new call for which a channel is being assigned. The new connection may also cause interference to existing calls on the same or in an adjacent radio channel. This is examined by determining the outgoing C/I for every potentially affected existing connection. The incoming and outgoing interference levels are estimated separately for downlink and uplink directions. The C/I estimation relies on the fact that the interference relations in a DFCA network are stable and predictable (because of inter-site synchronization) and controlled (use of cyclic frequency hopping). The C/I estimation is performed by combining information from the measurement reports, the interference matrix and a BSC radio resource table where the current status of radio resources occupation is stored.

In the C/I estimation process, the DFCA algorithm determines the interference that a new user would experience with different available time slots, MA list, and MAIO combinations.

This is done by searching all entries in the Interference Matrix to identify interference caused by the co-channel and adjacent channel connections. In case of adjacent channel interference, an adjacent channel protection is applied (usually 18 dB) to reflect the fact that adjacent channel interference is less harmful for the connection than the co-channel (as a result of Rx filter attenuation). For each channel the C/I value is estimated taking into account the strongest interferer to achieve the most pessimistic value. If the analyzed interferer had been reported in the latest MS measurement report, the C/I value is taken from that report, otherwise the C/I value is taken from IM table. The final C/I level is calculated by taking into account the power reduction applied to the interfering call to obtain a more realistic value. The difference between estimated C/I and the target C/I value (required C/I) is calculated. This difference is calculated for the incoming and outgoing C/I for both the UL and DL direction. Among these four C/I differences (calculated for the strongest interferer), the lowest is determined. This most restrictive C/I difference is then used in the channel selection procedure.

Having determined the most restrictive C/I difference for all the candidate channels, the channel with the highest C/I difference is checked against the soft-blocking limit. The soft-blocking limit is a configurable threshold in dB to ensure that the new connection would not generate or would not suffer from too strong interference. Any channel request that breaks the soft-blocking limit will be directed towards non-DFCA resources of the cell. Hence this soft-blocking check is an admission control mechanism in DFCA networks.

For the DFCA layer, DFCA also takes care of the training sequence code allocation. After a suitable channel has been found, the BSC determines the most suitable TSC by examining all the interfering connections. Specifically, the BSC determines the training sequence code that has been used by an interfering connection characterized by the highest C/I relation. This means that for the selected MA, MAIO and time slot combination, the C/I differences are checked TSC by TSC. The method of the training sequence code selection aims at avoiding the worst-case situation where a significant interference source uses the same training sequence code as the new connection. Such a conflict with a significant interferer would result in an incorrect channel estimation in the receiver and would consequently lead to a link level performance degradation. For that reason TSC used by the connection with the highest C/I difference is selected by DFCA algorithm, to guarantee the new connection will have the smallest chance for adverse effects from TSC cross-correlation.

The channel selection process above can be used for any service type (e.g. EFR, AMRHR, SDCCH, (E)GPRS) according to its service specific C/I target and C/I soft-blocking thresholds.

In case of an SDCCH allocation, the radio time slot and DFCA hopping parameters (MA and MAIO) with the highest C/I difference are selected. If several combinations have the same C/I value then the time slot having the most SDCCH connections is selected. If on the selected time slot there are already SDCCH sub-channels being used, then for the next SDCCH allocation on that time slot the same DFCA hopping parameters (as for the first sub-channel) must be used.

The DFCA algorithm may be involved also in the PS territory creation procedure which is triggered when an (E)GPRS service is being activated in a cell. In such a case, DFCA calculates the C/I for every time slot that is selected for the PS territory and selects the combination with the best radio quality. If on a DFCA TRX there is already an existing PS territory that needs to be upgraded, the DFCA will allocate the next time slots with the same DFCA hopping parameters as previously used by the PS territory (multislot connections require that the same hopping parameters are applied to all the assigned time slots). For these time slots, DFCA calculates the C/I and compares them against the soft-blocking threshold. If C/I soft-blocking occurs for any time slot, only time slots starting from the soft blocked ones towards the border of PS territory can be selected for an (E)GPRS upgrade. As the radio channel for the PS territory upgrade is selected by the DFCA algorithm from C/I estimates in real time, the maximum allowed interference level affecting each TCH and PDCH can be controlled by means of a C/I soft-blocking check. This C/I soft-blocking check indeed works as an admission control mechanism for the PS domain on DFCA layer.

10.5.3   Channel Allocation Based on Service-Specific C/I Requirements

DFCA controls the interference in the network and by this provides better quality for the radio channel. It also allows different QoS requirements to be set for different types of services (by means of different C/I target thresholds). This is very important, especially when the robustness of these services against interference varies significantly (e.g. due to different channel coding schemes). For example, an AMR connection can tolerate much higher interference level than an EFR connection while still maintaining comparable performance in terms of speech quality. As the least robust service determines the minimum frequency reuse distance, the AMR connection is often served in the hopping layer with 1/1 reuse while non-AMR connections are served by the BCCH layer (which usually offers better quality than hopping layer). For (E)GPRS services, higher C/I requirements are usually set for the benefit of higher modulation and coding schemes. Examples of service type specific C/I target and soft-blocking thresholds are given in Table 10.3.

The DFCA algorithm considers service type specific C/I requirements already during channel allocation and provides the channel with just the required quality. Furthermore, the same requirements are evaluated each time a new potentially interfering channel is allocated, thus aiming to keep the C/I above target for the whole duration of the connection. A new channel can be allocated by the DFCA only if it does not cause an ongoing connection to fall below the soft-blocking limit. Thus, DFCA automatically optimizes the usage of the frequency resource without the need for network planning changes, which is often a consequence of growing traffic.

For SAIC-capable mobiles, which provide sufficient quality at higher interference levels, a special configurable offset is applied to the connection type specific C/I target and C/I soft-blocking thresholds.

Table 10.3   Examples of service type specific C/I target and soft-blocking thresholds

Connection type	Target C/I [dB]	Soft blocking threshold [dB]
FR/EFR	15	10
HR	18	13
AMR FR	9	4
AMR HR	15	10
SDCCH	14	9
(E)GPRS	20	15

10.5.4 Forced Half-Rate Mode

The DFCA feature controls the interference and allocates the channel according to the required QoS. However, in high network load conditions, most resources are occupied and DFCA has less freedom during the channel selection procedure. It is therefore important to increase the half-rate usage in the high load scenario, even when there are sufficient hardware resources available in the cell. In this situation, a conventional load based channel rate selection algorithm between full-rate and half-rate will not be sufficient. Therefore a "forced Half-rate mode" has been introduced with DFCA. Forced HR mode is a way of performing channel rate selection between full-rate and half-rate, based on the average of the estimated incoming DL C/I values. The average DL C/I ratio provides a good benchmark of the load on the DFCA frequencies and can therefore be used to force HR allocation. Forced HR mode is triggered if the averaged C/I drops below a configurable threshold (Figure 10.13). When HR allocation is forced in the cell, it is triggered also in the interfering neighbor cells (to avoid a synchronized full-rate channel interfering with both corresponding half-rate channels in the serving cell). When the cell is in the forced HR mode, all new connections are assigned as HR channels. If during forced

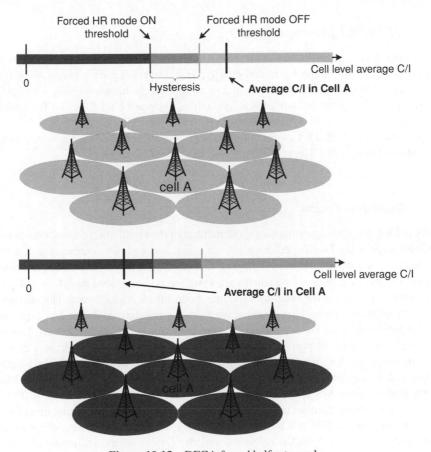

Figure 10.13 DFCA forced half-rate mode.

HR mode the DFCA algorithm is unable to provide a HR channel above the soft-blocking C/I threshold, the call is treated as soft-blocked and the resources on non-DFCA layer (e.g. BCCH) are searched for an allocation. If there are no resources on the non-DFCA layer, an FR connection is attempted on the DFCA TRX.

In general, triggering a DFCA forced HR mode with higher average C/I will lead to earlier use of half-rate which will improve the overall C/I conditions in the DFCA network. DFCA actually benefits from the greater degree of freedom during channel selection provided by the intense use of HR channels which can lead to improved C/I conditions and better overall speech quality (see Section 10.5.5.1). Forced half-rate mode allows DFCA to obtain optimum performance independently of load-based channel rate selection, which can be used conservatively to prevent unwanted half-rate usage in non-DFCA layers (Figure 10.13). In DFCA networks, the load criterion is checked first and if the condition for half-rate activation is not fulfilled, the C/I is compared against the C/I forced HR mode threshold. Thus, on the DFCA layer there are more possibilities to activate a channel as half-rate – HR mode can be chosen even if the load has not exceeded the threshold for a load-based channel rate selection.

10.5.5 DFCA Performance Results

By controlling interference, the DFCA functionality provides the best possibleway of assigning radio channels with the highest possible quality. This improves performance of both traffic and signaling (FACCH and SACCH) channels compared to the non-DFCA scenario. As a consequence, the key performance indicators such as Dropped Call Rate (DCR), uplink and downlink Rx Quality as well as Frame Erasure Rate (FER) improve.

The performance of the DFCA has been evaluated in different scenarios by network level simulations and field trials. The results are given in the next sections.

10.5.5.1 Simulation Results

A complete DFCA RRM algorithm was implemented in the simulator. In a real network-based simulation scenario, the DFCA performance was compared with a reference case that utilized a so-called ad-hoc random frequency hopping (an ad-hoc plan is a frequency plan that does not respect frequency groups – frequencies are assigned on a per-need basis).

Only voice capacity on the hopping layer has been taken into account. The real network scenario was used as it always has a varying degree of irregularities and it was expected that highly dynamic nature of DFCA should be able to adapt to the irregularities. The typical scenario was created by importing a network layout from a real network in a city center and its surroundings. An uneven traffic distribution was used to achieve the cell loads that were observed in the real network. As a result, the cell traffic was highly variable. Mobility was simulated with the MS speed of 3km/h. Mean call duration was fixed at 100 seconds. Power control and discontinuous transmission (DTX factor 50%) in uplink and downlink were enabled. The number of DFCA TRXs per cell ranged from one to five according to the real network configuration. In both the reference and the DFCA cases, 18 hopping frequencies were used.

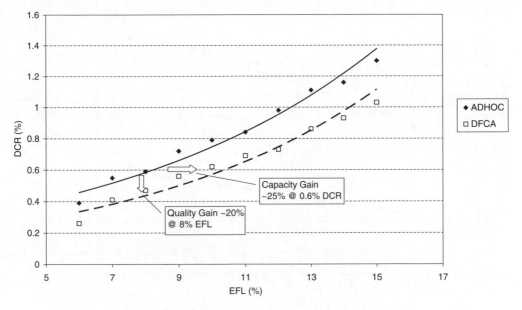

Figure 10.14   DFCA gain in terms of dropped call rate.

As shown in Figure 10.14, activation of the DFCA feature improves the DCR statistic. Dropped call rate decreases because calls, which in the non-DFCA mode were abnormally released due to interference, are now prevented from being dropped by the DFCA allocating a channel with a lower interference level.

Simulation results show a DCR reduction of approximately 20% at 8% of EFL. This performance improvement is called quality gain. At the same time DFCA reaches DCR of 0.6% at about 10% of EFL. This means that DFCA provides a capacity gain of 25%.

In addition to the improved dropped call rate, the simulations clearly indicate an improvement in terms of Rx Quality. At 8% of EFL the percentage of bad RxQual samples (here RxQual > 5) was reduced by approximately 60% in downlink and 75% in uplink (Figure 10.15). The number of measurement samples with bad Rx Quality is correlated with the level of interference that is reduced by DFCA which allocates a channel that affects existing calls in the least harmful way.

During the simulation campaign speech quality was evaluated in terms of Bad Quality Samples (the percentage of samples with FER higher than 4.2%). When compared to the reference scenario, DFCA considerably reduced the bad speech quality probability (FER > 4.2%) by 30% in the uplink and 40% in the downlink respectively (Figure 10.16).

An additional simulation campaign was performed for mixed PS and CS traffic in the same scenario as above but only 1 TRX was used with 6 hopping frequencies. In the reference case random frequency hopping with 1/1 reuse was applied. In the scenario with DFCA in use, the management of the PS territory (i.e. creating, upgrading and downgrading), was fully controlled by the DFCA algorithm. Two services were simultaneously simulated: voice with AMR HR (50% of all calls) and FTP download using (E)GPRS. The performance of the

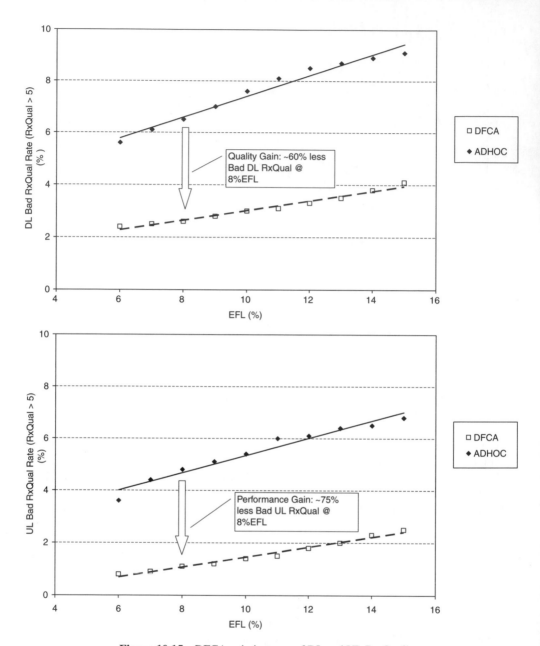

Figure 10.15   DFCA gain in terms of DL and UL Rx Quality.

CS calls was monitored in terms of dropped call rate and speech quality quantified by FER whereas for PS service, the average LLC throughput and delay statistics were evaluated.

The simulation results in Figures 10.17 and 10.18 show that the DFCA feature is able to improve the performance of CS users also if the CS and PS traffic is served on the same layer. Without DFCA the performance of the CS service would suffer from the disruptive impact

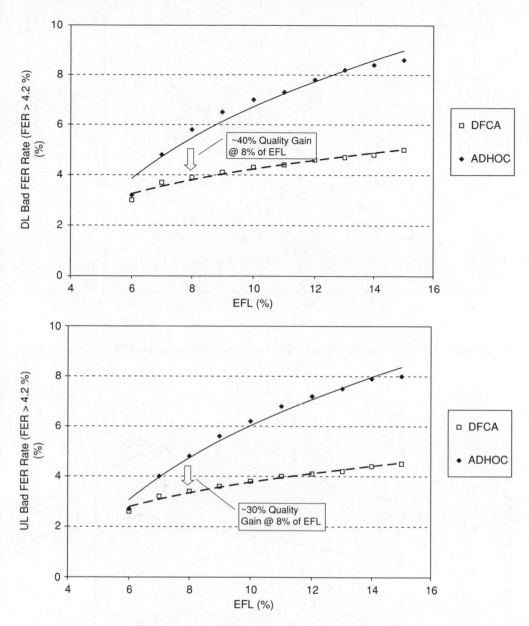

Figure 10.16   DFCA gain in terms of DL and UL FER.

of PS connections for which neither DL power control nor DTX is applicable. This negative effect is minimized in the DFCA network owing to the intelligent radio resource allocation scheme.

The activation of DFCA has also a positive influence on the performance of PS services. Indeed, DFCA provides better average C/I per time slot and therefore higher MCSs can be used. This, in turn, leads to higher average LLC throughputs and shorter delays as shown in

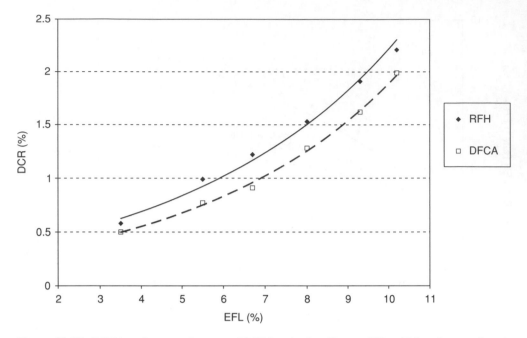

Figure 10.17   DFCA performance in terms of DCR in mixed traffic case (CS and PS on the same layer).

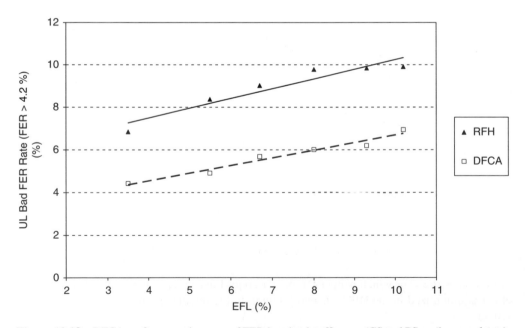

Figure 10.18   DFCA performance in terms of FER in mixed traffic case (CS and PS on the same layer).

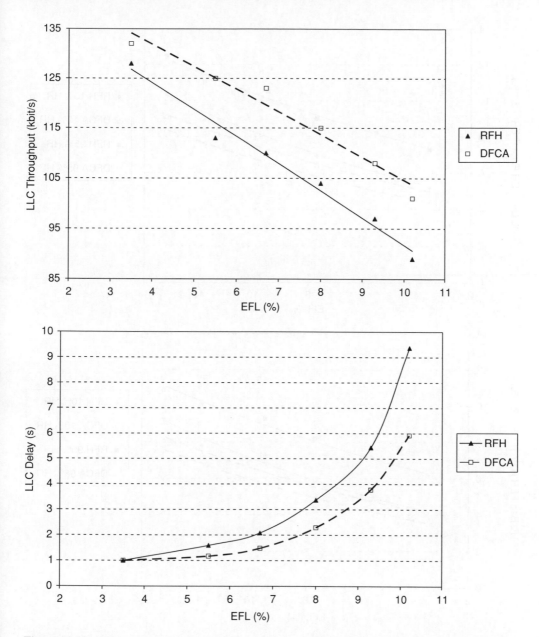

Figure 10.19   DFCA gain in terms of LLC throughput and LLC delay in mixed traffic case (CS and PS).

Figure 10.19. Generally, the presented simulation results show that DFCA increases spectral efficiency for both the CS and PS domains.

A separate simulation campaign was carried out to evaluate the impact of half-rate penetration on speech quality in DFCA networks (Figure 10.20). In the non-DFCA case, high HR usage degrades speech quality. In scenarios with low half-rate utilization, DFCA provides

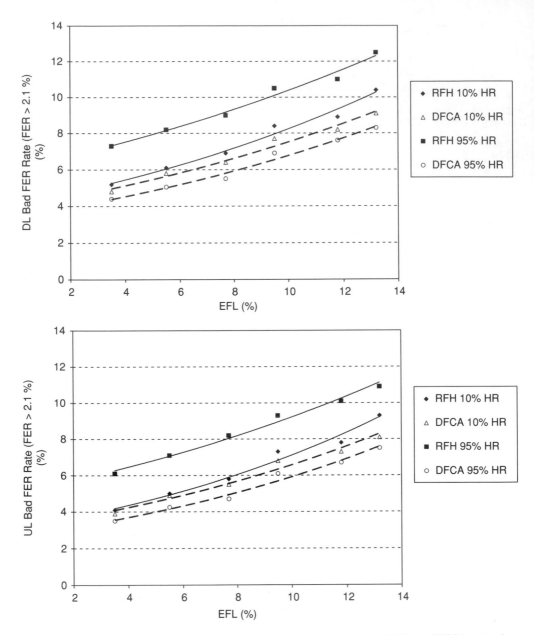

Figure 10.20   Impact of half-rate utilization on the speech quality in non-DFCA and DFCA scenarios.

slightly better speech performance than the non-DFCA case. What is interesting, is DFCA with very high (e.g. 95%) HR utilization offers similar speech quality as the non-DFCA network with low HR usage. This means DFCA with high half-rate utilization is possible while voice quality kept within acceptable levels. Higher HR utilization may be transformed into higher BTS hardware efficiency (as more traffic can be served with a lower number of TRXs) and thus

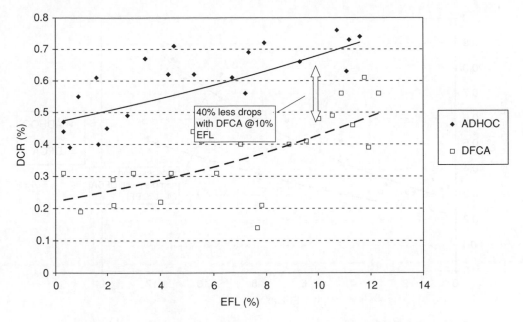

Figure 10.21   DFCA performance in terms of dropped call rate in the real network.

into reduced costs of hardware and transmission. It also helps in minimizing the combining losses improving DL link budget and hence increasing cell coverage.

10.5.5.2   DFCA Deployments in Live Networks

To verify the DFCA feature performance in real-life network scenarios, several field trials have been carried out. The main focus of the trials was to validate the gains revealed by the system level simulations, that is the impact of DFCA on the selected key performance indicators such as dropped call rate and speech quality. Furthermore, the efficiency of the DFCA in the narrowband scenario has been verified. The influence of high AMR HR penetration rate on DFCA network performance has been evaluated as well.

Figure 10.21 shows the performance of the DFCA in terms of dropped call rate observed in one of the trialled clusters. The benchmark case for the DFCA was random hopping with an ad-hoc frequency plan (6 MHz for hopping + 3.6 MHz for BCCH). It can be seen that for a given load operating point for example 10% of EFL, after DFCA activation, the dropped call rate was improved by approximately 40%. This is due to the better radio conditions having a positive influence on the performance on the signaling channels (FACCH and SACCH).

Apart from the DCR, the RxQuality statistics were also collected over the network where the trial took place (see Figure 10.22). It can be seen that DFCA provides approximately 45% less samples with bad Rx Quality (samples with RXQUAL equal to 5 and 6 were considered as bad quality ones) in DL and about 50% in UL at an EFL of 10%. The Rx Quality improvements are due to a reduction in interference provided by the DFCA functionality.

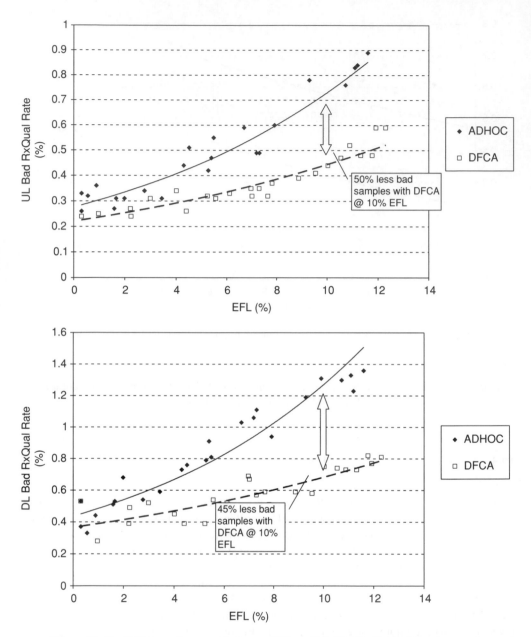

Figure 10.22   DFCA performance in terms of DL and UL Rx Quality in the real network.

By improving the interference conditions in the network, the DFCA significantly enhances the speech quality. The trial results show that not only frame erasure rate (Figure 10.23) but also mean opinion score (MOS) reflecting user perceived speech quality (Figure 10.24) are improved when DFCA is enabled.

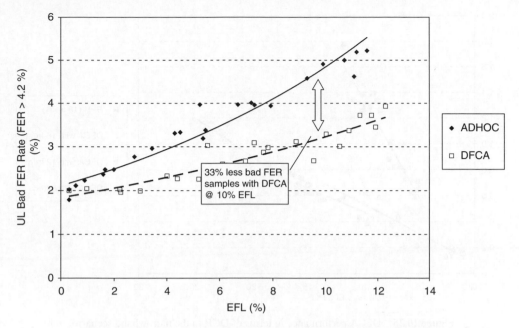

Figure 10.23   DFCA performance in terms of UL FER in the real network.

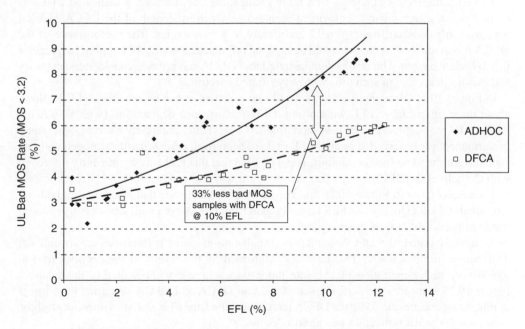

Figure 10.24   DFCA performance in terms of UL MOS in the real network.

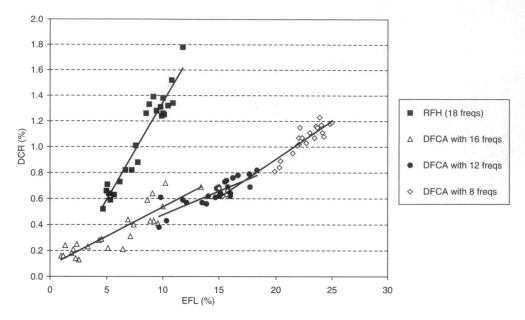

Figure 10.25   DFCA performance in terms of DCR in the narrowband scenario.

To verify the DFCA efficiency in a more challenging environment, a dedicated trial was carried out in a narrowband deployment scenario in which the length of the DFCA MA list was gradually reduced from 16 to 12 and finally to 8 frequencies. The performance of the DFCA functionality was benchmarked against the reference case with 18 hopping frequencies in a 1/1 reuse pattern. The results demonstrated that DFCA can provide considerable capacity and quality gains also in scenarios with very limited spectrum.

In Figure 10.25 depicting the dropped call rate statistics, it can be seen that DFCA allows about three times higher EFL without any DCR performance degradation (with DFCA the DCR of 1% was achieved at EFL of more than 20% while in the reference scenario the same performance was observed at around EFL of 7%). Moreover, comparing the dropped call rate statistics for the same load conditions, it can be seen that this KPI was significantly improved with the help of the DFCA even when the spectrum was reduced from 18 to 16 frequencies.

The results given in Figure 10.26 and Figure 10.27 demonstrate that not only dropped call rate but also the Rx Quality and FER statistics gain from DFCA even with significantly reduced bandwidth available for the hopping layer.

As already stated, the DFCA enables high half-rate usage as it improves the overall C/I distribution in the network. This has been confirmed by the results of the network trial in which very high penetration of half-rate mode was achieved without quality degradation. Figure 10.28 compares the performance of the non-DFCA and DFCA scenarios in terms of uplink frame erasure rate. With the DFCA feature activated, the FER statistics improve slightly, even though the HR utilization was almost doubled.

Summarizing the performance figures from simulations and field trials, one can conclude that the DFCA functionality provides significant spectral efficiency gains. On the one hand, higher network capacity can be achieved without compromising the most important KPIs,

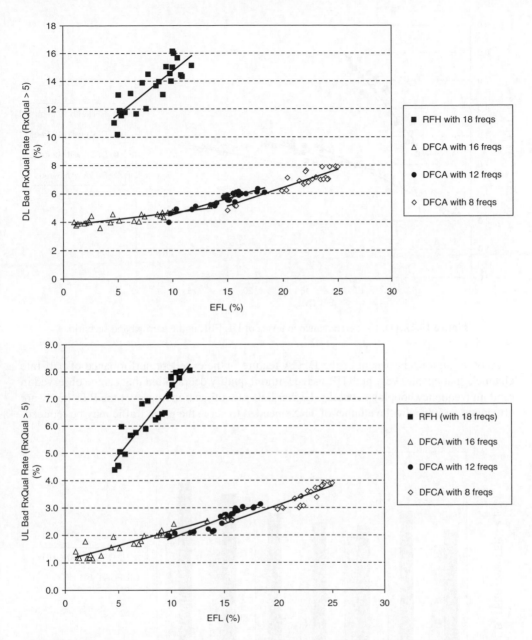

Figure 10.26 DFCA performance in terms of DL and UL Rx Quality in the narrowband scenario.

while on the other hand, DFCA allows operators to serve the same traffic with the reduced frequency spectrum while keeping quality at a satisfactory level. The DFCA functionality can also efficiently manage a mix of CS and PS traffic in one common frequency band enhancing both CS and PS services performance.

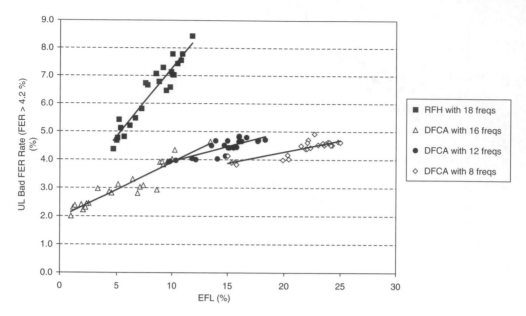

Figure 10.27   DFCA performance in terms of UL FER in the narrowband scenario.

Another significant benefit of the DFCA feature is the excellent performance of half-rate channels that enables very high HR usage without quality degradation that can be observed in random frequency hopping networks. Higher half-rate utilization improves the BTS hardware efficiency and thereby the number of TRXs needed to serve the given traffic may be reduced.

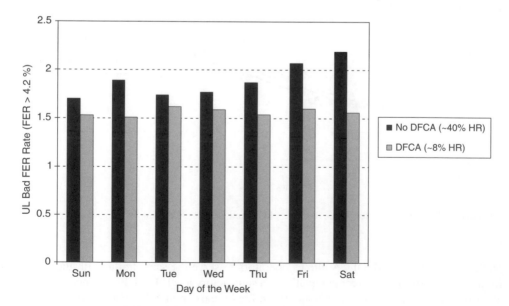

Figure 10.28   DFCA performance in high half-rate utilization scenario.

Moreover, DFCA simplifies the network planning. The DFCA TRXs do not have preassigned frequencies, frequency hopping parameters and training sequence codes since DFCA allocates the radio channel parameters "on the fly". This also simplifies the network capacity upgrades as new TRXs can be introduced with minimum planning effort.

References

[1] 3GPP TS 45.002 v8.0.0, "Multiplexing and Multiple Access on the Radio Path", December 2008.
[2] T. Halonen, J. Romero and J. Melero, *GSM, GPRS and EDGE Performance*, 2nd edition, New York: John Wiley & Sons, Ltd, 2003.

11

Advanced Admission and Quality Control Techniques

Sebastian Lasek, Krystian Krysmalski, Dariusz Tomeczko and
Sebastian Lysiak

11.1 Introduction

In recent years, EGPRS technology has been evolving towards a high speed system that should enable operators to provide new, more advanced and consequently more demanding services. Offering services characterized by stringent quality requirements in combination with huge increase in PS traffic volume has brought forward the challenge of how to manage network resources more efficiently. This has increased the demand for techniques that increase network capacity and consequently the operators' revenues while ensuring satisfactory quality for the end user. One way to face this challenge is the introduction of Quality of Service (QoS) management techniques that provide measures to secure sufficient performance for all services while maximizing the number of served subscribers. QoS management provides the mechanisms for differentiation between services and users, prioritization of premium subscribers and urgent traffic as well as guaranteeing certain quality levels for more demanding applications at the expense of lower priority ones.

The aim of this chapter is to introduce the principles of QoS management in GERAN networks. The main focus is Admission Control and Quality Control functionalities. However, for the sake of completeness, other complementary mechanisms such as Packet Flow Management, QoS-aware channel allocation algorithms and data flow scheduling are also discussed. Section 11.2 explains the fundamentals of QoS management and provides a concise description of the 3GPP QoS architecture including the concepts of the PDP Context and BSS Packet Flow Context. In Section 11.3 the procedures related to Admission Control are presented. Section 11.4 provides a detailed description of the Quality Control mechanisms including load balancing and its exemplary realization with Network Control Cell Reselection. In Section 11.5, the simulation results of a QoS-aware GERAN network are presented and analyzed.

GSM/EDGE: Evolution and Performance Edited by Mikko Säily, Guillaume Sébire and Eddie Riddington
© 2011 John Wiley & Sons, Ltd

Finally, Section 11.6 is dedicated to features aiming to improve the performance of EGPRS networks in the support of conversational traffic class services, namely Network Assisted Cell Change and Packet Switched Handover.

11.2 Quality of Service Management

Quality of Service, according to 3GPP, is the collective effect of service performance which determines the degree of satisfaction of a service user. It is characterized by a combination of factors such as service operability, accessibility, retention, integrity and other applicationspecific ones. To provide a mobile networks' subscribers with satisfactory quality, while maximizing network capacity, a set of mechanisms for efficient QoS management has been introduced. The QoS management in mobile networks is based on the fact that a user's expectations vary a lot depending on the application type and that different performance levels may be requested by users of different services.

The basic criteria for the service differentiation are: reliability (transmission link erroneousness), throughput and delay requirements. For example, an e-mail application is characterized by strict reliability and relatively relaxed delay requirements while a videoconferencing requires small delays but tolerates some transmission errors. Services having strict delay and loose reliability requirements are regarded as real-time services, for example videoconference, Pushto- Talk over Cellular, audio/video streaming or Voice over IP. Services characterized by high reliability and relaxed delay requirements are regarded as non-real-time services, for example e-mail applications, web browsing or file downloading. Furthermore, real-time services usually require a guaranteed bit rate that is implied in the use of a video or audio codec, while non-real-time services typically do not. This differentiation of the quality requirements of various applications is the essence of QoS management in EGPRS networks. By classifying and handling particular services according to their real needs, the QoS management allows efficient network resources utilization and maximization of the number of satisfied users. The concept of services differentiation is depicted in Figure 11.1. In this example, higher priority, in terms of bandwidth, is assigned to the video streaming in order to guarantee minimum delay and sufficient throughput, while web browsing and e-mail services, for which delay is less critical, are allocated less bandwidth. Since e-mail is a typical background application, lower priority is assigned to it compared to interactive web browsing.

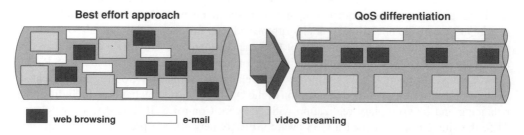

Figure 11.1   Services differentiation in QoS management [2]. Reproduced by permission of G. Gomes and R. Sanches, *End-to-end Quality of Service over Cellular Networks*, Figure 4.1, John Wiley & Sons, Ltd, 2005.

QoS management includes measures to negotiate the quality to be provided to the application, to prioritize particular connections and to guarantee certain performance for higherpriority data flows at the expense of quality degradation of lower-priority ones. Guaranteeing adequate performance for the most demanding applications requires mechanisms for the detailed estimation of expected quality of the incoming data connection and its impact on ongoing data flows. The so-called Admission Control (AC) functionality is in charge of blocking a new connection establishment should its quality be inadequate or else it might degrade the performance of already existing data flows to below acceptable limits. An efficient AC algorithm needs information on the current utilization of the Base Station Subsystem (BSS) resources. To make such information available already at the admission control stage, the Packet Flow Management feature has been introduced, which controls the QoS profile negotiations between SGSN and BSS and handles the BSS Packet Flow Contexts (see Section 11.2.2.1). The AC mechanism is supported also by the resource reservation functionality which, after having admitted a newconnection, keeps resources reserved for a given period of time. QoS management might also include an Allocation Retention Priority (ARP) feature which allows pre-emption of lower priority connections in order to free resources for incoming higher priority ones. The ARP mechanism also enables higher priority connections to retain their quality against lower priority ones in congested scenarios. To sustain the negotiated quality for the lifetime of the data flow, Quality Control (QC) functions have been introduced. The QC mechanisms aim to maintain the negotiated quality by detecting and reacting against potential QoS downgrades. Another important component of QoS management is in the QoS-aware selection of the transmission mode. For instance, some services with stringent delay requirements need to be realized in unacknowledged mode (since acknowledged mode could result in unacceptable delays), while those with high reliability prerequisites may require an acknowledged mode.

To realize an efficient QoS management, the above-mentioned functionalities need to be supported by a QoS-wise resource allocation algorithm and data flow scheduler as depicted in Figure 11.2. The QoS-aware data flow scheduler is aimed at ensuring that the high priority data flows experience at least the minimum specified quality even in highly loaded conditions by sharing resources between connections according to their priorities. The resource allocation

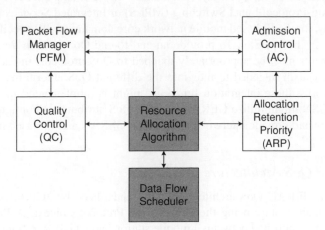

Figure 11.2 QoS management components.

algorithm aims to assign resources to particular data flows according to the requests coming from other QoS mechanisms and provides feedback information to certain QoS algorithms. Detailed description of QoS mechanisms and related parameters is provided in the next sections.

The QoS mechanisms allow operators to realize different strategies, for example maximization of the number of satisfied users equally for all the services, or treating some selected services as higher priority than others. The QoS management prioritization functions may be used by the operators to differentiate not only between services but also between subscribers. For example, high-priority packages with higher bit rates may be provided for premium users, and low-priority affordable packages for budget users. Furthermore, it is possible to manage user priorities dynamically, for example to downgrade the priorities of the users having exceeded their subscribed data volume limits.

The QoS mechanisms themselves are not a remedy for lack of network capacity. Rather, they help in well-dimensioned networks to minimize the effects of high traffic load, especially during busy hour congestion conditions, and thus improve the user's experience. QoS management may be regarded as a solution that ensures optimum usage of scarce network resources and as a prerequisite for GERAN networks offering high data rate and delay-sensitive services, such as video streaming or Voice over IP.

11.2.1   End-to-End QoS Architecture

In the context of end-to-end Quality of Service, the EGPRS network may be regarded as the sub-system that provides data connectivity between a user terminal (e.g. mobile station or laptop connected via mobile station) and an external IP network. The end-to-end transmission path, between a user terminal and a service provider, consists of three domains: the Radio Access Network, the Core Network and the external IP network (see Figure 11.3). Since different QoS mechanisms are used along the end-to-end path, it is necessary to coordinate the QoS management between all domains and this increases the complexity of the end-to-end QoS management.

In the external IP network, QoS may be realized with for example Differentiated Services (Diff- Serv), Multi-Protocol Label Switching (MPLS) or Integrated Services (IntServ) mechanisms. Similarly, for the IP-based mobile network core domain, any IP-specific QoS solution may be applied by the operator. To provide an end-to-end QoS in such a scenario, service quality requirements must be appropriately mapped to QoS protocols in each domain. Furthermore, a mechanism intended to integrate the different QoS algorithms will be needed. In EGPRS networks, these integration functions might be implemented in the GGSN. For example, the GGSN might utilize GERAN-specific QoS attributes to configure the DiffServ edge functionality and provide interworking between GERAN and IP-based networks [2, 8].

11.2.2   3GPP QoS Architecture in GERAN

The concept of a GERAN QoS architecture was standardized by 3GPP in Release 97 and Release 98, the main feature being the Packet Data Protocol Context. A PDP context is a logical connection established between a mobile station and a GGSN to transfer all IP traffic to and from the mobile station. The PDP context activation procedure makes the mobile station

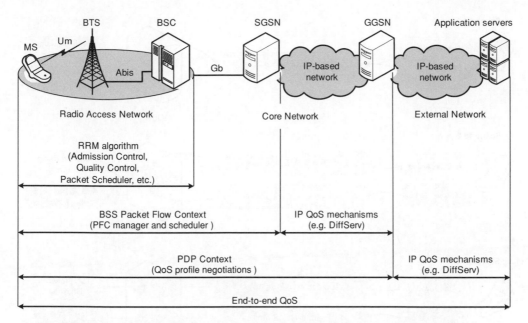

Figure 11.3   End-to-end QoS architecture [2]. Reproduced by permission of G. Gomes and R. Sanches, *End-to-end Quality of Service over Cellular Networks*, Figure 4.7, John Wiley & Sons, Ltd, 2005.

visible in the external IP network and allows aQoS profile for an incoming data transfer session to be negotiated between a mobile station and a network. While a PDP context activation is usually triggered by the application layer in a terminal, a PDP context activation may also be requested by a GGSN, although this would only be possible if a mobile station is assigned a static IP address (which is not that common).

The concept of the PDP context is the essence of services differentiation in GERAN. The basic idea is explained in Figure 11.4. QoS parameters are agreed on a per PDP context basis, that is all applications sharing one PDP context have the same QoS profile. For individual treatment of particular applications, separate PDP context must be activated. In 3GPP Release 97 and Release 98, each PDP context has its own IP address. In 3GPP Release 99, an en-hancement was introduced allowing several PDP contexts (with different QoS profiles) to be assigned to one common IP address. All the PDP contexts using the same IP address have to be connected to the same Access Point Name, where the APN identifies the access point to the external IP network and one APN may address one or several access points of the same GGSN. The mapping between services and APNs via services that are available is done by the operator and is usually based on QoS requirements as depicted in Figure 11.4. For example, it is possible to provide all services via the same APN (scenarios 1 and 2 in Figure 11.4) or to group services according to their QoS requirements and map them to different APNs (scenario 3 in Figure 11.4). If different services are available via different APNs, the service differentiation may require different PDP contexts connected to different APNs [1].

The PDP context activation request should ideally contain a detailed description of the QoS requirements based on a set of standardized QoS attributes. However, mobile station implementation depends on whether a PDP context activation request conveys explicit values

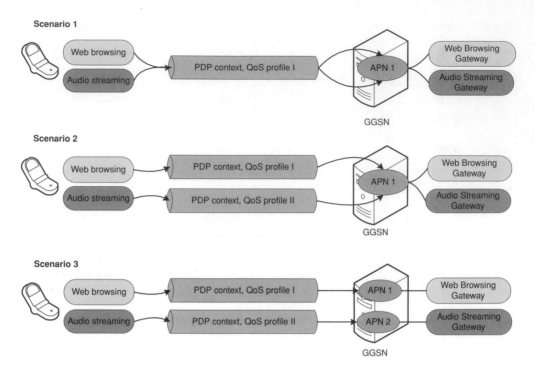

Figure 11.4   Service differentiation based on PDP context [1]. Reproduced by permission of T. Halonen, J. Romero and J. Melero, *GSM, GPRS and EDGE Performance*, 2nd edition, Figure 3.9, John Wiley & Sons, Ltd, 2003.

of certain QoS parameters or not. It is even possible that a PDP context activation request is empty. Given this, it is possible to divide terminals into two groups: QoS-aware and non-QoS-aware. Non-QoS-aware terminals (or application clients) are unable to indicate any QoS requirements, in which case all the necessary information is retrieved by an SGSN from the HLR (each subscriber has its own QoS profile created in the HLR). QoS-aware terminals are able to indicate to the network explicit values of certain QoS attributes. Upon receiving a PDP context activation request from a QoS-aware terminal, an SGSN checks with the HLR if the requested service is available for the subscriber and triggers the GGSN to establish a connection with an external IP network. If the requested QoS attributes exceed the upper limits defined in the HLR profile, the values from the HLR are used. If only some of the QoS attributes are explicitly indicated by a QoS-aware terminal, the missing information will be retrieved from the HLR. Having verified the QoS profile and set up the PDP context, the network indicates to a terminal that the PDP context has been activated with a PDP context activation accept message, which contains the agreed QoS profile and the assigned IP address to be used in communication with an external network. If the assigned QoS parameters are unacceptable, a mobile station may terminate the session by starting the PDP context deactivation procedure.

When requesting the PDP context activation, a terminal may indicate to the network (besides the QoS requirements), the APN via which the requested service should be realized. If different

Table 11.1   Release 97/98 QoS attributes [5, 6]

QoS attribute	Description
Precedence class	Indicates the relative priority of maintaining the service. There are three service precedence levels: • High precedence - guarantees service ahead all other precedence levels • Normal precedence - guarantees service ahead of low priority ones • Low precedence - receives service after commitments for high and normal precedence services have been fulfilled
Delay class	Defines end-to-end transfer delay through EGPRS network. There are three predictive delay classes for which maximum values of mean transfer delay and 95th percentile are determined. There is also one best-effort class for which no limits are specified.
Reliability class	Indicates the transmission characteristics that are required by an application. Data reliability is defined in terms of: • Probability of Service Data Unit (SDU) loss • Probability of SDU deliver out of sequence • Probability of duplicate SDU delivery and • Probability of corrupted SDU There are five different reliability classes.
Throughput class	Defines the bandwidth requested by a PDP context. User data throughput requirements are defined by two negotiable parameters: • Peak Throughput - specifies the expected maximum bit rate of data transfer across EGPRS network. There is no guarantee that this peak throughput can be achieved or sustained. • Mean Throughput - specifies the average bit rate of data transfer across EGPRS network (aggregated over the PDP context lifetime). There is no guarantee that this throughput can be achieved or sustained.

applications are mapped by the operator to different APNs (scenario 3 in Figure 11.4) and if for each subscriber a separate QoS profile was created for each APN, it is possible to realize the strategy of service differentiation even for non-QoS-aware mobile stations. The only prerequisite is that a non-QoS-aware terminal indicates, in the PDP context activation request, the appropriate APN. This requires proper terminal configuration. This kind of workaround is not necessary for QoS-aware terminals that are able to differentiate services by requesting service-specific QoS attributes during PDP context activation even towards the same APN (scenario 2 in Figure 11.4).

According to 3GPP Release 97 and Release 98, the service differentiation is based on four attributes: precedence class, delay class, retainability class and throughput class. Definitions of particular attributes can be found in Table 11.1[5,6].

The 3GPP Release 97 and Release 98QoS architecture supports only non-real-time services, such as e-mail, web browsing or file transfer. To give operators better control of QoS in their networks and to allow real-time services, the concept of BSS Packet Flow Context (PFC) was introduced in 3GPP Release 99. With this mechanism an SGSN checks with the BSS whether the requested QoS can be provided by the radio access network before a PDP context is admitted. Having verified the resource's availability, a BSS creates the Packet Flow Context

Table 11.2   3GPP QoS traffic classes [7]

Traffic class	Service type	Main characteristics	Exemplary applications
Background	• Non-real-time	• Minimized bit error rate • Destination does not expect data within a certain time	• E-mail • File downloading
Interactive	• Non-real-time	• Request-response scheme • Minimized bit error rate	• Web browsing
Streaming	• Real-time	• Unidirectional continuous stream • Minimized delay variation	• Audio streaming • Video streaming
Conversational	• Real-time	• Conversational pattern: stringent and low delay • Minimized delay variation	• Voice over IP • Video over IP

for the PDP context. The single BSS PFC can be shared by several PDP contexts with identical or similar negotiated QoS requirements. The set of QoS requirements of a PFC is called an Aggregate BSS QoS Profile (ABQP). The ABQP determines the quality that must be provided by a BSS to all the packet data flows established within the PFC. Radio as well as Abis and Gb interfaces are here considered. PFC QoS parameters are negotiated between an SGSN and a BSS based on the PDP context QoS attributes. More details on BSS PFC may be found in Section 11.2.2.1.

In addition to the BSS PFC, 3GPP Release 99 brought to GERAN the concept of QoS traffic classes and a new set of QoS parameters. These changes were introduced to keep the GERAN QoS management aligned with 3GPP Release 99 for UTRAN.

Four different traffic classes have been defined to distinguish between the various service characteristics: background, interactive, streaming and conversational. These classes are differentiated mainly by delay sensitivity, that is the conversational class is meant for the most delay-sensitive applications while the background class is the most delay-insensitive. The conversational and streaming classes are mainly dedicated to carry real-time traffic (see Table 11.2). The conversational class applies to the most delay-sensitive applications such as Voice over IP, video over IP or real-time gaming, while the streaming class applies to delay-sensitive applications such as audio and video streaming where the delay variations can be compensated (up to the certain extent) with the use of buffering mechanisms. The interactive and background classes are mainly intended to be used by applications with relaxed delay but more stringent reliability requirements. The interactive class is based on typical request-response communication scheme such as web browsing. In the case of backgroundtype services, the receiving side tolerates relatively longer response times and hence is less delay sensitive than interactive ones. Classic background applications are the e-mail and file downloading applications. Owing to different characteristics, particular traffic classes are handled by the QoS management algorithms differently. For example, for the streaming and conversational data flows, certain minimum throughputs and maximum delays may be guaranteed at the expense of background and interactive ones. Moreover, the interactive traffic class has a higher scheduling precedence than the background one. This means the background data flows have the least priority and

Table 11.3 Release 99 QoS attributes [7]

QoS attribute	Description
Traffic class	Type of application for which PDP context is to be established. There are four traffic classes: background, interactive, streaming and conversational.
Delivery order	Indicates whether in-sequence delivery of service data units (SDU) is required by the application or not.
Maximum SDU size (octets)	Defines the maximum allowed SDU size. Maximum value of 1500 octets.
SDU format information (bits)	List of possible exact sizes of SDUs for which UTRAN can support transparent RLC protocol mode. Not applicable to GERAN.
Delivery of erroneous SDUs	Indicates whether SDUs with detected errors shall be delivered to application layer or not.
Residual BER	Defines the undetected bit error ratio
SDU error ratio	Defines the ratio of lost or erroneous SDUs
Transfer delay	Determines the maximum delay for 95th percentile of the distribution of delay for all delivered SDUs during the PDP context lifetime. Delay is defined as a time from a request to transfer an SDU to its delivery to the destination point. It specifies the delay tolerated by the application.
Maximum bit rate	Defines the upper limit of bit rate expected by (and delivered to) the user or application. There is no guarantee that this maximum bit rate can be achieved or sustained.
Guaranteed bit rate	Defines bit rate which must be guaranteed to the user (or application) during whole PDP context lifetime.
Traffic handling priority	Determines the relative importance of SDUs belonging to the PDP context compared to SDUs of other PDP contexts. Applicable to interactive class only - allows differentiation within interactive class.
Allocation/retention priority	Determines the relative importance of allocation and retention of the PDP context comparing to other PDP contexts. This is a subscription parameter and is not negotiated by mobile station.

may only use resources when neither conversational nor streaming nor interactive applications need them.

Definitions of the other 3GPP Release 99 QoS parameters are presented in Table 11.3 [7]. Not all the attributes are relevant for all the traffic classes, that is the traffic handling priority attribute is applicable only to the interactive traffic class while the transfer delay and the guaranteed bit rate parameters are relevant only for the streaming and conversational traffic classes. Moreover, the EGPRS Base Station Subsystem depends on whether and which of the parameters are used in QoS management. Furthermore, some parameters, such as residual BER or SDU error ratio, are characterized by limited tuning flexibility [7]. In practice, only traffic class, maximum bit rate, guaranteed bit rate (for streaming and conversational traffic classes only), transfer delay, allocation/retention priority and traffic handling priority (interactive traffic class only) provide flexibility in QoS profiles parameterization.

Table 11.4   Mapping between QoS attributes of R97/98 and R99 [7]

R97/98 QoS attribute	R99 QoS attribute
Precedence class	Allocation/retention priority
Delay class	Interactive and background traffic classes and traffic handling priorities defined for interactive class
Reliability class	Combination of residual BER, SDU error ratio and delivery of erroneous SDUs
Peak throughput class	Maximum bit rate

To ensure backward compatibility with 3GPP Release 97 and Release 98 mobile stations, the mapping between Release 97/Release 98 and Release 99 attributes has been defined as shown in Table 11.4 [7].

11.2.2.1   Packet Flow Management

As already mentioned, in 3GPP Release 99, the concept of a BSS Packet Flow Context, known as Packet Flow Management (PFM), was defined to improve QoS handling in 3G and 2G networks. This mechanism allows exchange of the QoS information between an SGSN and a BSS already at the stage of QoS negotiations (PDP context activation). Figure 11.5 shows the message flow of the PDP context activation in the 3GPP Release 99 scenario [6].

In the 3GPP Release 97 and Release 98 scenarios, detailed information about radio resource occupation cannot be utilized by the AC since the PDP Context Activation procedure takes place between the MS and the SGSN, and it is totally transparent to a BSS.

In the 3GPP Release 99 scenario, when both the network and the MS indicate their support of the PFM, the information flow is extended, that is having agreed the PDP context with a

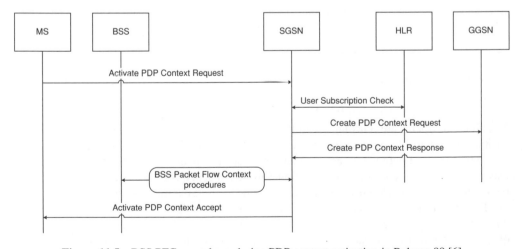

Figure 11.5   BSS PFC procedures during PDP context activation in Release 99 [6].

GGSN, an SGSN starts the BSS PFC set-up procedures. To this end, it recalculates some QoS parameters values (already accepted by the GGSN) to create the Aggregated BSS QoS Profile Information Element which defines the QoS requirements for the given Packet Flow Context. Some parameters such as transfer delay must be recalculated since the BSS can guarantee QoS only on the path between an MS and an SGSN – not across the whole delivery chain. An SGSN assigns to the PFC the Packet Flow Identifier (PFI) according to the QoS parameter values in the ABQP. Besides four predefined PFI values which correspond to signaling, SMS, TOM (Tunneling of Messages) and best-effort (these four do not require the BSS PFC Procedure), the PFI assignment algorithm is vendor-specific. However, the idea is that PFC with identical or similar requirements should get the same PFI. The ABQP information element complemented with the Allocation/Retention Priority parameter is sent to the BSS in a Create BSS Packet Flow Context Request message. Having checked during the Admission Control phase whether the available resources satisfy the QoS requirements, the BSS replies to the SGSN with a Create BSS Packet Flow Context Accept message. If the resources are insufficient to meet the QoS requirements, the BSS might restrict the requested QoS parameter values. The agreed QoS profile is then sent by the SGSN to the MS within the Activate PDP Context Accept message. If the assigned QoS parameters are not acceptable to the mobile station, it may reject the assigned PDP context [6].

The BSS PFC establishment procedure may be also triggered by a BSS being requested to transfer a LLC PDU for which no BSS PFC is available. A BSS will then trigger an SGSN (by sending the Download BSS Packet Flow Context Request message) to create a new BSS PFC and provide the required ABQP information element. The BSS PFC may be modified during its lifetime, for example if an earlier agreed QoS profile can no longer be sustained by the BSS. Finally, the BSS PFC may be deleted by an SGSN via the Delete BSS PFC Request message when for example the related PDP context is closed. In the case of UL TBF, the Packet Flow Timer (sent to the BSS together with the ABQP in the Create BSS Packet Flow Context Request) defines how long a PFC shall be stored in a BSS after the transmission of the last UL LLC frame.

A single mobile station may have several BSS PFCs created. They are stored in a BSS in the form of BSS Context. The logical relation between BSS Context and BSS PFCs is depicted in Figure 11.6. The PFC of a given MS can be assigned to one (MS1 PFI 15) or more applications (MS2 PFI 16) using the same PDP Context. On the other hand, the same PFI can be associated with more than one MS (PFI 20) if the users perceive the same quality. Note that PFM impacts transmission not only on the Gb but also on the Abis and radio interface.

The information stored in the BSS Contexts allows QoS differentiation on two different levels:

- between mobile stations, for example in the case of congestion, the performance of an MS with the lower priority PFC may be degraded prior to a performance deterioration of an MS with a PFC of higher priority;
- between PFCs of one mobile station, for example being aware of the number of active flows and being able to distinguish between them, a BSS can share the available radio resources in an optimum way (where the streaming and conversational PFCs will be preferred).

The packet flow schedulers, being vendor-specific solutions, are usually based on more or less complex variations of the Weighted Round Robin scheduling algorithms. To enable a BSS

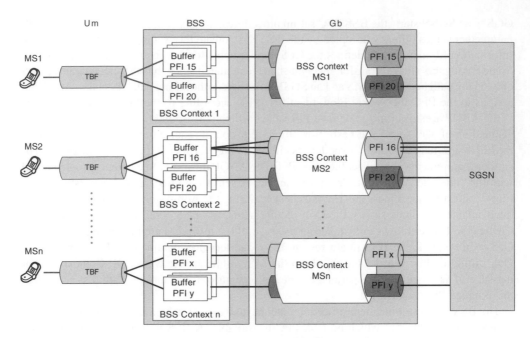

Figure 11.6   The concept of the BSS Context.

to schedule a particular LLC PDUs with proper priority, a BSS must be provided with the information on the QoS profile of all the PDUs. In 3GPP Release 97 and Release 98, this information, in the form of a QoS IE containing a subset of QoS parameters, is sent in each DL/UL-UNITDATA message (the BSSGP layer message containing LLC PDU) exchanged between an SGSN and a BSS. In the case of 3GPP Release 99, since all the QoS parameters are, in general, available in the BSS before the LLC PDUs appear, there is only a need to link the content of the UNITDATA message with the stored ABQP. This is realized with the help of the Packet Flow Indicator which is included in the UNIDATA message body as a new Release 99 Information Element.

Introduction of the Packet Flow Management functionality in 3GPP Release 99 also improved the performance of the Gb flow control algorithm, which aims to adjust the speed of transmission of DL LLC PDUs from the SGSN towards the BSS.

This can be used to limit the data rate in order to avoid BSS buffers' overflow and guarantee minimum data loss in case of cell reselection, or it can increase the data rate to guarantee a smooth transmission without delays and avoid the situation that BSS has nothing to transmit on the radio interface while data is kept in SGSN. The 3GPP Release 99 defines two levels of data transmission control between the SGSN and the BSS: cell[1] and mobile station levels [12]. As long as a BSS can control data flow transmission speed on the Gb interface to a resolution of a whole cell only, few mobiles with poor radio performance could fill up the BSSGP Virtual Connection (BVC) buffer, which in turn would force a slowdown in data transfer towards

---

[1] Formally it is BVC which stands for BSSGP Virtual Connection and in practice refers to the logical connection between a PCU and a single cell.

all mobiles in this cell. Control and buffering on the MS level avoid this phenomenon and decrease the total buffer size. Furthermore, 3GPP Release 5 has defined the third level of the Gb flow control: the Packet Flow Control level. Analogously, this change prevents the buffer in BSS dedicated to a given MS from being filled up by the data of the "slow" PFC with lower priority. For example, if streaming and background PFCs are used in parallel by a single user, preference on the air interface will be given to the streaming user. This may lead to protracted storage time of the background PFC PDUs in the BSS buffer and consequently its potential congestion. That is why it is better that the arrival rates of LLC PDUs of particular PFCs are controlled by the BSS.

11.3 Admission Control

11.3.1 Primary Functionalities of the Admission Control

The aim of theQoS-awareAdmission Control (AC) is to calculatewhich network resources are needed to satisfy the negotiatedQoS requirements, determine if these resources are available or not and reserve these resources [6].With Admission Control, access to the network is granted only if certain conditions related to satisfying the QoS requirements of the new Packet Flow Contexts are met. Moreover, the Admission Control algorithm has to assure that there will be no performance deterioration of the already established connections beyond the acceptable limits set by their respective PFC. The AC checking may be performed not only during the PFC creation but also due to the modification of a PFC or during a cell change. The efficient and strict Admission Control functionality is one of the main differentiators between the legacy "best effort" approach and the QoS-aware network management approach.

During the Admission Control process performed in the Packet Control Unit, availability of the resources across the whole transmission path is estimated, where it is assessed whether the radio interface, Abis interface, Gb interface associated with the PCU as well as PCU hardware (including but not limiting to available processing power, buffer sizes etc.) are each capable of handling a PFC. If the Admission Control fails, the creation or the modification of a PFC is declined or queued or alternatively the Aggregate BSS QoS Profile (ABQP) profile may be downgraded. Consequently, an efficient Quality of Service mechanism is always performing a trade-off between the number of connections served by the network and provided quality of the transmission. Performance of the Admission Control is presented in Section 11.5.

During the Admission Control phase, a detailed comparison has to be performed between the capacity needed to satisfy the QoS requirements of the PFC and available PS capacity [15]. In the case of the radio interface, the capacity figures could be defined as the number of radio blocks per second needed by the given ABQP and the number of radio blocks per second available (where the maximum capacity of the single radio time slot would be 50 radio blocks per second). The detailed procedure will depend on the QoS traffic class with the most important differentiation being between non-real-time and real-time services. In the next two sections, calculation of the needed and available capacity in the radio interface is presented (given that the radio interface is usually the bottleneck of the whole transmission path [1]).

11.3.2 Calculation of the Available Capacity

The number of time slots that could be used to serve PS traffic is one of the major inputs into the Admission Control and resource reservation algorithms. Generally, two types of radio time slot may be distinguished: time slots dedicated to PS traffic; and shared time slots that can be utilized by CS or PS traffic.

The assessment of available capacity could be based on information about the current PS load (e.g. the number of radio blocks used for all NRT and RT allocations across all the available time slots), or in the case of RT services, it could be directly calculated from the number of allocated resources. That way, temporarily inactive PFCs, that is the ones for which there is no ongoing TBF, will also be included in the number of occupied resources.While the inclusion of allocated RT resources improves the Admission Control performance, it could unnecessarily increase the blocking rate of the new PFCs. Hence, one could assume a certain overbooking might be allowed even for RT PFC. In this case, an appropriate trade-off in the actual implementation would have to be found. To avoid unnecessary transmission breaks caused by TBF establishment, longer TBF release times may be defined for real-time services (especially for conversational ones). Such an extension of a TBF lifetime could be achieved by utilizing Delayed TBF Release and Extended UL TBF functionalities. Moreover, even when transmission is not currently scheduled for a TBF, the necessary resources needed to satisfy QoS requirements could be treated as occupied during the Admission Control phase.

When calculating available capacity, the signaling messages sent on the radio interface that contribute to PACCH load (such as acknowledgements or packet assignments), should also be taken into consideration [2]. They reduce the amount of resources available for data transmission since PACCH is dynamically allocated (in both DL and UL direction) on the same physical channel as the application data traffic.

The active load, for example as measured by the scheduler algorithm, might not be sufficient to determine the influence of changes in traffic load and radio network conditions that may affect the Abis interface as well as the radio interface occupancy in highly dynamic manner. A sudden deterioration in interference in the network implies the use of lower coding schemes or a higher number of retransmissions which will directly affect the achievable throughput and may also negatively affect the delay factors. This will lead to an increase in the PS resources' occupancy time related to the given data volume which will result in a reduction in the resources available for other PFCs. This phenomenon should be taken into account with the application of carefully parameterized safety margins [15]. For example, a certain number of radio blocks shared on a demand basis could be restricted to existing TBFs and not made available for new TBFs. Moreover, in the calculations of the available capacity, an input from Quality Control entity could be taken into account as well, that is whether the already used resources for existing PFCs are indeed enough to satisfy their needs in terms of QoS requirements. The detailed set-up may be tuned with different levels of risk depending on the QoS traffic class [1].

11.3.3   Calculation of the Needed Capacity

In the assessment of needed capacity, for example for satisfying throughput or delay requirements, the calculations used in the Admission Control algorithm may be non-trivial in the sense that a main contributor might be the afore-mentioned dynamics in radio channel conditions. For example, the interference situation in the network, multipath propagation, constant changes in the load volume handled by the network etc. will all have an impact on the needed capacity. One approach is to use reference factors that tie the expected or measured interference situation in the network (e.g. in the single cell or the frequency layer, additionally differentiated by frequency reuse factor [2]) to the achievable throughput per radio time slot and delay values. The Link Adaptation restrictions (e.g. the set of available modulation and coding schemes for the given MS) could be taken into account as well. Moreover, BLER values could further limit the usage of certain coding schemes due to the delay requirements of the ABQP. For details about Link Adaptation and BLER factors, please refer to Section 11.4.2.

Another approach could be based on analyzing the relationship between throughput or delay requirements and resources needed for currently served PFCs, to predict the number of resources needed for a new PFC. Such an analysis may be performed either among PFCs of a given MS, or among PFCs of different MSs but having similar requirements.

Having estimated the throughput achievable per single radio time slot, the number of time slots needed to satisfy the QoS requirements of the given PFC could be assessed by dividing the requested guaranteed bit rate by the assessed time slot throughput. In this way, the needed number of radio blocks per second for the particular PFC could be provided as well.

A novel approach to the capacity calculation could also be based on utilization of the DFCA algorithm. Please refer to Section 11.3.5 for further details.

The use of the acknowledged or unacknowledged mode, determined according to the SDU error ratio attribute, affects this assessment as well. For example, in the former case, the potential retransmissions should also be included in the needed capacity calculation. When using acknowledged mode, the RLC Non-Persistent Mode feature may be applied, especially with conversational services where delay jitter should be avoided (for more details, see Chapter 2). Moreover, a mobile station's capability, such as its level of (E)GPRS support (whether e.g. it is EGPRS-2 capable or not), its multislot class, frequency bands the mobile could operate in and support for complementary PS features, such as for example Downlink Dual Carrier, have to be considered as well.

The guaranteed bit rate is foremost a requirement of real-time Packet Flow Contexts. However, it might also be beneficial to introduce a minimum bit rate threshold for nonreal-time services. The network would then try to maintain at least this minimum bit rate for non-real-time services (e.g. 1 radio block per second for the given TBF related to NRT PFC) in order to assure general PS service availability. This could be beneficial especially for mobiles that do not fully support QoS functionalities.

11.3.4   Multiplexing of the Resources

The PS radio resources may be shared by more than one TBF and more than one PFC. This fact directly influences the throughput calculations since the time slot capacity must be shared in such a case [2]. Certain restrictions related to the maximum number of TBFs/PFCs that utilize the same radio resources may be applied in order to reduce the transfer delay caused by multiplexing, even if the throughput requirements were met.

The channel allocation algorithm should aim to distribute the PS traffic evenly among the resources in order to minimize the average number of TBFs per PDCH, assuming the CS load volume allows for such a strategy. Taking into account bursty traffic generated by some services, it might be beneficial to multiplex, for example streaming with background services to utilize the common radio resources during the silence period of the TBF related to the streaming class [1]. Such a policy could be justified especially during the PS traffic peak. It should be also noted that due to statistical multiplexing gain [2], it is usually beneficial from the throughput point of view to assign the highest possible number of time slots for the MS (taking into account the split between DL and UL time slots and related limitations) even if the multiplexing factor were higher in this case. A certain trade-off between the throughput and delay constraints will have to be found.

The afore-mentioned multiplexing phenomena could affect achievable throughput for certain TBF/PFC in amore severeway ifPDCHsare shared between mobileswith different capabilities in terms of supported modulation schemes. In order to grant to the mobile the chance for UL transmission, an appropriate Uplink State Flag has to be sent in the downlink direction, unless a fixed allocationmethod is used. If this flag is dedicated to a GPRSmobile station, the downlink block within which USF is sent has to be GMSK modulated. This fact directly influences, for example the EGPRS DL bit rate if the GPRS and EGPRS TBFs are multiplexed onto the single PDCH. If USF GRANULARITY is set to 1 for uplink TBF (so-called USF Granularity 4 functionality), theUSF flag grants transmission for four successive blocks in the given time slot instead of one. Although in this case the impact of multiplexing is limited, this phenomenon should be taken into account during the Admission Control process. The problem becomes even more severe for EGPRS2 services, the performance of which could be significantly limited when multiplexing with legacy mobiles, especially GPRS mobiles. It should be noted that while USF Granularity 4 is an effective way of preventing DL throughput degradation, it negatively affects the performance of scheduling in the UL direction since it reduces the flexibility of the scheduler (four free consecutive radio blocks have to be found in order to grant the transmission turn).

In Figure 11.7, the exemplary influence of multiplexing of EGPRS2A mobiles with EG-PRS ones is presented at various EGPRS2A penetration figures (assuming there is no policy

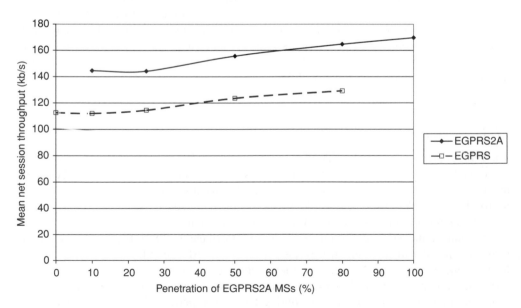

Figure 11.7   Influence of EGPRS and EGPRS2A multiplexing on net session throughput.

regarding routing of EGPRS and EGPRS2A to different radio resources). It can be seen that the higher the EGPRS penetration, the lower the mean EGPRS2A throughput. When the ratio of EGPRS2A to EGPRS mobiles is increased, the throughput as perceived by EGPRS mobiles can be seen to also increase.

This phenomenon could be observed since the resources' occupancy time for transmitting certain amount of data is lower for EGPRS2 than for EGPRS MSs. Hence, the freed resources could be utilized for EGPRS transmission. The simulations were performed for 1/3 frequency reuse pattern and for a heavy loaded network. USF Granularity 4 was applied.

Generally, one should avoid allocating MSs with different capabilities to the same radio resources. Moreover, it is likely to happen that only some part of the hardware resources in the radio access network (e.g. TRXs) are capable of transmitting and receiving in for example EGPRS- 2 mode. It is then reasonable to prevent (up to the certain extent) the allocation of GPRS or EGPRS connections to such TRXs. The policy with respect to applying such limitations should also be aligned with mobile station penetration figures related to the support of the given technology (like, e.g. EGPRS level).

11.3.5   Channel Allocation

Once the information about needed and available capacity is available, the Admission Control, together with the channel allocation algorithm, can search for the most applicable allocation so that an even distribution of the load across the time slots is achieved. Moreover. use of dedicated resources instead of shared ones [16] may be of special interest especially for demanding, real-time conversational services, since dedicated resources could not be occupied or pre-empted by the CS traffic in any case. Furthermore, scheduling weights that reflect the relative priority of the PFC in the scheduling phase could be taken into account during the allocation process in order to find best the possible resources for the new PFC. For further details, refer to Section 11.4.1.

The simplified example of used and free capacity assessment for the single TRX on a radio block basis together with a safety margin applied is presented in Figure 11.8. Unless Downlink Dual Carrier functionality is employed, it is not allowed to extend the allocation beyond one TRX. In the example provided, a new PFC could be admitted in the system since the assessed needed capacity is lower than the free capacity.

Usage of the Extended Dynamic Allocation[2] (EDA) deteriorates the scheduler flexibility in UL direction and hence should be taken into account during Admission Control and channel allocation procedures, especially for UL RT allocations. This potential deterioration could be caused by:

- a short break (of the length of one UL transmission turn) as a result of decreasing the number of scheduled slots for UL (applicable to mobiles that are unable to monitor the assigned USF in the DL PDCH corresponding to the lowest numbered uplink PDCH of the shortened scheduled allocation [10], as depicted in Figure 11.9);

---

[2] In Extended Dynamic Allocation, a USF flag grants transmission on the time slot on which the USF was received and all higher numbered time slots belonging to the UL allocation.

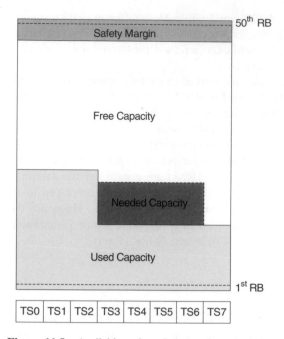

Figure 11.8 Available and needed capacity calculation.

- the fact that among the scheduled slots for EDA, the one with the highest number is always included;
- UL PACCH signaling load related to DL TBF that limits the EDA usage in UL.

Moreover, it has to be taken into account that all PACCH messages in DL are transmitted on the PDCH corresponding to lowest numbered time slot in the UL EDA assignment. If the Shifted USF feature [10, 11] is employed, PDCH which corresponds to second lowest numbered time slot in the UL assignment is used instead. These limitations should also be taken into account in the Admission Control process since it affects the signaling load distribution. In the case of concurrent TBF allocation, however, PACCH messages could be transmitted by the network on any of the common time slot in the DL/UL assignment.

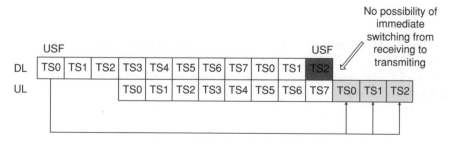

Figure 11.9 Scheduling limitations in case of EDA.

Another limitation that has to be considered in the capacity assessment is the UL power reduction for multislotULallocation applied not only according to the used modulation scheme and frequency band but also due to MS Multislot Power Profile (in order to avoid possible overheating in the transmitting circuitry [3]). However, in some cases use of Extended Dynamic Allocation instead of Dynamic Allocation could be the only way to ensure that requested UL throughput demanding QoS requirements are met.

Even if most PS applications have a asymmetrical DL-oriented character, it is essential that both directions (DL and UL) should be evaluated jointly during Admission Control phase and a new PFC should be admitted only if it passes both UL and DL requirements.

If Admission Control grants access to the network, the PS resources may be reserved at once even if TBF is not yet created. Such a shift from common TBF-oriented to PFC-oriented allocation policy is not necessary for the non-real-time services but improves the probability that real-time PFC will be handled properly. If immediate PS resources "pre-allocation" are used, this has to take place across the whole delivery chain in order to make this solution fully reliable.

It is clear that the Admission Control algorithm is very closely connected with the channel allocation process (being a part of RRM) and in fact, in the QoS-aware network it is not possible to treat these two algorithms independently [6]. The DFCA algorithm usability for the Admission Control and QoS differentiation could be here especially marked. With the help of the DFCA, it is possible (as presented in detail in Chapter 10) to find the radio channel allocation that meets a given service quality requirement (both CS and PS) and at the same time minimizes any impact to the already established connections. The PS service differentiation could be performed taking into account the respective Packet Flow Contexts [1]. Moreover, the control in interference level experienced by a particular user could be tuned according to different QoS needs. More precisely, one of the targets of the QoS-aware DFCA algorithm could be, for example to ensure high and constant CIR levels for the streaming connections while for the non-real-time services the CIR levels could be compromised. In such a way, real-time PFC requirements with respect to throughput or delay could be met even in a highly loaded network.

If it finally happens that there are no resources available for establishing a new PFC, the Admission Control procedure may trigger pre-emption procedures related to the already established PFC depending on, for example QoS traffic class and priorities (e.g. THP together with ARP) in order to grant access for higher priority PFCs [1] or alternatively the renegotiation of QoS profile could take place.

11.3.6   Admission Control during PFC Runtime

An efficient Admission Control also plays an important role during Packet Switched Handover or Network Controlled Cell Reselection (NCCR) procedures. The NCCR functionality can be seen as an intermediate step between the conventional mobile-based cell reselection and PS handover. With the help of PS Handover or NCCR, it is possible to undertake efficient PS load management. Performance of such a management could be further improved by means of carefully configured Admission Control process. Before triggering a cell change (or handover), it could be checked whether a new flow in the target cell would satisfy the QoS requirements and simultaneously that the performance of the already active PFCs will not be affected in the target cell. For details about PS Handover and NCCR functionality, refer to Sections 11.4.3.1 and 11.6.2 respectively.

Another aspect that has to be considered in the Admission Control phase is the priority that is normally given in the GSM networks to CS traffic. It could be reflected, for example in the handling of shared PS/CS time slots where CS traffic takes precedence over packet allocations and under heavy voice load may downgrade the number of PS time slots. The reorientation of this policy may be considered during QoS introduction in the network. It might be reasonable to avoid downgrading if the quality of the established PFCs were affected and there are no means of reallocating them. Such a strict policy against CS-triggered pre-emption of PS resources could be applicable only for some PFCs, for example for the real-time conversational ones.

11.4   Quality Control

11.4.1   Scheduler

The introduction of an Admission Control network is to ensure that certain QoS requirements related to Packet Flow Contexts can be met. Hence the scheduler, which is responsible for handling the transmission turns for the different Packet Flow Contexts, needs also to be QoSaware. The scheduling algorithm takes into account distribution of TBFs across the available PDCHs. In the downlink direction, the scheduler needs to be PFC-oriented (rather than TBoriented), since more than one PFC could be handled within one TBF, as depicted in Figure 11.10 (the Multiple TBF feature is assumed not employed here).

The BSS can control the DL data transfer for a given MS on a PFC basis via the Gb interface by utilizing the PFC Flow Control functionality (see Section 11.2.2.1). Moreover, data flows related to non-real-time PFCs could be restricted if there is a risk that a real-time PFC would not satisfy its QoS requirements.

An exemplary utilization of the Multiple TBF feature is depicted in Figure 11.11. Support for the Multiple TBF functionality is essential, especially if different RLC modes are required for various TBFs related to a single MS.

In the uplink direction, the scheduler is responsible only for distributing the permissions to access the physical radio resource on a TBF basis. Moreover, sequence of TBFs may utilize the same PFC in both DL and UL directions.

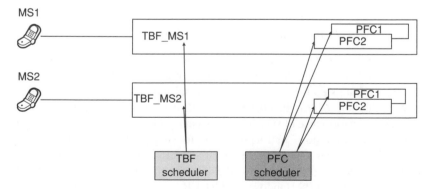

Figure 11.10   TBF and PFC schedulers in non-Multiple TBF case.

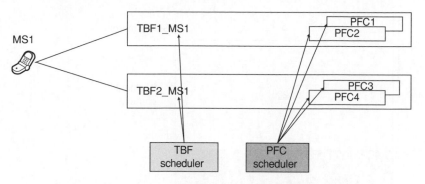

Figure 11.11   TBF and PFC schedulers in Multiple TBF case.

The scheduler has to prioritize the blocks according to the QoS class related to the given Packet Flow Contexts and simultaneously utilize the capacity left after satisfying the guaranteed bit rate requirements in the most efficient way. This could be performed while handling in a reasonable way the multiplexing effects described in the Section 11.3.4 as part of Admission Control and by utilizing an efficient polling algorithm to minimize the influence on the actual traffic performance. For the non-real-time services that are without throughput and delay requirements, the scheduling weights need to reflect the relative priority of the various PFCs and could be assigned according to, for example traffic class, ARP, traffic handling priority, etc. In the case of real-time services, the scheduling weights should be derived first and foremost from throughput and delay requirements. This process could be handled in a similar way as for the calculation of needed capacity during the Admission Control phase (refer to Section 11.3.3). These weights could be used, for example in a deficit round robin [15] algorithm implemented in the scheduler where each PFC is served in a cyclic way and the serving time is assigned according to each of the PFC's traffic needs. PFCs are not handled in a uniform way (as in the simplest round robin scheduling algorithm), instead a certain prioritization is applied according to ABQP. The scheduler should prioritize blocks corresponding to real-time services. However, in order to leave room for non-real-time services during the scheduling phase, a minimum bit rate (to be understood as a non-binding target to be achieved) may also be introduced. This could even be essential, especially when the same resources (e.g. time slots) are shared between non-real-time and real-time services. It should be also noted that the scheduler has to respect the maximum bit rate limits set by the PFC, and the PFC throughput should not exceed the guaranteed limit unless it will not affect the performance of the other PFCs (including NRT ones). In the first iteration of the scheduling algorithm, real-time related TBFs could be given transmission turns while in the second iteration, non-real-time TBFs could be included (as long as there is any residual capacity left after the first phase).

An example of handling three different TBFs in a scheduler: one RT and two NRTs is given in Figure 11.12. In this example, the residual capacity left after assigning transmission turns for the RT and NRT services is shared among the three TBFs according to their relative weights (e.g. three times more radio blocks are assigned to TBF3 compared to the TBF2). Radio blocks allocated as a safety margin could also be utilized by both the RT and NRT scheduler.

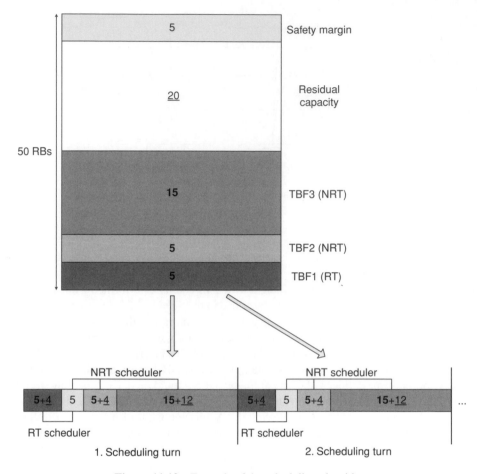

Figure 11.12 Example of the scheduling algorithm.

In general, the RT service performance should be preserved by the Quality Control functionality, even in highly loaded network conditions as depicted in Figure 11.13, where, during congestion, the RT services have been protected at the expense of delaying the packets related to non-real-time services. If there were no QoS functionality in the network, then all services could be negatively impacted.

Moreover, higher scheduling weights for the signaling in downlink could be assigned than for the traffic channels, for example when a Packet Uplink Ack/Nack message has to be sent and an UL radio block has to be assigned for polling purposes. The same rule is also applicable in uplink for example when a polling response is needed.

11.4.2   Link Adaptation

Among the various other mechanisms that help to maintain an acceptable quality of service, the Link Adaptation procedure can also be considered a part of QoS-aware network management.

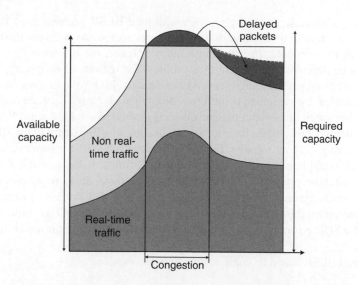

Figure 11.13   RT and non-RT traffic handling by Quality Control during congestion.

This algorithm needs to be carefully parameterized to ensure not only the maximum possible throughput under changing radio conditions but also to satisfy the delay requirements as well. Usage of Incremental Redundancy is clearly favored in terms of throughput when less robust coding schemes are used, even in imperfect radio conditions, but simultaneously in such circumstances, multiple retransmissions may lead to degradation in the transfer delay [1]. In Figure 11.14, the throughput performance of MCS6 and MCS9 is shown both with

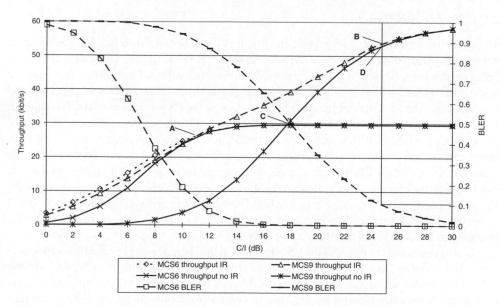

Figure 11.14   Throughput and BLER with and without IR for MCS6 and MCS9 coding scheme.

and without Incremental Redundancy together with their BLER performance. The boundary of when a deterioration in throughput or delay factor is acceptable is determined by the QoS requirements of the particular Packet Flow Context. Hence, the optimum Link Adaptation algorithm may be different for different QoS profiles. For reasons of simplicity, two EGPRS coding schemes are shown in Figure 11.14. Assuming ideal Link Adaptation with Incremental Redundancy enabled, the switching point A may be applied to change between coding schemes for the case where only throughput maximization is desired. Switching point B may be used where certain delay requirements have to be met, leading to a limitation in the BLER (in this case to 10%) during the transmission. For acknowledged mode, the transfer delay parameter from the ABQP could be mapped to acceptable BLER limits. If Incremental Redundancy is not enabled, switching points C and D for the cases without delay requirements and with some BLER constraints may be applied respectively. Note that in case of unacknowledged mode where Incremental Redundancy is not applicable, BLER limitations could be calculated according to the SDU error ratio – the more stringent the requirement, the lower the acceptable BLER.

The presented BLER could be computed as follows:

$$BLER = 1 - \frac{Ncorrect\_blocks}{Nneeded\_transmissions}$$

Ncorrect_blocks is the number of RLC data blocks received correctly in the given period and Nneeded_transmissions is the number of transmissions needed for correct reception. The general formula for BLER calculation is also valid for the cases where Incremental Redundancy is used.

11.4.3   Load Balancing

To prevent a degradation to PS service quality, it is often beneficial to balance the PS load: between the time slots inside a given TRX; between TRXs within the cell and (if possible) between the cells. This is not always possible during the Admission Control and resource reservation phase, which is why a mechanism performing some periodical checking and distributing of the load may be required. Reallocation of the PS services may also be required, for example to optimize the throughput of the non-real-time services. Hence, the use of Network Controlled Cell Reselection or Packet Switched Handover may be beneficial or in some cases essential. Moreover, Network Controlled Cell Reselection in conjunction with Network Assisted Cell Change can decrease the outage time during cell change, and hence can improve the quality and performance of the service as perceived by the end user [2]. PS Handover can reduce this outage even further. The NCCR feature is described in more detail in Section 11.4.3.1. Further information on the NACC and PS Handover functionalities is also provided in Sections 11.6.1 and 11.6.2 respectively.

It is also beneficial to compress the CS allocation in the shared resources as much as possible unless it deteriorates quality of CS calls. Such a policy, which will result in a CS and PS traffic split [1], provides additional room for PS load balancing and helps in admitting more users during the Admission Control phase. Due to certain limitations introduced by a mobile's multislot class, contiguous PS allocation across the time slots is usually preferred. Moreover, it should be emphasized that unless the PS Handover feature is utilized in the network, it is usually easier and more efficient to move the CS calls, for example to other radio cells with the help of load-based handovers, rather than reallocate PS traffic. Furthermore, prioritization of certain layers (either BCCH or non-BCCH [13]) for PS traffic may be beneficial as well.

11.4.3.1 Network Control Cell Reselection

A cell reselection procedure allows a mobile station to change cell either in standby or in ready state. In the standby state (when an MS is GPRS attached but has no radio resources allocated for packet transfer and thus is in packet idle mode), the cell reselection is performed autonomously by the MS – so-called mobile controlled cell re-selection. However, in the ready state (when packet transfer is ongoing or has just ended but the ready timer has not yet expired), the cell reselection can be initiated either by the MS or by the network. In the latter case, the network orders the MS to send measurement reports, commands the MS to change cell and also decides to which cell the MS shall change. This mode is supported when the Network Controlled Cell Reselection functionality is enabled in the network.When the network makes the decision to trigger a cell reselection, it sends a Packet Cell Change Order message to force the MS to switch to another cell. There might be several criteria evaluated by the Network Controlled Cell Reselection algorithm that triggers a cell change and selects the target cell for the cell change. For instance, a power budget criterion, which is based on measurement reports containing information about the carrier level of the serving and adjacent cells, might trigger the MS to make cell reselection towards a cell with a lower path loss. Furthermore, by means of the budget criterion, some neighboring cells or layers (e.g. a less interfered 1800 MHz layer) can be made more attractive to an MS in packet transfer mode than to an MS in GMM idle mode. Neighbor cell capabilities can also be considered to determine a cell's attractiveness, for example the NCCR algorithm could force an EGPRS MSs to select an EGPRS capable cell and keep the GPRS only MSs in a non-EGPRS capable cell in order to minimize allocation of MSs with different capabilities on the same radio resources. As another criterion, the load situation in the serving and neighboring cells can be taken into account by the network to trigger a cell reselection. In such a case, the network may try to move the MSs in packet transfer mode from the overloaded cell to a neighboring cell that offers acceptable radio conditions and has enough free resources to accommodate the additional traffic.

Normally, in networks without any QoS management, the channel allocation algorithm allocates packet data channels as long as there are available resources in the cell. This may lead to a situation where some cells are congested, while their neighboring cells are not. Network controlled cell reselection due to traffic reasons is, in principle very similar to handover due to traffic reasons in the CS domain and can optimize the use of hardware resources and thereby boost the network capacity.

The basic idea of load balancing is outlined in Figure 11.15. When a certain load in the cell is exceeded that is the PS traffic in the cell exceeds a predefined threshold (THR1) reflecting the percentage of busy PS resources, the terminals in the serving cell involved in data transmission can be subjected to a reallocation to a neighboring cell. The goal of the algorithm is to decrease the PS channel utilization below a value defined by another threshold (THR2). Only neighboring cells that have enough free PS capacity, that is their load is below a certain threshold (THR3) can be considered as target cells for reselection due to traffic reasons.

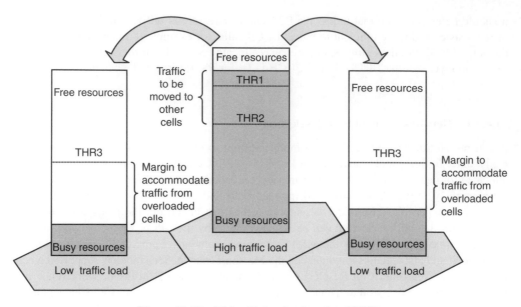

Figure 11.15 PS load balancing based on NCCR.

More complex algorithms based on real-time and non-real-time traffic load and cell capacity information may estimate the actual load level that every candidate cell would reach if it is selected as the target cell for a particular ongoing PS connection. Of course, the target cells considered by the load balancing algorithm need also to fulfill the criteria relating tominimum radio conditions. These are determined based on for example measurement reports sent to the network by theMS. There might be several different implementations as regards the procedure for the selection of terminals to be transferred to the neighboring cells. Some basic solutions are based on releasing the load of the serving cell by forcing a certain number of connections to be transferred to neighboring cells according to the power budget criteria. Some more advanced solutions are aware of the QoS profile of the ongoing PFCs. For instance, the terminals having PFCs established for a low delay-sensitive service class, such as, for example background, can be moved first to the neighboring cells. If this does not help to decrease the load to below the target level, the MSs with high QoS connections may become candidates for transfer to adjacent cells in the second step.

The Network Controlled Cell Reselection may also be triggered by a Quality Control function as a result of a QoS degradation as described in Section 11.4.4.

11.4.4 Countermeasures against Quality Degradation

Due to dynamics in the radio system such as the interference situation in the network or the influence of other factors such as possible hardware failures or CS traffic peaks that could degrade the PS traffic performance, the quality of the transmission has to be monitored constantly against the negotiated QoS. The Quality Control mechanism may then attempt to maintain connection quality by detecting degradation in service quality and reacting to it. Quality Control is in fact responsible for addressing all these problems that could not be addressed in an earlier phase, for example during the Admission Control or scheduling phase.

Quality Control (like Admission Control) should take care of all parts of the delivery chain, since only then can it be reliable. However, when considering Quality Control, the radio interface should be assumed to be the most critical part [1].

The Block Error Rate (BLER) of the TBF is one of the most critical performance indicators, for example BLER of an ongoing TBF could fall below a certain threshold even if the signal level is relatively high (for details of BLER calculations, refer to Section 11.4.2). Delay values can be assessed in an indirect way, for example with the help of constant BLER monitoring. In some cases, it is the variance of delay that is of interest and not the delay as such. Throughput as perceived by the user also has to be constantly controlled and a certain interaction with the scheduling algorithm may needed to be detected, for example a degradation in throughput. In the DL, throughput should be monitored in the BSS domain to ensure it does not drop below the guaranteed bit rate or actual incoming throughput from SGSN (if it is lower than the guaranteed bit rate [15]). For the UL, the situation is not as straightforward since the incoming throughput is not known by the network (actually only the UL PFC is known). For this reason, a throughput lower than the guaranteed bit rate does not necessarily mean there are bandwidth limitations across the delivery chain. An unacceptable degradation could therefore be detected, for example by comparing available throughput related to the UL PFC on the radio interface (assessed with the help of BLER evaluation and throughput recalculation for the particular coding scheme) and UL PFC throughput measured on the Gb interface. Assuming the radio interface is the bottleneck, then if these two throughput values are equal and lower than guaranteed bit rate, then it can be assumed the available bit rate in the UL is insufficient. On the other hand, if the available throughput is higher than the throughput on the Gb interface, it could be assumed the QoS parameters are satisfied. Other factors could be monitored as well, such as the mean bit rate per radio block (to determine the actual quality of the TBF which could be compared against the theoretical radio block capacity), or information about signal strength, etc.

If Quality Control faces a degradation in the bit rate or delay factors for one or more of the Packet Flow Contexts, then there are many possible countermeasures that can be applied, such as:

- reallocation of the TBF into different resources (e.g. TRX);
- network-controlled cell reselection and Inter-System NCCR;
- PS Handover together with Inter-System PS Handover;
- renegotiation of the QoS (including downgrading of the QoS traffic class, e.g. from streaming to interactive);
- dropping the PFC(s) (if there are no ways of ensuring that QoS requirements could be satisfied).

During the corrective actions performed by the Quality Control, the Admission Control functionality should also be utilized in order to assess the influence of allocating a PFC on a different resource.

It is evident that even though the Admission Control and Quality Control functions are distinct parts of the QoS-aware network, they should also be designed to be complementary to each other.

11.5   Performance of the QoS-aware GERAN Networks

A QoS-aware GERAN network controls the access to the resources by taking into account the QoS requirements of the incoming Packet Flow Context and all other Packet Flow Contexts that have already been established according to the rules presented in the previous subsections. This section presents system-level simulations that have been performed to show the performance of the typical QoS-aware RRM algorithms including Admission Control and Quality Control procedures in various network scenarios. To simplify the analysis, a regular network without CS traffic was simulated and only the downlink direction was taken into account. A hopping layer with 1/1 frequency reuse pattern (2 TRXs/cell in 900-MHz band) was analyzed.

To show the performance of QoS service differentiation procedures, the following three services were simulated in various network load conditions:

- H.263 video streaming based on real-time streaming class;
- WWW service based on non-real-time interactive class;
- FTP download based on non-real-time background class.

The guaranteed bit rate for the streaming service was set to 64 kbit/s – sufficient for the considered application. Two different scenarios were analyzed. In the first, an even distribution of the service penetration across the users was assumed. In the second scenario, the proportion of background, interactive and streaming application users was set to 1:1:3. Net session throughput was calculated as the mean quotient of the received data to the time where MS has a TBF (or the TBF was in the establishment phase). Figure 11.16 shows that even in a

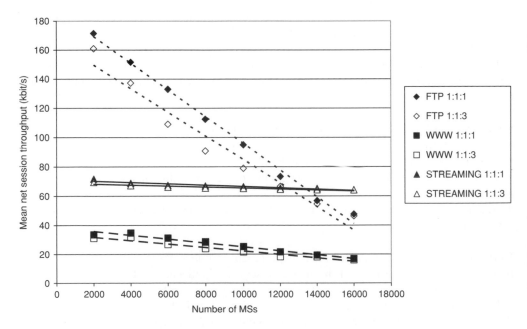

Figure 11.16   Throughput differentiation in QoS-aware network.

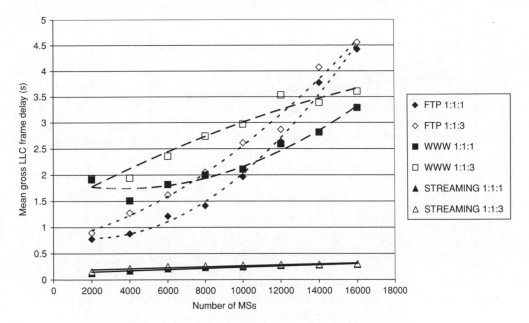

Figure 11.17   Delay differentiation in QoS-aware network.

highly loaded network, the mean throughput for the admitted streaming Packet Flow Contexts does not degrade below the guaranteed bit rate. The performance of the streaming service is sustained at the expense of a throughput deterioration experienced by the non-real-time users. Furthermore, the performance degradation for the background service can be seen to be significantly worse as a result of being handled with a lower priority than the interactive service. The influence of the network load on delay values is presented in Figure 11.17, where mean gross LLC frame delay is shown. It is defined as the time difference between storing an LLC frame in the PCU buffer and its correct reception by the MS. The lowest delay is perceived by the streaming application being served with the highest preference. Furthermore, the delay degradation with increased traffic load is barely visible for the streaming class while the delay values of the non-real-time services deteriorate significantly. The background service handled by the network with lowest priority suffers most from higher delay degradation when compared to the interactive one.

Moreover, it can be seen that the performance of the non-real-time services degrades more (in terms of both throughput and delay) for the scenario where the streaming application contributed more to the total offered load (1:1:3 vs. 1:1:1). This reflects the inherent behaviour of a QoS-aware network that prioritizes the real-time PFCs over non-real-time ones.

In Figures 11.18 and 11.19, the performance of the streaming service in a non-QoS-aware network is shown. It can be seen that without QoS management, the network is unable to provide satisfactory performance to the application even at moderate offered loads. One of the reasons for this phenomenon is the lack of AC and QC functionalities. Furthermore, in QoS-aware networks, the PFC throughput cannot exceed the guaranteed one if it affects the

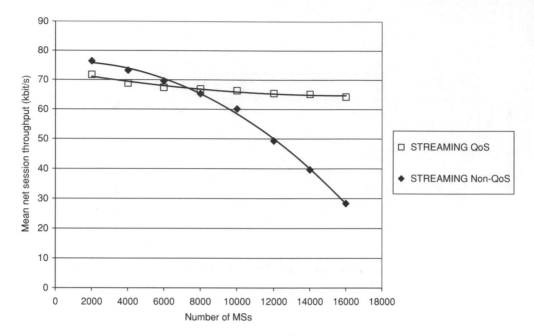

Figure 11.18   Throughput in non-QoS-aware network.

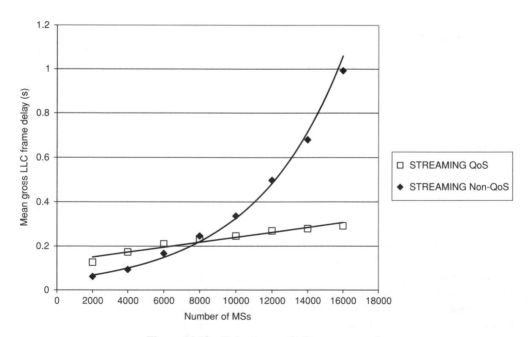

Figure 11.19   Delay in non-QoS-aware network.

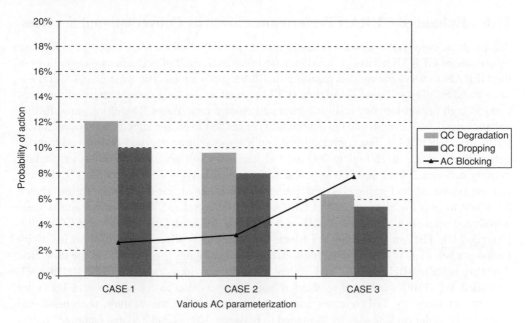

Figure 11.20 Relationship between Admission Control and Quality Control.

performance of other PFCs. In non-QoS-aware networks, there is no such mechanism and hence, higher throughputs and lower delays may be possible but only at low offered loads.

To achieve the expected network performance in terms of throughput and delay values, efficient Admission Control and Quality Control mechanisms are needed. Moreover, there is a clear relationship between the probability of blocking caused by the Admission Control and the probability of dropping or degradation performed by Quality Control. Prior to dropping or degradation, other countermeasures such as reallocation may be triggered by the QC. In the case of streaming services, degradation means that the PFC is downgraded to an interactive one. The actual Admission Control algorithm may be tuned according to different strategies. If the Admission Control is strict, the probability of service degradation or dropping during PFC runtime is lower. This could be achieved, for example with the help of applying certain safety margins. On the other hand, the number of admitted users is lower in comparison with more relaxed approaches towards Admission Control tuning. The effect of three different AC parameterization strategies in case of streaming services is depicted in Figure 11.20. Case 1 corresponds to the least strict AC parameterization, while in case 3 the most stringent approach was applied. The results were obtained for a heavy loaded network where the effect of Admission Control and Quality Control performance is especially visible.

The presented simulation results show that QoS-aware RRM is needed in order to achieve efficient service differentiation and to guarantee satisfactory performance for real-time applications in GERAN. Therefore modern, more advanced services may be offered by GSM operators. At the same time, the trade-off between quality and capacity should also be considered.

11.6   Enhanced GERAN Performance towards Conversational Services

Although the conversational traffic class is defined in 3GPP Release 99, its support in Release 99 compliant GERAN networks failed to materialize as a result of performance limitations in the GERAN to serve the most demanding real-time applications. The main issues were: too high transfer delay and round- trip time (RTT) as well as too long service interruption times and too high packet loss rate resulting from cell change procedures. Therefore, some further enhancements on the radio and Gb interfaces were needed to enable support of conversational traffic class in GERAN. The conversational applications require one-way end-to-end transfer delays in the range of 150 ms to 200 ms and packet loss of about 1–3%. However, delays ranging between 300 ms and 500 ms could still be considered as acceptable [14, 17]. Aiming to meet conversational traffic class requirements, a number of features have been introduced by 3GPP to improve performance in GERAN. In 3GPP Release 4, two new features were introduced, namely Extended UL TBF and Network Assisted Cell Change (NACC). With Extended UL TBF, an UL Temporary Block Flow may be kept active during short inactivity periods (when there is no data to be transferred). The analogous functionality in the downlink direction is called Delayed DL TBF Release, which was introduced in 3GPP Release 97. Extended UL TBF is intended to shorten both one-way end-to-end delays and RTTs by avoiding unnecessary TBF releases and establishments. With this feature, the end-to-end delay for IP audio packets may be shortened to between 320 ms and 370 ms. Detailed studies on the Extended UL TBF feature may be found in [14]. The NACC feature was introduced to improve cell change mechanisms and shorten service outage time during this procedure. Further details on this feature may be found in Section 11.6.1. However, service interruption times offered by GERAN in Gb mode with the NACC mechanism, although improved, are still unacceptable for conversational traffic class. In order to avoid Gb mode limitations and support conversational class services in GERAN, the Iu-mode for GERAN was standardized in 3GPP Release 5. The Iu-PS interface connects the 2G BSS to the 3G core network via 3G SGSN. With this architecture, GERAN subscribers can be offered the same services as UTRAN ones. This solution, however, requires Iu-mode capable mobile stations.

To further reduce the service outage time during cell change in GERAN Gb mode and to enable real-time services with stringent delay requirements, the PS handover feature was introduced in the 3GPP Release 6. The PS handover is intended to improve the performance of the cell change procedure by reusing the CS handover principle, that is by allocating resources to a mobile station in a target cell before a mobile station is moved to that cell. Although the PS handover is intended for real-time services, especially conversational traffic class, it is applicable to all PS services offered by GERAN Gb mode. Further details on the PS handover functionality can be found in Section 11.6.2.

3GPP Release 7 brought further enhancements aiming to improve QoS in EGPRS networks. Besides features such as Downlink Dual Carrier, EGPRS2 and MSRD, aimed at increasing the user perceived throughput (see Chapter 2 for further details), a set of latency-reducing mechanisms have been designed, namely: Reduced Transmission Time Interval (RTTI), RLC Non-Persistent Mode, Fast Ack/Nack Reporting and TBF pre-allocation [17]. In Reduced Transmission Time Interval, the transmit time of the single radio block is reduced from 20 ms to 10 ms by transmitting four bursts (comprising one radio block) in parallel onto two different time slots and hence using only two consecutive TDMA frames. By halving the transmission time, a significant reduction is provided in delay and round-trip time. In Fast

Ack/Nack Reporting, the bandwidth needed for feedback messages (Packet Uplink/Downlink Ack/Nack) is minimized to allow frequent acknowledgements reporting. This is realized by a very short Piggy-backed Ack/Nack report embedded in a normal RLC data block instead of typical PUAN/PDAN messages in dedicated radio blocks. In RLC Non-Persistent Mode, delays are reduced compared to typical acknowledged mode of operation, especially in severe radio conditions. With RLC acknowledged mode, correctly received data blocks cannot be relayed to the upper layer until all preceding ones have been successfully retransmitted. In RLC nonpersistent mode, if a configurable timer expires, the non-correctly received blocks are replaced with dummy information bits and together with all subsequent correctly received ones are forwarded to the upper layer. A detailed description of the RTTI, FANR and RLC NPM features can be found in Chapter 2. In TBF pre-allocation, a TBF may be established before the user data is ready for transmission and kept open by the network via the Extended UL TBF feature. Thanks to TBF pre-allocation, the TBF set-up time is removed from the service initial access time. The combination of all the delay and latency-reducing features enables GERAN to fulfill the QoS requirements of the most demanding real-time services and to support all the traffic classes, defined in the 3GPP Release 99, including the conversational one.

11.6.1   Network Assisted Cell Change

It has already been mentioned that the "conventional" cell reselection algorithm is unsuitable for services that require seamless cell change operation, such as real-time services. This is because "conventional" cell reselection is performed by the terminal without notification to the network. Furthermore, the mobile cannot make a random access to restart the data transfer in the new cell until a consistent set of system information for that cell has been correctly received. In other words, an MS that performs cell reselection must first acquire a defined minimum set of (P)SI in the target cell before resuming the TBF in the new cell. The service interruption time that arises during cell reselection is therefore dependent on how quickly the MS can acquire the minimum set of system information in the new cell. This procedure can take a couple of seconds, which makes "conventional" cell reselection unsuitable for the more delay-sensitive service classes like, for example streaming.

To support these services in GERAN, enhancements to the existing cell change mechanism are necessary. To this end, the Network Assisted Cell Change feature was introduced in 3GPP Release 4 for MSs, performing cell changes between cells belonging to the same BSC. The feature can reduce the service outage time for an MS in packet transfer mode from a couple of seconds down to even below one second by giving the network the possibility of assisting an MS before and during the cell change. The assistance is given both by the sending of neighboring cell system information and with introduction of new procedures that will be briefly described in the later part of this section.[3] The NACC feature introduced in 3GPP

---

[3] This sentence and the previous one are a paraphrase of 3GPP TS 44.901 v8.0.0, section 4.2 sentence "The feature reduced the service outage time for an MS in packet transfer mode from a couple of seconds down to 300–700 ms by giving the network a possibility to assist the MS before and during the cell change. The assistance is given both as sending of neighbouring cell system information and with introduction of new procedures." ©2008. 3GPP[TM] TSs and TRs are the property of ARIB, ATIS, CCSA, ETSI, TTA who jointly own the copyright on them. They are subject to further modifications and are therefore provided to you "as is" for information purposes only. Further use is strictly prohibited.

Release 4 covers only the case of intra-BSC cell reselection. In order to support NACC in the case of inter-BSC cell changes, the RAN Information Management (RIM) protocol was introduced in 3GPP Release 5. The task for RIM in the context of the NACC feature is to support the retrieval of the relevant system information data of cells managed by neighboring BSCs and to provide this information for the NACC supported cell change procedure [4, 12].

The principle of the Network Assisted Cell Change feature is to minimize the service outage time for all QoS traffic classes when an MS in packet transfer mode moves between GSM cells. The outage time reduction is achieved by giving the network the possibility to assist an MS before and during cell reselection.[4] As mentioned before, there are two independent mechanisms specified for NACC.

The first, called Neighbor Cell System Information Distribution, allows the network to assist an MS by providing the neighboring cell system information required for initial access after a cell change. This information is sent to the mobile station on the PACCH in the source cell. An MS in packet transfer mode that receives this information stores it for 30 seconds. It may then use this information when accessing a target cell and hence avoid scanning the BCCH channel for the necessary system information. Having switched to a new cell, the MS may directly initiate the access procedure via RACH and then ask for the missing system information on the PACCH by using the Packet SI Status procedures.

The second mechanism provides the network with the means to order mobiles of 3GPP Release 4 or later to enter so-called Cell Change Notification (CCN) mode before reselecting a cell. In CCN mode, an MS notifies the network about when cell reselection criteria are fulfilled and delays the execution of cell reselection so that the network can respond with neighbor cell system information. Having entered the CCN mode (Figure 11.21 depicts the message flow in CCN mode), the MS informs the network with a Packet Cell Change Notification (PCCN) message that it wants to change a cell. After sending the message to the network, the mobile station continues the ongoing packet transfer for either a maximum time of about 1 second or until the network responds with a Packet Cell Change Continue or Packet Cell Change Order message. The Packet Cell Change Order message might be sent to indicate a different target cell from the one proposed by the mobile station (e.g. when the target cell proposed by MS is in a congestion status), or the Packet Cell Change Continue message might be sent to force the MS to continue the cell reselection procedure to the target cell proposed by the MS. Upon reception of the PCCN message, the network may send the required system information about the proposed or any other cell selected by the network in a Packet Neighbor Cell Data message to allow the MS to perform an immediate initial access in the target cell when the re-selection has been performed.

By providing (P)SI information on the target cell still on a dedicated link in the source cell, the service interruption time due to cell reselection can be significantly reduced. There are a few different performance indicators providing information about the service interruption time caused by the cell reselection. Figure 11.22 shows one of them, namely Data Outage

---

[4] This sentence and the previous one are a paraphrase of 3GPP TS 44.901 v8.0.0, section 4.3 sentence, "the Network Assisted Cell Change feature introduced in Release 4 is a tool to minimize the service outage time for all QoS classes when a GPRS MS in packet transfer mode moves between GSM cells belonging to the same BSC." © 2008. 3GPP™ TSs and TRs are the property of ARIB, ATIS, CCSA, ETSI, TTA who jointly own the copyright on them. They are subject to further modifications and are therefore provided to you "as is" for information purposes only. Further use is strictly prohibited.

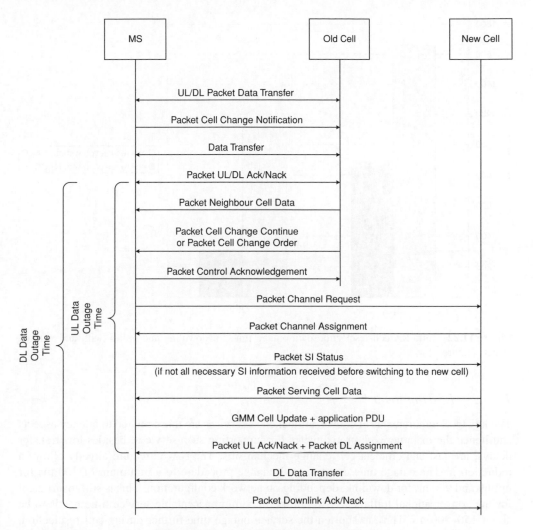

Figure 11.21 Intra-RA NACC message flow.

Time defined as the period between reception of the last Packet Downlink/Uplink Ack/Nack message in the old cell and reception of the first Packet Downlink/Uplink Ack/Nack message in the new cell. In this figure, the Data Outage Time is given for an intra-Routing Area scenario. Note that in case of inter-RA cell change, the service interruption times may get longer owing to the RAU procedure during which an application data transfer is not possible.

As depicted in Figure 11.22, the NACC feature can significantly reduce the data outage time arising from the cell change procedure compared to "conventional" cell reselections. Consequently, more delay-sensitive applications, like streaming ones, may be enabled in GERAN. However, to meet the delay requirements of conversational-like services, the service interruption times during cell change procedures need to be further reduced.

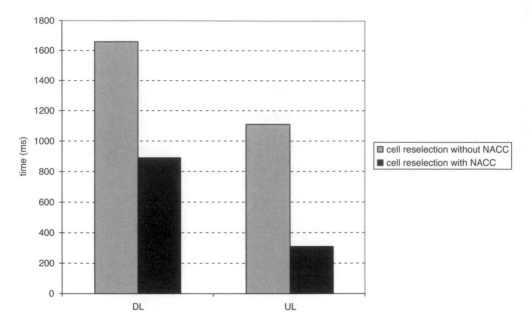

Figure 11.22 Intra-RA cell reselection data outage time for downlink and uplink with and without NACC feature.

11.6.2   PS Handover

The key QoS requirements of conversational class services are those related to latency aspects and hence the main network prerequisites for supporting such services, besides low transfer delays, are fast and efficient cell change mechanisms. The NACC feature already allows a reduction in data outage time during a cell change procedure to a minimum of 300 ms for uplink and 900 ms for downlink, depending on network configuration. This is still insufficient for the conversational traffic class where the maximum acceptable service time needs to be about 150–200 ms. Thus, to shorten the service outage time further during cell reselection, Packet Switched Handover for GERAN A/Gb mode was introduced in 3GPP Release 6. Although introduction of this feature was driven by real-time services, it is applicable to all types of PS applications. PS Handover improves the performance of the packet cell reselection procedure towards the performance of circuit-switched handover by allocating resources in the target cell before the mobile station leaves the old cell. New mechanisms supporting handover of BSS Packet Flow Context between source and target cells have also been introduced on the Gb interface. In this case, new BSSGP signalling messages that carry information related to BSS PFC being handed over, including corresponding QoS profiles, have been defined. The PS Handover functionality also provides mechanisms for data forwarding which reduce packet losses during a cell change.

The PS Handover procedure can move an MS with one or more packet flows from a source cell to a target cell, where the source and target cells can be located either within the same BSS (Intra-BSS HO), between different BSSs within the same SGSN (Intra-SGSN HO), or between different SGSNs (Inter-SGSN HO), or even between different radio access technologies

(Inter-RAT HO). In the case of the PS Handover, the whole cell change procedure is controlled by the network, with the decision about whether and towards which cell a terminal should be handed over being based on for example measurement reports received from a mobile station. Before an MS is ordered to change cell, the network needs to reserve the appropriate resources in a target cell and depending on whether a target cell belongs to the same or different BSS or SGSN, then the network may involve different procedures in the HO preparation phase [9]. The exemplary message flow for the intra-SGSN inter-BSS PS Handover procedure is depicted in Figure 11.23.

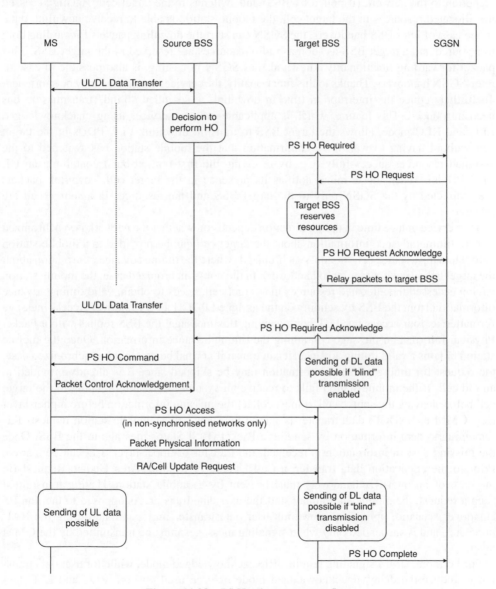

Figure 11.23   PS Handover message flow.

As soon as radio resources in a target cell are allocated, the mobile station is commanded to make a cell change (PS Handover Command message), where the network provides the MS with a description of the allocated resources and with the system information needed for access in a target cell. The provisioning of part of the target cell system information before the cell change allows the mobile station to access the assigned resources without having to perform the signaling phase on common control channels. This significantly reduces service outage time (in the order of couple of seconds compared to legacy cell change procedure). On receiving the PS Handover Command message, the mobile station acknowledges its correct reception to the network (if polled by BSS) and switches to the target cell. Having accessed the allocated resources in the target cell, the mobile station is able to receive downlink data. In the case of inter-BSS handovers, the SGSN can start the downlink packets forwarding both to a source and a target BSS as soon as radio resources are reserved in the target BSS. This packet forwarding functionality, known also as SGSN bi-casting, is also possible in case of inter-SGSN handovers. Thanks to this functionality, the service outage time may be shortened. To further reduce the interruption time in downlink, a so-called "blind" transmission has been introduced. This feature, which is applicable only to services using unacknowledged LLC and RLC mode, allows the target BSS to transmit downlink LLC PDUs in the target cell without having to wait for a confirmation that the mobile station has switched to the assigned resources successfully (e.g. by receiving the first normal burst containing an UL LLC PDUs). When the terminal notifies its presence in the target cell, downlink packets are redirected by the SGSN only to the target BSS and the resources in a source cell are released.

The service outage time in uplink direction depends on whether the network is synchronized (where timing advance information about the target cell can be provided to a mobile station before leaving a source cell) or non-synchronized (where the timing advance information about the target cell is unknown prior to handover). In the non-synchronized case, the mobile station having accessed the allocated resources in a target cell, needs to obtain valid timing advance information from the BSS by sending on the assigned PDCH a PS Handover Access message formatted as four access bursts. Upon receiving this message, the BSS replies with a Packet Physical Information message containing the timing advance information. Once the mobile station obtains a valid timing advance, it can transmit normal bursts. In the synchronized case, the request for timing advance information may be skipped since it is already provided in an old cell. If the mobile station fails to receive all system information relating to the target cell before leaving the source cell, it must collect the missing information before higher-layer (e.g. GMM or SNDCP) data transfer is started in a new cell. A mobile station requests this remaining system information by sending a Packet (P)SI Status message to the BSS. Once the missing system information is received, the transfer of upper-layer data can be started. Prior to the application data transfer, a Cell Update or Routing Area Update Request (in the case of an inter-RA handover) must be sent by a mobile station. Having transmitted such a request, the mobile station may start the user data transfer. As opposed to the non-PS Handover scenario, uplink and downlink user data transfer may continue during the RAU procedure and is suspended only when signaling messages must be transmitted by the GMM layer.

Furthermore, GMM signaling requires RLC acknowledged mode, while for user data transfer, to reduce the delay, unacknowledged mode may be used on both RLC and LLC layers. Without the Multiple-TBF feature (allowing an MS to have more than one TBF active simultaneously), the RLC mode would in this case need to be changed from unacknowledged to acknowledged mode. This means the ongoing TBF must be released and a new one established which would lead to an impact on the service interruption time. Considering that in case of an ongoing the DL TBF, an UL TBF establishment may only be requested when sending a DL TBF polling response, the service outage time would be dependent on downlink polling algorithm.

In the following system-level simulation results [18], PS handover performance (Figure 11.23) is given in terms of service interruption time for a conversational type of application (VoIP). The used network simulator modelled all of the relevant details affecting the service outage time starting from the physical layer up to radio access network and core network level (cell deployment, frequency reuse, mobility, Routing Area Update, etc.). A simple model of VoIP speech is used as the traffic model (where the data frame is generated every 20 ms, the size of the frame is matched to the predefined bit rate of 8kb/s and the call arrival rate is 5 calls/hour/terminal). Only one time slot in UL and one time slot in DL were allocated to a mobile station having an active call. No multiplexing was allowed, so only one mobile is allocated on each time slot. Extended UL TBF mode and Delayed DL TBF Release were applied, but not Multiple-TBF. The service outage time is measured at LLC layer. To measure the most significant delay that may occur during PS Handover, two different definitions were used depending on whether the intra-RA or inter-RA scenario was analyzed. In the case of intra-RA PS handover, the largest service interruption time occurs when switching between the source and the target cells and this was defined as the period between the reception of the last correct LLC data block in the old cell and the reception of the first correct LLC data block in the new cell. In the case of inter-RA PS handover, the most significant service outage time is when GMM signaling messages are transmitted in the target cell and this was defined as the time between the reception of the last LLC data block before the first mobility management message is sent and the reception of the first LLC data block after the mobility management message is received. The first GMM message is the ROUTING AREA UPDATE REQUEST in uplink and the ROUTING AREA UPDATE ACCEPT in downlink.

The service interruption time introduced by PS handover on the LLC layer for downlink and uplink LLC frames for the intra-RA scenario is depicted in Figure 11.24.

When SGSN bi-casting is enabled, the typical service interruption time is about 20–60 ms in downlink and 60–80 ms in uplink. These service interruption times are significantly improved over those introduced by the NACC procedures and fully satisfy the delay requirements of conversational traffic class services. The presented results are for a synchronized network scenario with SGSN bi-casting enabled. Without SGSN bi-casting, the outage time in the downlink might extend to 80–140 ms for the synchronized network and 120–180 ms for the non-synchronized network, while uplink service interruption times in the non-synchronized network extend to 80–120 ms [18].

The service interruption time introduced by Routing Area Update procedures on the LLC layer for downlink and uplink LLC frames for the case of inter-RA PS handover is depicted in Figure 11.25.

In the case of an inter-RA PS handover, the service interruption times are longer when compared to the intra-RA scenario and are about 400 ms in downlink and 700 ms in uplink. This is caused by the need to change to RLC acknowledged mode to send the GMM message. The presented results for uplink were obtained with a polling algorithm optimized for fast uplink

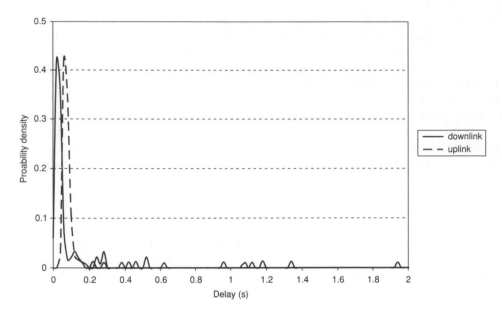

Figure 11.24 Intra-RA PS Handover service interruption time for downlink and uplink LLC frames transfer – synchronized network with SGSN bi-casting. Reproduced by permission of Rexhpi V., Bohaty D., Harniti S., Sebire G., "Handover of packet-switched in GERAN A/Gb mode" Global Telecommunication Conference 2005, GLOBECOM'05, IEEE. © 2005.

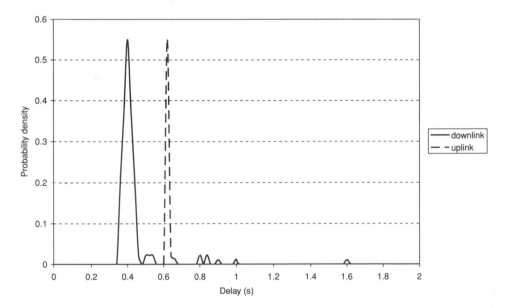

Figure 11.25 Inter-RA PS Handover service interruption time for downlink and uplink LLC frames transfer – optimized polling algorithm. Reproduced by permission of Rexhpi V., Bohaty D., Harniti S., Sebire G., "Handover of packet-switched in GERAN A/Gb mode", Global Telecommunication Conference 2005, GLOBECOM'05, IEEE. © 2005.

TBF re-establishment. These service interruption times are not sufficient for the conversational traffic class services, even though they represent a significant improvement compared to the inter-RA cell reselection procedures even with the NACC feature (where typical outage time is in the order of few seconds). To reduce inter-RA PS handover service interruption times to acceptable levels for conversational traffic class services, the Multiple-TBF feature which avoids a TBF mode change for GMM signaling transfer is necessary [18].

The results show that the PS Handover feature can reduce the service interruption time significantly compared to a pre-Release 6 GERAN A/Gb mode cell change mechanism. For this reason, the introduction of PS Handover functionality is a major step towards support of the conversational traffic class services in GERAN networks.

References

[1] T. Halonen, J. Romero and J. Melero, *GSM, GPRS and EDGE Performance*, 2nd edition, Chichester: John Wiley & Sons, Ltd, 2003.

[2] G. Gomes and R. Sanches, *End-to-end Quality of Service over Cellular Networks*, Chichester: John Wiley & Sons, Ltd, 2005.

[3] 3GPP GP-032231 "Introduction of Mobile Station Multislot Power Classes", August 2003.

[4] 3GPP TR 44.901 v8.0.0, "External Network Assisted Cell Change (NACC)", December 2008.

[5] 3GPP TS 22.060 v3.5.0, "General Packet Radio System (GPRS)", Service Description, Stage 1, October 2000.

[6] 3GPP TS 23.060 v3.17.0, "General Packet Radio System (GPRS)", Service Description, Stage 2, December 2006.

[7] 3GPP TS 23.107 v3.9.0, "Quality of Service (QoS) Concept and Architecture (Release 1999)", September 2002.

[8] 3GPP TS 23.207 v5.10.0, "End-to-end Quality of Service (QoS) Concept and Architecture (Release 5)", September 2005.

[9] 3GPP TS 43.129 v8.1.0, "Packet-switched Handover for GERAN A/Gb Mode", Stage 2, February 2009.

[10] 3GPP TS 44.060 v8.5.0 "Radio Link Control/Medium Access Control (RLC/MAC) Protocol", May 2009.

[11] 3GPP TS 45.002 v8.0.0 "Multiplexing and Multiple Access on the Radio Path", December 2008.

[12] 3GPP TS 48.018 v8.3.0, "BSS GPRS Protocol (Release 8)", May 2009.

[13] C.F. Ball., K. Ivanov and F. Treml, " Contrasting GPRS and EDGE over TCP/IP on BCCH and non-BCCH Carriers", in *IEEE VTC*, 2003.

[14] C.F. Ball, C. Masseroni and R. Trivisonno, "Introducing 3G-like Conversational Services in GERAN Packet Data Networks", in *VTC*, 2005, Spring, 30 May–1 June 2005.

[15] D. Fernandez and H. Montes, "An Enhanced Quality of Service Method for Guaranteed Bitrate Services over Shared Channels in EGPRS Systems", in *IEEE, VTC*, 2002.

[16] K. Ivanov, C.F. Ball and F. Treml, "GPRS/EDGE Performance on Reserved and Shared Packet Data Channels", in *IEEE*, 2003.

[17] F. Lironi, C. Masseroni, S. Parolari and E. Vutukuri, "Provision of Conversational Services over the GERAN: Technical Solution and Performance", in 18th Annual IEEE International Symposium on Personal, Indoor and Mobile Radio Communication, PIMRC, 2007.

[18] V. Rexhepi, D. Bohaty, S. Harniti, and G. Sébire, "Handover of Packet-Switched Services in GERAN A/Gb Mode", in *Global Telecommunication Conference GLOBECOM'05*, IEEE, 2005.

12

Capacity Enhancements for GSM

Kolio Ivanov, Carsten Ball, Robert Müllner, Hubert Winkler,
Kurt Kremnitzer, David Gallegos, Jari Hulkkonen, Krystian Majchrowicz,
Sebastian Lasek and Marcin Grygiel

12.1 Introduction

In this chapter, we introduce some of the proprietary solutions that are available to improve
the performance and capacity of a GSM/EDGE network.

First in 12.2, we describe Progressive Power Control, a technique that enhances power
control in AMR by adapting the thresholds that govern it resulting in lower interference and
a reduced dropped call rate. Then in 12.3 we describe Temporary Overpower, a technique
that targets the associated control channels in order to increase retainability in the network.
Retainability is also enhanced by minimising the size of the signalling commands at call
set-up and during handover. This is described in 12.4. Other optimizations are also described
in 12.5 for the Radio Link Timeout parameter, in 12.6 on the criteria for performing intra-cell
handovers and in 12.7 on service dependent channel allocation based on a multiple frequency
reuse scheme.

Finally in 12.8, proprietary solutions to the GERAN transport system are described, namely
efficient IP transport techniques that contribute to savings in terms of bandwidth and cost.
Different migration paths for GERAN transport evolution are also outlined.

12.2 Progressive Power Control for AMR

Progressive Power Control (PPC) is a technique that improves both the quality and capacity
of an AMR network by minimizing the use of high transmission power, while at the same
time maximizing the use of the high bit-rate codec modes. PPC can also be used to address
the imbalance in quality between the control and AMR traffic channels.

In this section, we provide an overview of the PPC technique and provide an evaluation of
its gains with the help of system level simulations and laboratory test measurements.

GSM/EDGE: Evolution and Performance   Edited by Mikko Säily, Guillaume Sébire and Eddie Riddington
© 2011 John Wiley & Sons, Ltd

12.2.1   Introduction

Power Control (PC) reduces interference in a network by adjusting transmission power levels based on radio link measurements. For example, PC might be used to optimize the transmission power in order to maintain the signal level such that a quality criterion for the connection is only just met. In [1], GSM power control is studied together with DTX in a variety of network configurations and it is clearly evident that PC improves network performance both with or without DTX or frequency hopping.

When applied to GSM voice, power control algorithms are commonly based on thresholds to the reported RxQual and RxLev parameters.

For example, high and low RxQual thresholds are often used to define a "target" RxQual (corresponding to a value between the high and low thresholds). Provided the maximum or minimum transmission power is not exceeded, the quality of a connection is kept to within the target quality criterion.

AMR link adaptation adapts the codec mode according to a signal quality criterion and as a consequence, the AMR codec mode usage should correlate to the targetted RxQual. For example, Figure 12.1 shows an example of AMR codec mode usage versus RxQual, based on results obtained from a test measurement (whose set-up is described below). When RxQual exceeds 4, the most robust AMR mode (AFS 4.75) is used; AFS 5.9 is used when RxQual is within the range 2 to 4, and the higher speech bit rates (AFS 7.4 and AFS 12.2) are used in the case of best two quality classes. In this configuration, a target RxQual in excess of 4 would push the AMR codec to the most robust modes, while a target RxQual less than 2 would favor the high bit-rate codec modes.

Progressive Power Control provides added flexibility to the basic power control method. This flexibility can be used to maximize the potential gains of AMR. Connections that use the highest power levels cause interference other users hence, from a system performance point of view, it is better to push the connections in poor radio conditions towards the robust

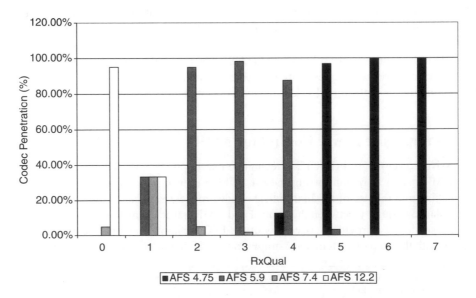

Figure 12.1   RxQual vs. AMR codec mode usage in an example test measurement.

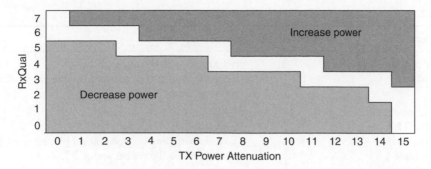

Figure 12.2 RxQual threshold adaptation.

modes of AMR, instead of increasing the transmission power to the highest levels. On the other hand, connections in very good radio conditions should be allowed to achieve maximum voice quality by using the highest speech bit rate modes because the highest power levels are not required. With PPC, this is done by adapting the RxQual thresholds in the power control algorithm. For example in Figure 12.2, the high and low RxQual thresholds change dynamically depending on the transmission power level. If high power is already in use, the quality threshold is tightened. Thus, connections requiring the highest power levels are pushed to the most robust modes of AMR. The target RxQual thresholds are then lowered, step by step as a function of transmission power until users in good radio conditions are permitted to use a transmission power level to achieve the highest quality speech.

PPC performance was studied using laboratory tests and system level simulations. The laboratory test scenario in Figure 12.3 depicts a 2G network where an Abis protocol analyzer and drive test tools collect appropriated log and traces from measurements results and where a signal generator provides the interference source to emulate the CIR conditions that trigger the power control algorithm thresholds.

The test case configuration consisted of two TRXs and frequency hopping was disabled. The reference PC RxQual thresholds were: AFS (4, 2) and AHS (2, 0), where the first and second number in the parenthesis corresponds to low and high quality thresholds respectively. PPC adapted the thresholds in the AFS case from (6, 4) to (2, 0) and in the case of AHS from (4, 2) to (2, 0). The AMR codec set consisted of the modes: AMR12.2, AMR7.4, AMR5.9 and AMR4.75 (except for the AMR12.2 mode, which is available only for AFS).

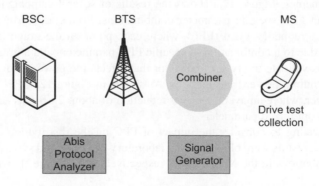

Figure 12.3 Laboratory test configuration for Progressive power control.

Table 12.1   Essential parameters of simulation model

Parameter	Value
Number of cells	>100 cells with 65 degree antennas
Re-use pattern	customized ad-hoc frequency planning (each sector with specific MA List)
Available spectrum	7 MHz
Number of TRX per cell	approximately 4TRX per sector
Radio conditions	taken into account by tuned model propagation
Traffic volume	realistic traffic figures per sector
AMR penetration	100%
Channel mode	TCH/AFS, TCH/AHS
Active Codec Set	TCH/AFS12.2, TCH/AFS7.4, TCH/AFS5.9, TCH/AFS4.75, TCH/AHS7.4, TCH/AHS5.9, TCH/AHS4.75
Link adaptation activated	Activated
PC RxQual lower thresholds for AFS	4
PC RxQual lower thresholds for AHS	2
PC RxQual upper thresholds for AFS	2
PC RxQual upper thresholds for AHS	0
PPC parameters:	RxQual classes max. 2 steps up and down compared to reference PC set-up

A dynamic, time-driven network level simulator for GSM/EDGE was utilized to evaluate network level capacity and quality gains (further details about the simulator environment can be found in Chapter 5). The simulated scenario was an irregular network in an interference limited macro environment. The essential parameters of the simulation are in Table 12.1.

12.2.2   PPC Impact on Power Levels

Power level distributions change when PPC is applied such that high power transmissions are avoided by adapting the RxQual thresholds that govern the power control mechanism. Instead of powering up, the robust modes are used leading to a reduction in interference. Use of maximum output power is expected to be less frequent than those ones observed in normal power control schemes. Figure 12.4 shows the results of several campaigns of simulations based on different PPC specific parameter combinations. Points are vertically grouped for a given Effective Frequency Load (EFL), where each point on the graph is the result of a given simulation due to a combination of specific PPC parameters and a given network load. These simulations validate the type of behavior that affects the power level distribution and the final average output power transmissions. As shown in Figure 12.4, PPC resulted in less transmission power, where an average power reduction of about 2 dB was achieveable in the case of PPC with optimized parameters.

It is also interesting to consider the impact of PPC on the distribution of output power levels. In this case, results were obtained from laboratory test trials and system simulations for power level distributions in the MS and BTS respectively. Results are shown in Figure 12.5 and Figure 12.6.

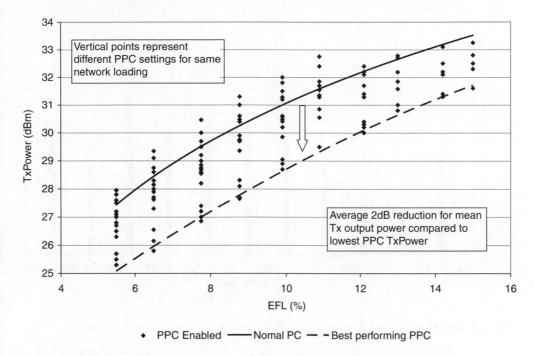

Figure 12.4 Mean TX Power figures comparison: Reference Case and multiple PPC settings.

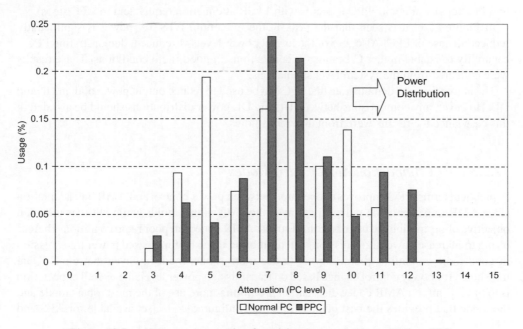

Figure 12.5 MS Power Attenuation from laboratory tests: Normal PC vs. PPC.

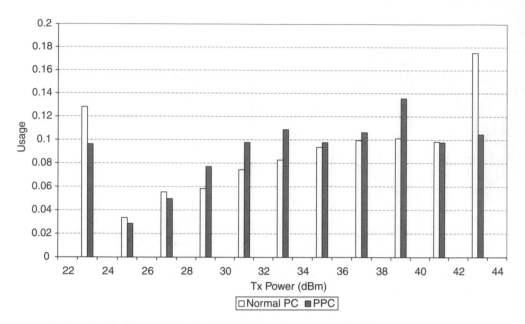

Figure 12.6   Normal PC vs. PPC BTS Transmit power distribution from simulations.

In Figure 12.5, it is seen that with PPC there is less utilization of high power levels. With normal PC, the highest peak of power transmission occurs with 10 dB attenuation (equivalent to 5 PC steps) while with PPC it occurs with 14 dB attenuation (equivalent to 7 PC steps).

In Figure 12.6, it is seen that the peak in the maximum BTS TX power is significantly reduced in case of PPC. Also, use of the lowest power levels is reduced, demonstrating PPC's capability to enable higher C/I ratios for connections in good radio conditions. These results were collected from the system level simulations.

The above findings demonstrate that PPC can be used to reduce output power both in UL and DL. However, any comparisons between UL and DL power distributions should be avoided in this case since the results were from different studies.

12.2.3   PPC Impact on AMR Codec Modes

In practical terms, PPC improves cooperation between power control and AMR link adaptation since AMR link adaptation will adapt to meet the RxQual thresholds while PPC has the main objective of promoting the use of the most robust AMR codecs in poor radio conditions (before trying to increase power transmission) and promoting the use of increased power transmission in good radio conditions (before trying to use more robust modes). Given that the RxQual thresholds for power control change based on transmission power level, the overall expectation is to have a shift on AMR codec distribution promoting more use of the most robust mode and the mode that provides the best quality from the configured codec set available for AFS and AHS.

Figure 12.7 depicts AMR codec usage taken from the field trial measurements. The trail in this case was from a high interference scenario and therefore useage of the most robust AMR

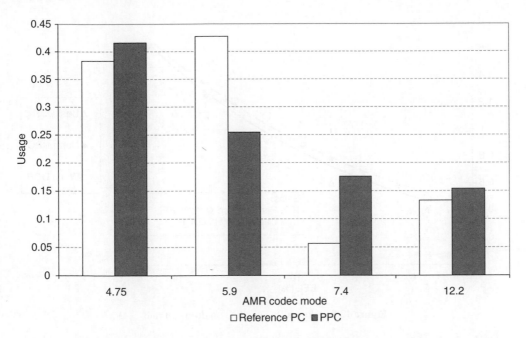

Figure 12.7   An example AMR Codec Distribution comparison between reference PC and PPC cases.

codec mode is relatively high. It can be seen that the use of the most robust mode is slightly increased. Then, there is clear increment in the best two speech quality AMR codec modes. This is a good example of the basic principle of PPC, how it can be used to push connections to more robust modes, and, at the same time favor use of the highest codec mode in good radio conditions.

12.2.4   Quality and Capacity Gains

This section depicts the potential capacity and performance gains obtained by progressive power control. Network capacity and quality have been evaluated using DCR and FER (measured as bad quality FER samples) respectively. Since the goal is to reduce high power transmission and thus overall interference, the benefits of PPC should also be evident on FACCH and SACCH performance behavior. Specifically, a reduction in the drop call rate due to the handover process and due to the radio failure perceived by radio link timeout is expected.

Figures 12.8 and 12.9 depict the DCR and FER performance for the real network simulation scenario described in Section 12.2.1 when configured for different combinations of PPC parameter settings. Shown are the results obtained for the best and worse PPC settings and for normal PC (shown for reference). For the best combination, a DCR quality gain of 19% or a capacity gain of 12% was achieved (Figure 12.8). These gains were obtained due to a reduction of interference derived from less utilization of maximum output power. The impact to the AMR codec mode distribution resulted in an impact to the FER samples, where a slight degradation can be seen for the best PPC settings (Figure 12.9). This demonstrates that the PPC algorithm can be used to balance quality difference between control and traffic channels, and same time increase network capacity.

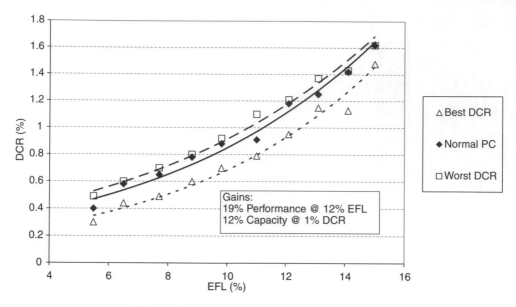

Figure 12.8   Normal PC vs. PPC on drop call rate.

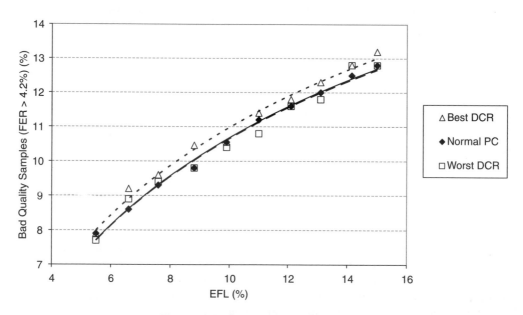

Figure 12.9   Normal PC vs. PPC on FER.

12.3 Temporary Overpower[1]

In the past number of years, a series of voice quality and capacity enhancement features have been introduced in narrowband deployments of GERAN (GSM/EDGE Radio Access Networks) mobile radio networks such as the Adaptive Multi-Rate (AMR) speech codec. Findings from field measurements have revealed that the Call Drop Rate (CDR) is the major limiting factor even at medium load in networks exploiting AMR with tight frequency reuse, due to excessive failures on the fast and slow associated control channels (FACCH/SACCH). The reason for this phenomenon which is typically observed at Effective Frequency Loads (EFLs) beyond 20%, is the significant physical layer performance imbalance of up to 6 dB in terms of carrier to interference ratio (C/I) between the robust TCH/AFS4.75 / 5.90 kbps speech codecs and the signaling channels. A novel strategy using temporary overpower on the control channels was introduced as a practical and fully backwards compatible option to improve the probability of successful decoding of the FACCH/SACCH frames. The evaluation of the novel approach by system level simulations shows a significant soft capacity gain by substantially reducing the CDR, enabling a system load above 30% EFL in homogeneous hexagonal networks. The temporary overpower strategy is especially attractive when extremely high network load situations temporarily occur.

In such conditions, a speech quality degradation down to a still acceptable level is often tolerated when providing the required capacity during busy hours load peaks. Moreover, the temporary overpower on FACCH/SACCH can be efficiently combined with the 3GPP Release 6 standardized features Repeated Downlink FACCH and Repeated SACCH.

12.3.1 Introduction

Originally in GSM the robustness of the signaling FACCH/SACCH channels was properly designed to match the error protection schemes of the standard full-rate and half-rate speech traffic channels (TCH/FS, TCH/HS). Signaling performance was sufficient to cope with the interference levels in early GSM deployments utilizing standard or enhanced full-rate speech (TCH/EFS) codecs even in tight 1/3 or 1/1 frequency reuse at a system load appropriate for those codecs [2–6]. Under poor radio conditions connections were terminated due to perceived bad speech quality by one of the parties before the call was dropped due to signaling errors such as for example radio link timeout on SACCH.

The robustness of the GSM traffic channels has been substantially increased by the introduction of AMR speech codecs capable of adapting the relative amount of speech and channel coding to the prevailing radio conditions. Robust AMR codec modes in full-rate, for example TCH/AFS4.75 and TCH/AFS5.90, provide sufficient speech quality even at very low C/I levels.

Today's state-of-the-art GERAN networks exploiting the robustness of AMR are deployed in aggressive 1 × 1 frequency reuse, offering high capacity. Since the robustness of the control channels FACCH and SACCH, in the following referred to as ACCH, has not been improved until recently, the high interference levels in such types of networks substantially degrade the

---

[1] This section is based on "Breaking through AMR Voice Capacity Limits Due to Dropped Calls by Control Channel Improvements in GERAN Networks", by Kolio Ivanov, Carsten F. Ball, Robert Müllner, Hubert Winkler, René Perl and Kurt Kremnitzer which appeared in *EEE PIMRC*, Berlin © 2005.

signaling performance especially in downlink (DL). In the uplink (UL) direction interference cancellation is used to improve also the performance of the signaling channels. While the speech quality is still acceptable, important control information for timing advance, power control (PC) and handover (HO) is not available. Even at reasonable speech quality provided by TCH/AFS4.75 codec mode, both SACCH and FACCH show a significant block erasure rate (BLER): radio link failure (RLF) is often observed due to successively bad decoding of SACCH blocks. The HO command sent on FACCH does not reach the mobile station (MS) in the serving cell. The physical layer information (containing the timing advance) sent on DL FACCH does not reach the mobile station in the target cell. All those events are the main reason for a dramatic increase in the number of dropped calls, limiting the network capacity well below the soft capacity limit for TCH. The analysis of the call drop statistics in real networks planned in tight reuse operated at medium to high fractional load also shows that the vulnerability of ACCH is the limiting factor for the network capacity.

There are two approaches to address ACCH performance enhancements in DL:

1. By improving the error protection schemes for FACCH and SACCH, implying changes in the 3GPP GERAN standard.
2. By improving the C/I on the ACCHs, applying temporarily higher transmission power on FACCH and SACCH bursts.

Regarding the first approach, the optional features called "Repeated Downlink FACCH" and "Repeated SACCH" were defined in the 3GPP GERAN specifications [6–8] for Release 6 compliant MS, capable of performing soft combining on two successively repeated FACCH/SACCH blocks. The MS indicates the support of these features via the MS Classmark 3 (CM3) Information Element called Repeated ACCH Capability (1 bit field) as defined in [9]. This permits their early implementation for pre-Release 6 compliant MSs. The repeated DL FACCH/SACCH procedure foresees a serial repetition of the FACCH/SACCH blocks. The MS can then enhance its link performance by attempting a soft decision combining of the repeated FACCH/SACCH blocks. The performance requirements for repeated FACCH/SACCH are defined in [10] with a reference performance of FER $\leq$ 5%. According to link-level simulation results, the expected gain in decoding performance is about 3–5 dB. This link-level improvement will reduce the number of lost FACCH/SACCH blocks and hence the number of dropped calls due to failed handovers and RLF will be minimized. More information about the Repeated Downlink FACCH and Repeated SACCH can be found in Chapter 7 on Control Channel Performance.

Regarding the second approach, the performance in DL of both FACCH and SACCH has been improved by introducing a novel strategy called Temporary Overpower (TOP). TOP improves the C/I on FACCH/SACCH bursts by temporarily increasing their DL transmit power with respect to the currently used power control level on speech bursts (on hopping TRXs with Static Power Reduction (SPR)) by for example 2 or 4 dB for mobiles in very poor radio conditions. As a result, the balance between TCH/AFS and ACCH link performance can be re-established. Consequently the CDR can be reduced leading to a significant increase in the GERAN network capacity for speech services. A particular benefit from the second approach is its applicability to both legacy as well as 3GPP Release 6 compliant MS.

12.3.2 Dropped Calls in GERAN Networks

12.3.2.1 Overview

In real deployments, the network-initiated curtailment of a call may originate from various reasons, for example radio interface failures, failures from HO, Abis interface, A interface, user actions, failures in the Base Transceiver Station (BTS), Base Station Controller (BSC) and at the mobile side. An MS malfunction has an especially strong impact on the measured CDR in real networks. In this simulation study, only dropped calls related to failures on the air interface are considered. Such failures are caused by excessive errors on the SACCH and FACCH signaling channels, mainly due to high interference or insufficient signal level in capacity and coverage limited scenarios, respectively. In the following, call drop performance has been addressed for interference limited deployments based on the RLF criterion as defined in [11] as well as on HO failures encountered during HO execution.

12.3.2.2 Dropped Calls Due to Radio Link Failure

The objective of prematurely terminating the call by the radio access network is to remove calls experiencing such a bad radio link quality that retaining the connection is useless. Excessive errors on the air interface not only cause unacceptable speech quality but they also may completely compromise the transmission of signaling messages required to maintain the connection in a mobile environment. In GSM, the RLF criterion has been defined as a main trigger mechanism to drop speech calls suffering from excessive bad quality. The aim of detecting an RLF in both the MS and BTS was to ensure that connections with unacceptable speech quality, which cannot be improved either by RF power control or HO, are either released or re-established in a defined manner. The RLF criterion is based on the radio link counter S. If the MS/BTS is unable to decode a SACCH message, S is decreased by one. In the case of a successful reception of a SACCH message, S is increased by two. In any case, S shall not exceed the value defined by the parameter RADIO LINK TIMEOUT set in the data base, for example at 24, and broadcast via SYS INFO messages. If S reaches zero, an RLF due to radio link timeout is declared and the call is dropped.

12.3.2.3 Dropped Calls Due to HO Failures

HO is required to maintain a call in progress as the MS passes from one cell coverage area to another and may also be employed to meet network management requirements, for example for congestion relief. HO may occur during a call from one TCH to another TCH. It may also occur from SDCCH to SDCCH or from SDCCH to TCH (directed retry) during the initial signaling period at call set-up. The HO may be either from a channel on one cell to another channel on a surrounding cell (Inter-Cell HO) or between channels on the same cell (Intra-Cell HO). In general, the HO process comprises two phases [7]:

- HO decision: detecting the necessity of an Inter-Cell HO for an ongoing call based on several criteria, for example quality, level, power budget, etc. as well as generating and evaluating the target cell list, or the necessity of an Intra-Cell HO due to bad quality or for channel mode adaptation reasons.

- HO execution: requiring the exchange of signaling messages on FACCH to transmit the HO or assignment (ASS) command to the MS and to perform the access on the new channel in the target cell in case of Inter-Cell HO or on the new channel in the serving cell in case of Intra-Cell HO.

12.3.2.4 Assignment and Handover Command Transmission

On a HO decision after receiving the ASS or HO command message from the BSC, the BTS of the serving cell uses the FACCH on the TCH to send these messages to the MS over the air interface. The HO command is a Layer-3 (L3) message with variable length depending on the type of coding used. If the HO command is too long, it is segmented by LAPDm into multiple L2 information (I)-frames with a payload of 20 bytes each. One I-frame is transmitted on the air interface by L1 using one FACCH frame on the TCH. The ASS command occupies a single I-frame. The FACCH steals one speech frame in the case of a full-rate (FR) or two speech frames in the case of a half-rate (HR) channel. Since the LAPDm window size is limited to one, the transmission of one segment must be acknowledged before the next one can be transmitted. For this purpose the timer T200 and the counter N200 are used. When the BTS transmits the first burst of the FACCH frame, the BTS starts timer T200 and waits for the acknowledgement frame from theMS, in this case the MS responds with a receive ready (RR) frame on the UL FACCH. If the acknowledgement from the MS is not received within T200, the FACCH is repeated and T200 restarted. The maximum number of FACCH repetitions is controlled by the counter N200, a fixed value of for example 34, specified by GSM for FACCH FR. If the transmission of the first or of a successive (if necessary) I-frame fails, that is after N200 retransmissions the FACCH frame has still not been successfully decoded, the BTS sends an error indication (ERR IND) message with cause "T200 expired (N200 + 1) times" to the BSC which releases the resources: a dropped call due to T200 expiry is the result.

12.3.2.5 Access Procedure on the New Channel of the Target Cell

After successful reception of the HO command, the MS will access the target cell by transmitting an unspecified number of HO access bursts on the new traffic channel. At the time of transmitting the first HO access burst, the MS starts timer T3124. After receipt of a HO access burst, the BTS starts transmitting the PHYS INFO message (a single I-frame) on FACCH to the MS and starts timer T3105. If the MS correctly receives the PHYS INFO before T3124 has expired, the MS stops T3124 and sends the Set Asynchronous Balance Mode (SABM) frame on the UL FACCH to establish the L2 connection for the submission of the HO complete message. If T3105 expires before the BTS receives the SABM frame, it repeats the PHYS INFO FACCH and restarts T3105. The maximum number of FACCH repetitions is determined by the counter NY1. If the PHYS INFO is not correctly received before the expiry of T3124 (which is fixed at 320 ms), the MS returns to the old channel of the originating BTS, sends a HO FAILURE message on the UL FACCH and tries to resume the connection. The SABM frame has to be correctly acknowledged by an Unnumbered Acknowledgement (UA) FACCH from the BTS. If the MS does not receive the UA frame, it retransmits the SABM. The time between two UA retransmissions is 34 TDMA frames in case of FR speech and the total number of transmissions is restricted to 5. If the UA frame is correctly received, the MS will send the HO complete message which concludes the non-synchronized Inter-cell HO execution procedure. If the MS was not able to correctly decode the UA frame (after it has been transmitted 5 times as described above), the MS returns to the old channel of the originating BTS, sends a HO FAILURE message on the UL FACCH and tries to resume the connection.

Exactly the same procedure is accomplished in the case of Intra-Cell HO except that the PHYS INFO message is omitted. In other words, after successful reception of the ASS command, the MS transmits on the new channel the SABM frame on the UL FACCH to establish the L2 connection for the submission of the ASS complete message. The latter has to be correctly acknowledged by the BTS, using the UA frame on the DL FACCH. If the MS does not receive the UA frame, it retransmits the SABM. The time between two UA retransmissions is 34 TDMA frames in the case of FR speech and the total number of transmissions is restricted to 5. If the UA frame is correctly received, the MS will send the ASS complete message which concludes the Intra-Cell HO execution procedure. If the MS was not able to correctly decode the UA frame (after it has been transmitted five times), the MS returns to the old channel, sends an ASS FAILURE message on the UL FACCH and tries to resume the connection.

12.3.3   Performance Improvements on the Associated Control Channels

12.3.3.1   Temporary Overpower Strategy

Figure 12.10 shows the link-level performance of AMR TCH/AFS4.75, TCH/EFS, FACCH and SACCH with ideal frequency hopping on a TU3 channel. When measured at the 1% FER/BLER limit, there is a performance gap of around 6 dB between TCH/AFS4.75 and the associated signaling channels. Recall that the latter have been properly designed to support TCH/FS and TCH/EFS at C/I operating point of 8–10 dB with FER = 1%.

As mentioned above, one option to reduce the C/I gap and improve the probability of successful decoding of the FACCH/SACCH frames at the MS side is to use TOP on these channels in DL. Overpower is suggested to increase the C/I exclusively for particular radio links suffering from poor radio conditions. Under normal radio conditions, both FACCH and SACCH are transmitted using the current power settings due to PC for the TCH. In critical radio conditions, however, based on a new trigger mechanism, FACCH and SACCH shall be transmitted independently of each other with 2 dB or 4 dB higher power (overpower) with respect to the currently used TCH transmission power level commanded by the DL PC procedure. To avoid the adverse effect of permanent overpower on the interference level experienced on TCH channels, overpower on ACCH is only applied if the measured radio link quality in DL of a particular radio link is below a certain bad quality threshold. For example, if the averaged RXQUAL in DL reported by the mobile is worse than 6.5, corresponding to a C/I of less than 4 dB. Note that according to Figure 12.10, TCH/AFS4.75 at C/I = 4 dB still provides sufficient speech quality with mean FER well below 1% whereas both FACCH and SACCH show a BLER of about 30%. TOP requires a static power reduction of the same amount on the TCH TRX providing headroom for the ACCH overpower. The resulting final transmission power is not allowed to exceed the maximum physical output power of the TRX. That means, for example in the case of a static 2 dB power reduction

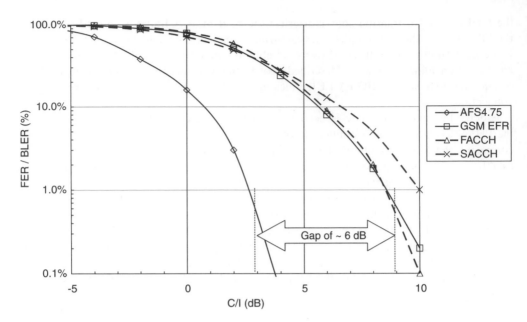

Figure 12.10 Link-level performance of TCH/AFS4.75, TCH/EFS and ACCH for TU3 with ideal Frequency Hopping.

and exhausted PC, an overpower command of 4 dB leads to a maximum 2 dB final power increase. TOP is applicable in interference limited high capacity scenarios implying no coverage limitations.

12.3.4   Network Simulation Model

To study the effect of TOP on the GERAN speech service performance in terms of capacity and quality, system level simulations have been performed in a tight 1/1 frequency reuse scenario with $3 + 3 + 3$ up to $6 + 6 + 6$ configurations (i.e. 3-sectored sites with 3 up to 6 TCH-TRX per sector) depending on the offered cell load.

Table 12.2 gives an overview of the essential parameter settings used in the system simulation model. An urban network deployment with slow moving subscribers (3 km/h) has been assumed. A hexagonal cell layout with a site-to-site distance of 1500 m, a propagation index of 3.76 and a slow fading standard deviation of 6 dB with correlation distance of 110 m have been used. A narrowband scenario has been assumed with 5 MHz spectrum availability in the 900 MHz frequency band. A BCCH planning in a typical 4/3 reuse results in 2.4 MHz (12 frequencies) left for the hopping layer and one guard channel.

Both power budget (better cell) Inter-Cell HO with HO-margin of 3 dB between cells of different sites (0 dB between cells on the same site) and HO due to quality with a quality threshold of 2 dB are modeled very close to the real system implementation based on signal level and quality measurements. Since in tight (1/1) reuse due to the high interference levels on the serving TCH especially at high traffic load, the power budget HO is strongly delayed

Table 12.2 Essential parameters of the system level simulation model

Parameter	Value
GSM band	900 MHz
Cell layout	hexagonal, regular cell layout
Number of sites	61 (183 cells)
Site-to-site distance	1500 m (cell radius: 500 m)
Re-use pattern	BCCH: 4/3 and TCH: 1/1
TCH frequencies per cell	12 for 1/1 re-use
TCH-TRX per cell	up to 6, depending on load
Path loss slope	37.6 dB per decade
Slow fading standard deviation	6 dB
Correlation distance	110 m
Frequency Hopping	Random over 12 frequencies on the TCH-TRX
HO-margin for HO due to Power Budget	3 dB (0 dB intra-site);
HO quality threshold in terms of C/I	2 dB (for both Inter-Cell and Intra-Cell HO)
Intra-Cell HO level threshold	-75 dBm
Inter-Cell HO level threshold	-100 dBm
Radio Link Timeout	24
Temporary Overpower on ACCH	2 dB, 4 dB with FACCH repetition
Adjacent channel suppression	14 dB
Channel mode	TCH/AFS
Active Codec Set	TCH/AFS12.2, TCH/AFS7.95, TCH/AFS5.90, TCH/AFS4.75
Power control	level and quality based
Fast fading profile	TU 3 (slow moving subscribers at a speed of 3 km/h)
Mean call duration	90 s (minimum call duration 10 s)
Voice activity factor (DTX)	0.6
FERmax	2.5%

causing dramatic degradation in signal quality, a HO due to quality is required to achieve appropriate network performance. An Intra-Cell HO is initiated if in addition to the bad quality (C/I < 2 dB), the measured received signal level exceeds -75 dBm.

Dropped calls resulting from decoding errors on FACCH and SACCH have been considered. For this purpose, the HO execution procedure including transmission of ASS and HO commands, physical information and acknowledgements over FACCH along with the transmission of power control and timing advance commands over SACCH have been modeled according to Section 12.3.1. However, in the case of assignment and HO failure, the call is deemed dropped instead of returning to the old channel. The Radio Link Timeout parameter has been set to 24.

TOP is implemented assuming 4 dB static power reduction on the TCH-TRXs and applying 2 dB overpower on both FACCH and SACCH bursts for a call suffering from bad channel quality (measured DL RXQUAL is worse than 6.5), and adding 2 dB more overpower whenever FACCH retransmissions are required. Adjacent channel interference has been considered assuming an adjacent channel protection factor of pessimistic 14 dB.

Since GPRS/EDGE and signaling channels in operational networks are mainly configured on the BCCH carrier, only speech service capacity on the TCH-TRXs has been taken into account. Simulations have been performed for DL only since it has been assumed that the DL is the limiting link due to interference cancellation technique introduced in UL. AMR codec mode adaptation based on C/I thresholds comparison has been implemented close to the real system, automatically switching between four AMR FR codec modes included in the active codec set. Latency in the link adaptation as well as in the PC algorithm has been modeled accurately. A mobile receiver with decent interference reference performance has been selected while advanced receiver algorithms in DL like SAIC [13] or sophisticated radio resource management [14] have not been taken into account. On the network side, TX diversity obtained for example by antenna hopping has been switched off.

The speech quality in terms of FER is measured over a sampling period of 1.92 seconds (i.e. 4 SACCH periods). The speech quality in the network is quantified by the percentage of speech samples with FER higher than a specific threshold for example FERmax = 2.5% (see Table 12.2). Since such speech samples indicate a bad speech quality, the probability of their occurrence is termed bad quality probability (BQP). The network capacity is determined by the EFL at which BQP does not exceed a certain threshold, for example 5% also called outage probability. Correspondingly the EFL at the outage BQP is termed soft capacity limit. Assuming that all eight time slots on a TRX are available for speech traffic using FR speech codecs, EFL is given by [9]:

$$EFL[\%] = \frac{Erl_{BH}}{8 \cdot N_{totalfreq}} \cdot 100[\%],$$

where Erl_{BH} is the busy hour served traffic in Erlangs and $N_{totalfreq}$ is the number of TCH frequencies in the system (e.g. 12 according to Table 12.1).

12.3.5   Simulation Results

12.3.5.1   Preserving AMR Capacity and Speech Quality

The EFL at which the speech quality in the network becomes unacceptable due to excessive interference is referred to as the soft capacity or soft-blocking limit.

Speech quality is defined in the network by BQP, where the speech quality in the network is deemed satisfactory when the BQP is less than 5% for FERmax at 2.5%. Hence, the BQP outage probability defines a soft capacity limit allowing network operators to adjust the QoS within a tolerable range according to their needs. In established high capacity GERAN networks an outage probability of 5–10% is widely accepted in extremely high load traffic situations emerging temporarily during busy hours.

Figure 12.11 depicts the trade-off between outage probability in terms of BQP and the network soft capacity in terms of EFL for a high capacity GERAN network in a tight 1/1 re-use with 12 hopping frequencies (2.4 MHz bandwidth for TCH). Two scenarios have been investigated: one without improvements on ACCH denoted in the legend by TOP-off, and another one featuring the TOP strategy on ACCH denoted by TOP-on. Both scenarios have been studied assuming a segmentation of the HO-command message into a single I-frame. The BQP graph with TOP-on reveals a performance degradation at increasing system load since calls which have been dropped in the TOP-off scenario due to excessive errors on the signaling channels are now maintained contributing a significant portion of bad speech

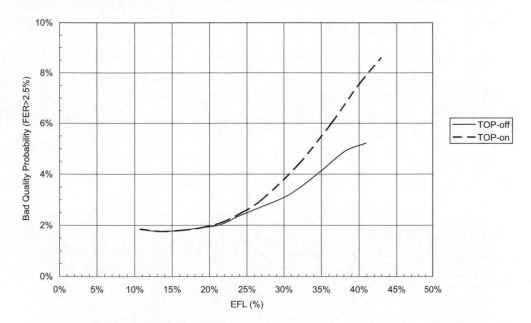

Figure 12.11 BQP vs. EFL for GERAN deployment in tight 1/1 reuse with and without improvements on the control channels SACCH and FACCH.

samples. According to Figure 12.11, using TCH/AFS codecs with TOP, a soft capacity of 34% and 40% EFL has been achieved at BQP outage of 5% and 7.5%, respectively. Note that with maximum installed 6 TCH-TRX per cell (see Table 1) a system load beyond 40% EFL causes a hard blocking of more than 2% (hard blocking of 2% is widely tolerated in matured mobile networks).

12.3.5.2 Boosting GERAN Capacity by Call Drop Suppression

Considering a CDR of 2% as the upper limit for Grade of Service (GoS) widely accepted in commercial networks, as shown in Figure 12.12, the maximum EFL supported in a network without ACCH improvements (TOP-off) is about 24%. In other words, the network capacity limit of almost 40% EFL at BQP outage of 5% depicted in Figure 12.11 (TOP-off) is captured by the CDR due to bad ACCH performance.

The applied TOP strategy proves to be able to shift the capacity limitation due to CDR to about 31% EFL harmonizing the GoS and QoS defined in terms of BQP. This corresponds to a network capacity gain of nearly 30% achieved by TOP.

It should be noted that the TOP strategy is an efficient solution for interference-limited scenarios whereas the improvements based on the 3GPP standard are suitable for both interference and coverage limited deployments. A combination of both strategies provides an additive gain, not presented in this study.

The impact of the second I-frame (in case of HO-command segmentation) on the CDR is in the range of 1–2% EFL. However, it should be noted that with increasing users' speed, the

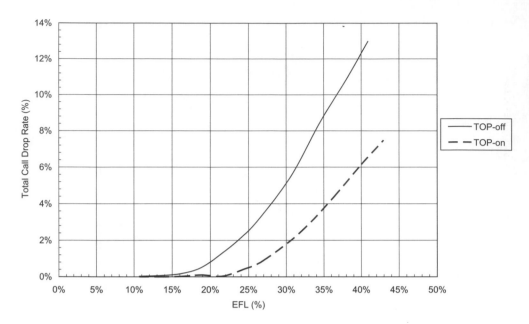

Figure 12.12   Total Call Drop Rate due to RLF and failed HOs vs. EFL with and without improvements on the control channels SACCH and FACCH.

effect of the delay caused by the second I-frame leads to a higher probability of losing the HO-command and consequently to a dropped call.

12.3.5.3   Impact of Radio Link Timeout and Handover Failure on Call Drop Rate

The main contributors to dropped calls on the radio interface are Radio Link Timeout (RLT) on SACCH and failures during HO execution on FACCH. The comparison of the results for the scenario TOP-off shown in Figure 12.13 clearly reveals FACCH failures to be the dominant contributor to dropped calls observed in highly loaded GERAN networks in tight 1/1 reuse accounting for about 90% of the total CDR. For example, at 25% EFL, the CDR due to RLT in a network without TOP is 0.1–0.2%, which is about 10% of the total CDR of 2% according to Figure 12.13 for the same scenario.

The overwhelming impact of failed FACCH transmissions on the CDR is even more visible in terms of CDR per terminated call, see Figure 12.14. While RLFs account for less than 1% of the dropped calls even at the highest load, the CDR due to failures on FACCH observed during Intra-Cell HO execution exponentially increases with system load. This phenomenon is explained by the dramatic growth in the number of Intra-Cell HO due to bad quality as the interference level in the network increases (see Figure 12.15).

TOP proves to be an appropriate technique to substantially reduce the number of lost calls caused by unsuccessful FACCH delivery. While the number of Intra-Cell HO per call is almost doubled with TOP (see Figure 12.15) the CDR per terminated call is almost halved by TOP (see Figure 12.14).

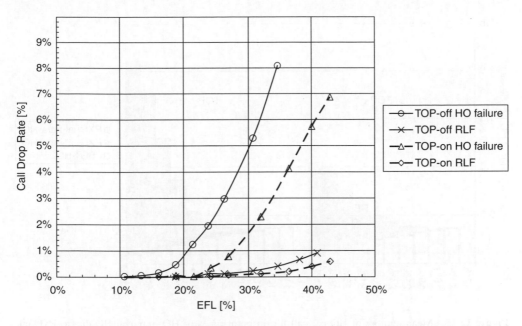

Figure 12.13 Call Drop Rate due to RLT on SACCH and HO failures on FACCH with and without TOP.

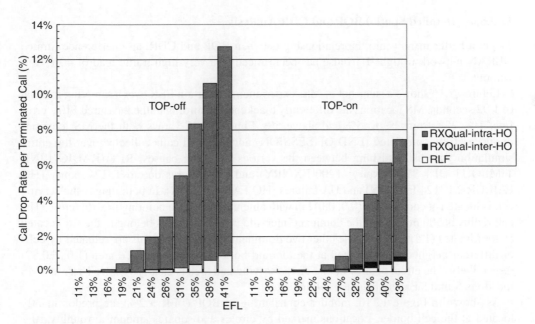

Figure 12.14 Call Drop Rate per terminated call with and without TOP due to RLF and FACCH failures on different types of HO: intracell HO due to RxQual, intercell HO due to RxQual and power budget HO (PBGT).

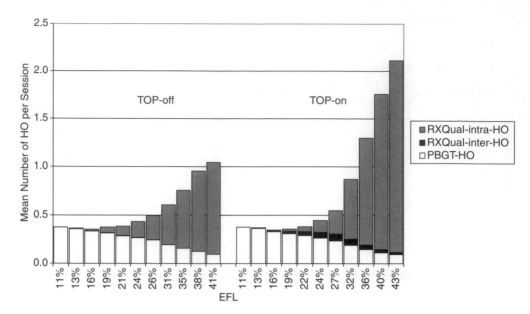

Figure 12.15 Mean number of HO per call for different types of HO: intracell HO due to RxQual, intercell HO due to RxQual and power budget HO (PBGT).

12.3.5.4 Detailed Spatial BQP and CDR Analysis

To get a better insight into the relationship between BQP and CDR, an interference limited GERAN network in tight 1/1 reuse has been traced at a very high traffic load of 40% EFL without TOP.

Figure 12.16 shows a snapshot of MSs experiencing a certain FER over a sampling interval of 1.92 seconds. MS locations are differently blackened according to the measured FER: error free, 1%, 2%, from 2% to 5% and above 5%. Figure 12.17 shows both the MS locations of successfully completed (END OF SESSION) and dropped calls collected over the entire simulation run distinguishing between the various dropping causes: RLT (RADIO LINK TIMEOUT DL), T200 expiry (T200 EXPIRY) and HO failures due to T3124 expiry (HO FAILURE T3124 EXPIRY) and UA failures (HO FAILURE UA). In both figures the serving cell is located at coordinates (0.0; 0.0) km with antenna main lobe pointing in x-direction. The cell radius is 500 m. Two of the dominant inter-site interfering cells belong to the site located at coordinates (1.5; 0.0) km. The other two dominant interfering signals are radiated by cells of different neighbor sites located at the top and bottom of the displayed area (1.0, ±0.87) respectively. The effect of slow fading on the serving cell area is clearly visible. There are MS locations found far away from the serving site.

As shown in Figure 12.16, mobiles contributing to BQP (FER > 2%) are predominantly located at the cell border. Locations marked by circles and squares amount to roughly 10% related to 7.5% BQP at 40% EFL, referring to Figure 12.11. According to Figure 12.17, mobiles suffering from a dropped call are also predominantly located at the cell border. This proves the strong correlation between bad link quality and call drop. Obviously the portion of calls dropped due to failed HO is overwhelming, as already illustrated in Figure 12.13.

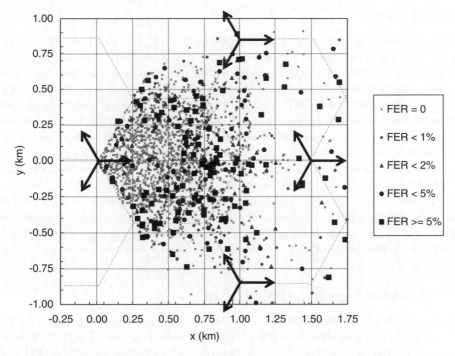

Figure 12.16   MS locations with different FER measured over four SACCH frames.

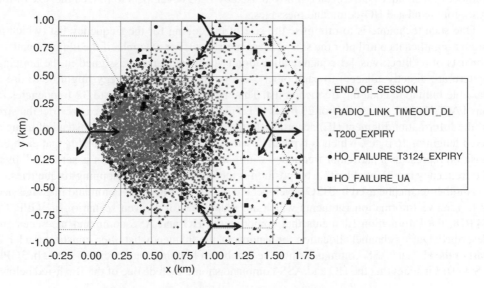

Figure 12.17   Locations of completed vs. dropped calls.

12.3.6   Summary

Field measurements in high capacity GERAN deployments in 1/1 reuse reveal dropped calls to be the major factor limiting the EFL to well below 30% at 2% CDR. This fact has been confirmed in this simulation study. Speech quality in those networks is still acceptable for EFL of up to 40%, assuming BQP outage of up to 5% while failures on the signaling channels limit the network capacity down to 24% EFL. To improve performance, a novel temporary overpower technique for the control channels FACCH and SACCH has been implemented in interference-limited scenarios. This technique can be interpreted as a separate/independent quality-based PC for signaling and traffic channels. A prerequisite for this approach is a static power reduction of the base station transceivers in the range of 2 dB to 4 dB. The simulation results reveal that the TOP strategy is the proper answer to the call drop issue, boosting GERAN AMR speech service capacity beyond 30% EFL in homogeneous deployments without coverage limitation. In particular, TOP supports both legacy and 3GPP Release 6 compliant handsets, providing an additive capacity gain for the latter in combination with the 3GPP optional features Repeated DL FACCH and Repeated SACCH.

12.4   Handover Signaling Optimization

In Chapter 7, it has been demonstrated that the HO success rate (and consequently the CDR) will be proportional to number of I-frames required to convey the layer 3 (L3) HO-command message. Likewise, the size of the L3 ASS-command (which is used during the TCH assignment phase at call set-up and during intracell HO) will have an effect on the call set-up success rate and the intracell handover success rate (and consequently the CDR). For these reasons, a number of techniques have been employed used by BSC developers to reduce the size of the ASS-command and HO-command messages.

One such technique is to minimise the size of the coding for the frequency list (which is often a significant contributor the size of these messages). For example, if the hopping pattern consists of a "block-wise" frequency allocation (where frequencies assigned in the hopping pattern are directly adjacent to each other), an optimium coding strategy might be to use a variable bitmap to code the frequency list. This method can support up to 32 frequencies in one FACCH block (consisting of 20 bytes of L3 payload). Hence the BSC can restrict the size of the Information Element (IE) according to the size of the frequency list instead of using a fixed length of 16 bytes, which is typically used. The principle of a "bitmap" is that each bit position represents a particular frequency within the frequency list. If a bit is set to "1", then the frequency represented by this bit position is included in the list of hopping frequencies.

Another technqiue used by BSC developers to reduce the HO/ASS-command message size is to remove Information Elements which are redundant for a given scenario. In 3GPP TS 44.018, the Information Elements that comprise the HO and ASS-command messages are described. During channel allocation (or handover), the frequency list often constitutes a large part of the HO and ASS-command when the number of hopping frequencies is large. In 3GPP TS 44.018 it states that the HO and ASS-command should include one of the IEs listed below:

- Frequency Channel Sequence (10 octets).
- Frequency list (4–131 octets).
- Frequency Short List (10 octets).
- Mobile Allocation (3–10 octets).

Each IE has certain restrictions on when and how they may be used. For example, the Frequency Channel Sequence IE may be used only if all frequencies are within the P-GSM band, while if the Mobile Allocation IE is present, the 17 octet Cell Channel Description IE should also be present.

In a system with a 1/1 frequency reuse, the number of hopping frequencies is typically large and the allocation is often identical between neighboring cells. Consequently, a lot of signaling is wasted by repeating the same frequency allocation over and over again.

A simple but efficient reduction of the ASS/HO command message size employed by some BSC manufacturers is to make the above-mentioned IEs optional in the message. If none of the IEs are present, the MS will assume the same hopping frequency allocation as in the old cell. For a BSC that uses a Mobile Allocation IE together with a Cell Channel Description IE in the HO/ASS command, the removal of the Cell Channel Description IE which consists of 17 octets, means with a high probability that the message length can be reduced by one I frame. This significantly speeds up handover and TCH assignment and improves performance in terms of call set-up and HO success rate – leading to a reduction in the call drop rate.

The technique of HO command and ASS command message length reduction has been trialled in a real network and the results of the trial showed a significant performance improvement brought by these HO signaling optimization techniques. In the reference scenario (AMR network with 1/1 reuse and 16 consecutive hopping frequencies in the MA list), the ASS command and HO-command messages were typically 37 bytes in length. After enabling HO signaling optimization, this was reduced to about 20 bytes. This reduced the number of I frames required to convey these L3 messages from 2 frames to a single frame. The technique had a significant impact on retainability: CDR at EFL of 15% was reduced by 25% as shown in Figure 12.18 and both ASS command and HO command message length reductions contributed to this retainability improvement. Drops during intercell HO at an EFL of 15% were reduced by 37% as depicted in Figure 12.19, while during intracell HO at an EFL of 15%, a performance gain of 39% was seen as depicted in Figure 12.20.

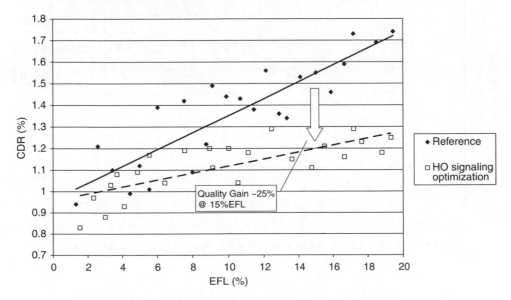

Figure 12.18 Call Drop Rate with and without HO signaling optimization.

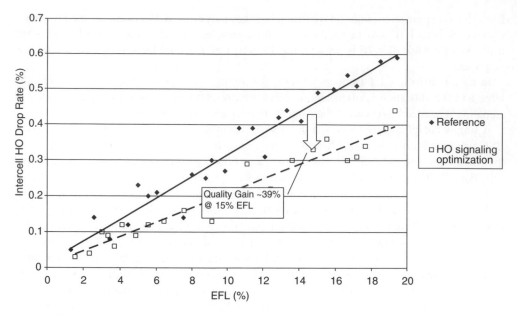

Figure 12.19   Intercell HO Drop Rate with and without HO signaling optimization.

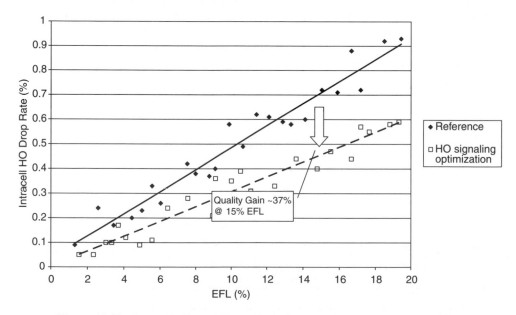

Figure 12.20   Intracell HO Drop Rate with and without HO signaling optimization.

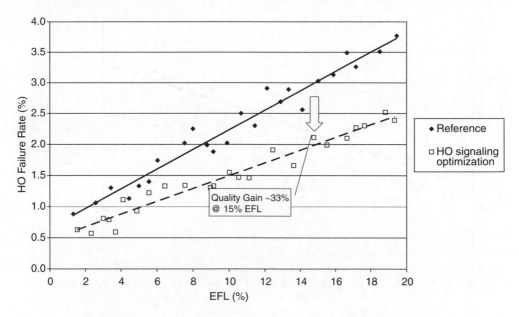

Figure 12.21 HO Failure Rate with and without HO signaling optimization.

As expected, HO signaling optimization improved HO performance, where a 33% reduction in the HO Failure Rate was seen (Figure 12.21). Additionally, a positive impact on accessibility was observed where the Call Set-up Failure Rate improved slightly as the consequence of the shorter ASS command message length as depicted in Figure 12.22.

12.5 Separate Radio Link Timeout Value for AMR

The purpose of Radio Link Timeout (RLT) is to control a forced release so that the forced call release does not occur until the call has degraded to a quality beyond which the majority of subscribers would have manually released the call (this is described in more detail in 12.3.2). Given that AMR calls are more robust than EFR calls and provide acceptable speech quality at much lower C/I ratios, the specification of a common RLT value for AMR and non AMR calls may cause the AMR calls to be unnecessarily dropped. For this reason, separate RLT thresholds for AMR and non-AMR calls have been investigated. The findings indicate that the RLT threshold for AMR should be at a much higher value than for EFR. Trials in a real network with AMR penetration of 70% have shown that when the RLT threshold is increased by more than 50% (e.g. from 20 for EFR to 36 for AMR), the overall CDR was reduced by 25% (Figure 12.23), while keeping the speech quality at the acceptable level (Figure 12.24).

12.6 AMR HR to AMR FR Handover Optimization

An intra-cell handover from AMR FR to the AMR HR, also known as a packing or compression HO, is triggered when the load in a cell is high and the quality is sufficient for the HR

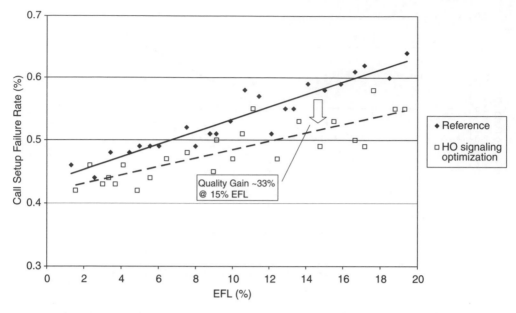

Figure 12.22   Call Set-up Failure Rate with and without HO signaling optimization.

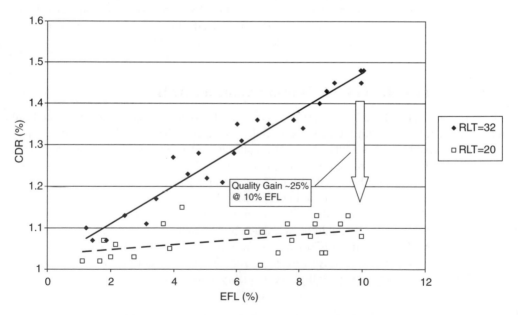

Figure 12.23   Call Drop Rate for different RLT values.

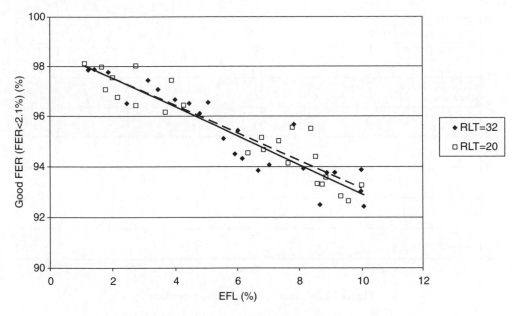

Figure 12.24   Good UL FER for different RLT values.

channel mode. When the quality drops to below a predefined threshold for the HR channel mode, a handover from AMR HR to AMR FR is triggered (also known as an unpacking or decompression HO). This is depicted in Figure 12.25.

When a rapid field drop occurs, a drop in quality will also typically occur. For an AMR HR call, this means triggering an unpacking handover (to AMR FR). If the signal strength

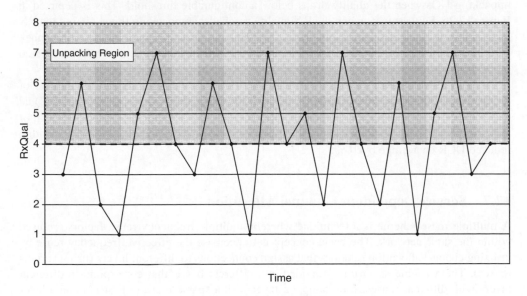

Figure 12.25   Standard unpacking procedure.

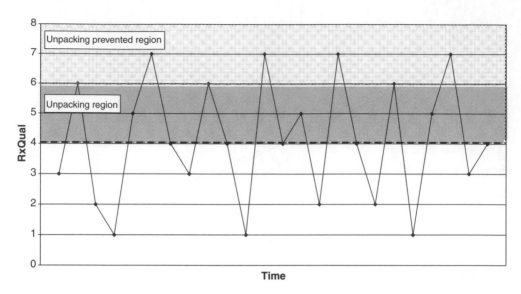

Figure 12.26 Improved unpacking procedure.

drops too quickly, a call drop might occur during the handover procedure. The same could also happen when the MS moves around a corner. Its own RX level does not change but the interferer level rapidly increases. This can cause the DL quality to drop to the very low level (e.g. RxQual = 7) and consequently to trigger a quality-based handover to AMR FR. Since it would occur in very poor radio conditions, there is a high chance the handover will fail and the call will be dropped. For that reason the handover algorithm can be improved by preventing unpacking HOs when the quality drops below a configurable threshold. This is depicted in Figure 12.26. Furthermore, when the RX quality reaches the unpacking threshold but the RX level is low, it might also be reasonable to avoid HO and instead remain on the HR channel. With this improvement, the HO algorithm prevents AMR unpacking and interference based intracell handovers when the RX level and/or the RX quality conditions are very poor.

Trials in a live network have proved that these HO improvements reduce the number of dropped calls during handover. Handover to AMR FR was in this case triggered when RxQual was 6 or above, while HO was prevented when RxQual was 7 and RxLev was −95 dBm or less. Figure 12.27 shows the results of the trial where the improved unpacking HO algorithm reduced DCR on average by 10%.

12.7   Service Dependent Channel Allocation

A multiple reuse scheme is a technique whereby multiple frequency reuse factors are used within the same network. The basic concept is to decrease the effective frequency reuse by utilizing channel allocation schemes that assign connections to different layers (using various reuses). The available spectrum is divided into different bands that correspond to different layers with different reuses, for example a layer with a sparse reuse (e.g. 4/12) and a layer with a tight frequency reuse like 1/1.

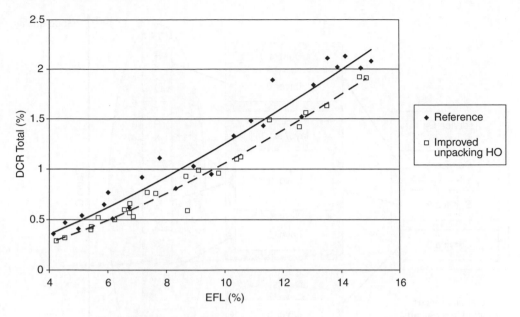

Figure 12.27   DCR with and without improved unpacking algorithm.

An example of a multiple reuse scheme could be 4/12 reuse for the BCCH TRX together with 3/9 and 1/1 reuse patterns allocated to the first and second TCH transceiver respectively. Nine frequencies are reserved for the second transceiver, thus the second TRX will have a 9 reuse. These 9 frequencies are never used by the first or third transceivers in any of the cells in the system. In the same way, the third TRX in all cells use one of the frequencies from the third set which consists of six frequencies in reuse 1/1. Altogether 33 frequencies are needed (12 for BCCH, 9 for reuse 3/9 and 12 for reuse 1/1). This method allows for gradual tightening of the reuse as more TRXs are installed (e.g. one TRX can be added to the layer with reuse 1/1). Furthermore, it also takes advantage of the fact that the traffic is not homogeneously distributed within the system.

Modern radio resource algorithms allow the allocation of resources from different layers based on service type. Now a days, a huge variety of available services with different C/I requirements need to be supported in the GERAN network. Some of them like (E)GPRS requires good radio conditions (high C/I) while other services like AMR are robust against interference and can be served in relatively low C/I. Service Dependent Channel Allocation (SDCA) is a technique that allows a radio channel to be allocated based on the service type. The basic concept is to allocate a radio channel for the specific service from a layer defined in the list of preferred layers. With this technique, TRXs are divided into layers, hence a layer represents a TRX or a group of TRXs which offer the same radio quality (as a result of belonging to the same frequency plan). Different layers have different qualities since they are planned with different reuse factors. Every service has its own list of preferred layers. Only resources included in that list can be allocated to the service. The list can contain one or more layers and the order of the layers could represent the order of preference. Figure 12.28 depicts the concept of channel allocation based on service type. In this example, a cell is defined

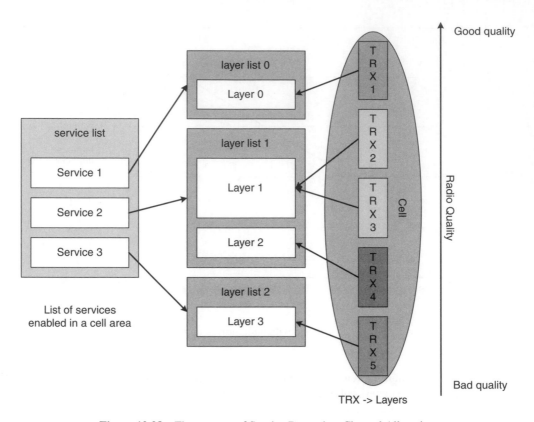

Figure 12.28   The concept of Service Dependent Channel Allocation.

with 5 TRXs and 3 layers: layer 0 represents a BCCH TRX planned with a 4/12 reuse pattern and offering superior quality; layer 1 represents TRX2 and TRX3 planned with a 3/9 pattern and offering good quality; layer 2 represents TRX 4 planned with a 1/3 pattern and provides medium quality; and the worst quality layer with 1/1 reuse pattern is layer 3 consisting of TRX5. Since in this example there are 3 services and each of them requires its own list of preferred layers there are 3 lists dedicated to each service type. Service 1 (e.g. EGPRS) is the service that has the highest C/I requirement. Therefore the Service Dependent Channel Allocation algorithm for EGPRS assigns resources from layer 0 only (which provides highest quality). Service 2 (e.g. EFR) requires good or medium quality, therefore in the associated list of layers, layer 1 with good quality is preferred unless there are no resources available, in which case layer 2 (which provides medium quality) is selected. Finally service 3 (e.g. AMR) has the lowest C/I requirements, therefore a channel can be allocated from layer 3 which is planned with the tight 1/1 frequency reuse and offering the poorest quality.

Service Dependent Channel Allocation allows efficient utilization of the frequency band. A channel is allocated according to the required quality and the list of layers which comes as the result of frequency planning. Furthermore, the parameters used to configure radio link control functions such as power control and handover can be differentiated for each service instead of being fixed for the whole cell. This might be needed since for different services, different strategies may need to be applied. For example, AMR is more resistant to interference than other services such as EFR and so a more aggressive strategy could be applied. As a result, parameters dedicated to the AMR service can be applied so that a power increase for AMR calls might be triggered later than for the EFR calls. Likewise, the handover parameters can be adjusted according to the C/I requirements of the service. A consequence of aggressive parameter settings might be a lowering of the level of interference leading to positive impacts to retainability and speech quality.

The concept of Service Dependent Channel Allocation provides higher flexibility when assigning resources to voice/data services. Each layer can be tailored to handle a specific service, whose quality requirements determine the adopted reuse scheme. The concept of layers also allows an additional degree of freedom in the link control functions, which in turn can lead to greater mobile network efficiency.

A potential benefit of SDCA which is most relevant is when the performance enhancement is transformed into the capacity gain. A simulation campaign was performed to estimate the potential capacity gain from a multiple reuse scheme that utilizes a Service Dependent Channel Allocation scheme. The performance of multiple layers together with the SDCA scheme were compared with a reference scenario, which consisted of a single baseband hopping layer within which and BCCH TRX was included. In total, 27 frequencies were included in the hopping sequence. Power control and DTX were activated in uplink and downlink. Three service types were used: EFR, AMR and EGPRS. In the SDCA case, the BCCH layer was planned with reuse 4/12 and served by EGPRS only. For EFR 1 TRX with synthesized random hopping over 6 frequencies was used. Finally AMR traffic was served by one layer consisting of 2 TRXs using synthesized random hopping with 9 frequencies in 1/1 reuse. In this campaign traffic was simulated using the following traffic mix: 50% AMR, 25% EFR and 25% EGPRS.

The results in Figure 12.29 show the trend in the overall drop call rate (for AMR and EFR) and suggest that the multiple reuse scheme and SDCA provide an additional gain

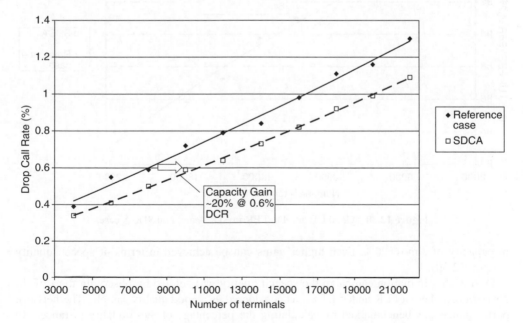

Figure 12.29 Drop Call Rate in SDCA and non-SDCA case.

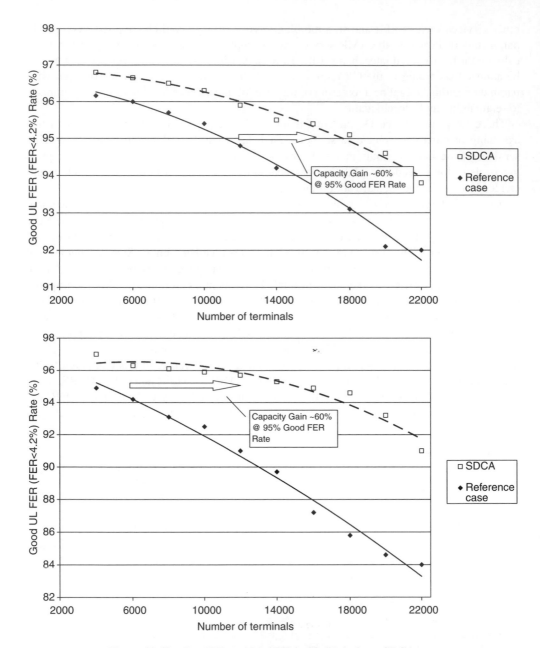

Figure 12.30   Good UL and DL FER in SDCA and non SDCA case.

in capacity of almost 20%. Even higher gains can be achieved in terms of speech quality (Figure 12.30).

During the simulation campaign the speech quality was studied by observing the FER in 2 s periods, where a FER under 4.2% was considered as a good quality sample. The network performance was benchmarked by calculating the percentage of these quality instances. In

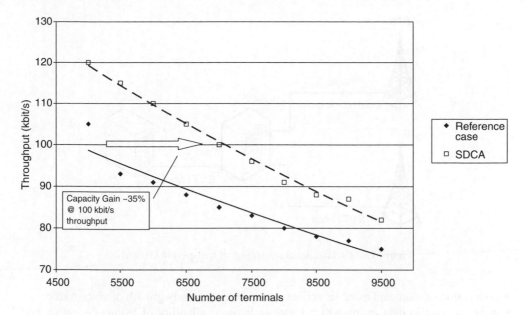

Figure 12.31 LLC throughput in SDCA and non-SDCA case.

both uplink and downlink, the percentage of good speech samples increased significantly with SDCA when compared to the reference scenario.

Finally, EGPRS throughput was studied. In the reference scenario, PS data was served using baseband hopping while SDCA allowed EGPRS to be allocated to the dedicated BCCH layer that offers excellent C/I conditions. This had the impact of increasing the throughput in the SDCA case, as depicted in Figure 12.31.

The results demonstrate that SDCA has the potential to provide significant performance enhancements and capacity gains when compared to a reference network that utilized baseband hopping. However, SDCA requires careful frequency planning and parameter optimization (adjustment for each service type) and the gains are dependent on network load and traffic distribution.

12.8 Advanced Abis Solutions

GERAN network entities (Base Station, BSC, TRAU) are connected to each other by means of E1/T1 TDM links which convey traffic related to a user as well as control and management planes over Abis and Ater/A interfaces (Figure 12.32).

Transmission to and from base stations accounts for a large part of total cost of ownership (TCO) in GERAN: radio access networks are non-homogeneous in terms of topologies and traffic volume per base station; base stations are numerous and often cover large areas; transport bandwidth is limited; and its scalability is stepwise in large steps of for example 2 Mbit/s in case of E1.

Moreover, GERAN just like other mobile communication networks is currently facing traffic growth challenges of a twofold nature:

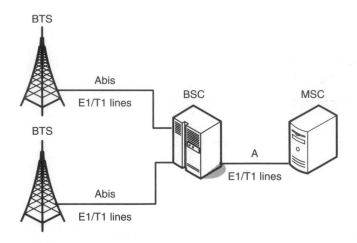

Figure 12.32 Traditional realization of transport in GERAN.

- connection of more and more subscribers contributes to steady growth of voice traffic;
- huge increase in data traffic which is driven by an availability of features enabling high throughput rates on a radio interface (EGPRS, DLDC, EGPRS2, . . .) and also by competitive flat rates.

The combination of these two effects results in a traffic "explosion" both in emerging and mature markets which additionally coincides with a change of traffic characteristics from voice-dominated to data-dominated as most of the innovations – both in terms of devices and in terms of end-user services – are conceived for the packet world. The traffic explosion and an inevitable migration towards data dominated mobile networks definitely increase bandwidth demand but do not necessarily bring proportional revenues for network operators as the flat rate trend for PS services leads to greater data traffic in the network but not to additional ARPU.

Therefore, to keep the business profitable, the costs should be uncoupled from capacity which means migrating from a TDM-based transport to an IP-based one. Thus the revolution in the GERAN transport aiming at the introduction of IP transport, which was supposed to begin soon anyway, has even been brought forward and has actually now already started.

12.8.1   Basic Principles for Saving Bandwidth and Costs

In a traditional Abis realization, a TDM signal is transmitted over E1/T1 links using a permanent mapping between the radio and Abis interface. Typically, on top of these permanent resources on the Abis interface, some dynamic capacity (depending on PS traffic volume to be served) is made available to enable high data throughput rates. The essential drawback of such a set-up is the blocking of transmission capacity even if there is no information to be sent (for example, during transmission of idle or silent frames). In such a case dummy bits are inserted to preserve a constant bit rate. Therefore, in the traditional realization of the transport in GERAN, Abis bandwidth is determined by a configuration of a base station: nearly the same bandwidth is reserved for base stations with the given configuration regardless of traffic load. For these reasons the traditional TDM-based transport that is adequate to handle voice and data with basic GPRS coding schemes CS1/CS2 (which were the only user plane traffic types present in GSM networks until EGPRS deployment) is becoming inefficient for a transmission of the bursty traffic nowadays dominating in GERAN networks.

Migration to the IP-based transport means the introduction of a new concept in which Abis frames conveying traffic and signalling information between a base station and an associated controller are subject to conversion prior to sending them to a transmission path. As a result of this conversion, the informative bits retrieved from incoming frames are used to form IP packets which are eventually sent over the Abis interface. Such realization of the Abis transport is referred to as "packet Abis" hereafter. A packetization process described above brings in a number of effects which contribute to bandwidth savings which are achievable with the Packet Abis and unattainable for the traditional Abis realization. The most important are the following:

- only informative bits are transmitted over Packet Abis, all other bits are not needed (for example, those used to preserve a constant bit rate in the legacy Abis implementation) and thus are not sent;
- the number of IP packets that are produced and transmitted corresponds exactly to the actual traffic load of a base station; moreover, further bandwidth optimization is observed due to the statistical multiplexing effect caused by an unbalanced load in sectors of the base station; bandwidth is used efficiently as it is allocated according to the actual needs, that is a common bandwidth is reused by different traffic types, a codec mode used on the radio interface determines a payload size, no resources are dedicated permanently, no "empty" packets are sent.

Obviously, the involvement of transport protocols leads to a certain transmission overhead but the savings from the packetization process more than compensates for the bandwidth consumption for the transport of the protocol headers introduced by the Packet Abis. In consequence, unlike the traditional Abis, the required transport bandwidth mainly depends on a load and a traffic profile observed in a base station and varies over time.

12.8.2 Transport Aspects

An additional advantage of the Packet Abis is that the above-mentioned principles are applicable to different realizations of a physical layer, namely the currently installed TDM links as well as Ethernet links (i.e. Packet Switched Networks) can transmit Packet Abis traffic.

The former allows reuse of an existing transport infrastructure and significant reduction of bandwidth usage on Abis lines at the same time (i.e. due to optimization effects brought in by Packet Abis, more TRXs can be handled by given amount of TDM links). Both are helpful to protect the installed base of TDM networks which is especially appreciated by those operators who have recently made big investments in leased lines.

The latter leads to phasing out the TDM transport and to the introduction of Packet Switched Networks (PSN) instead. The process affects the existing network topology but it paves the way to a common, fully integrated transport network that can be freely shared between different radio access technologies as well as by fixed networks which is beneficial in terms of costs and operability.

The Ethernet-based IP transport is more cost efficient in comparison to the TDM-based one but, on the other hand, it introduces some complexity that is irrelevant to the TDM-based transport and special features are needed to cope with this. The most crucial issues are related to:

- transport impairments;
- synchronization distribution;
- higher security requirements;
- lack of guaranteed bandwidth.

The grade of service must not be reduced after migration from TDM-based transport to the Packet Switched Networks (PSN) where a certain packet loss rate, packet delay and packet delay variations are inevitably present. Although all these transport impairments are unavoidable in PSN, it is possible to achieve "TDM-like" performance and to preserve voice quality and data throughputs at the level before migration by keeping the packet loss rate, the packet delay and the packet delay variation below definite thresholds. Therefore, stringent requirements concerning the packet loss rate, the packet delay and the delay variation are put on a PSN, and satisfying the required service level is a prerequisite for the selection of the PSN as a transport network.

Delivery of a synchronization signal from a master (e.g. base station controller) to a slave entity (i.e. base station in case of GERAN) with an expected time and frequency accuracy is a fundamental requirement in the GERAN networks (Figure 12.33). For the traditional transport realization, the synchronization signal is reliably distributed among the master and slave entities by means of TDM lines connecting these entities, while in a PSN the synchronization

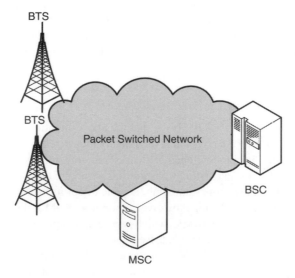

Figure 12.33   Usage of packet switched network for GERAN transport.

chain is broken and in consequence isolated physical clocks cannot synchronize with each other over time. Thus several techniques are worked out and made available to guarantee the distribution of the synchronization signal between distant nodes for various network configurations, topologies and customers' preferences. Available methods can be classified into one of the following categories:

- external synchronization input, for example a GPS receiver;
- physical layer, for example Synchronous Ethernet [13];
- clock recovery algorithms [14].

A GPS receiver installed at a base station is the source of a timing signal which can guarantee a required accuracy and stability. However, in this method the installation of the GPS receiver is necessary at each node, which in some cases can be unacceptable, for instance, due to costs. Synchronous Ethernet makes use of a concept known from for example PDH/SDH networks where an extremely accurate reference signal is transmitted over a physical layer and all other clocks in the network can adjust their timing via a dedicated port. In this method the reference timing signal is delivered to the Ethernet switches using a synchronization port. The method maintains a stable and a precise clock signal regardless of a network load when it is supported at every hop along the chain of nodes between the source of reference clock (e.g. BSC) and the base station. Precision Time protocol (defined by [14]) is a solution in which a reference source generates timing packets transmitted to a single distinct node. In the case of GERAN, this is the base station site which the timing packet is produced for. The slave function installed in the base station uses complex algorithms to interpret a reception rate of time-stamped packets and retrieve a clock signal.

Use of a public Ethernet or IP networks for GERAN transport purposes raises security concerns because it makes the connections between base stations and controllers more vulnerable to possible attacks than in the case of the traditional TDM-based transport. The security aspects are especially crucial in the case of large PS networks which might be shared between various systems and managed by different operators. Such networks often offer global connectivity which is conducive to increasing the risk of attacks. Thus a number of countermeasures have been worked out to reduce the potential security weaknesses of PS networks and, in particular, to improve the security level for Packet Abis. The key security features are:

- access control based on control lists or identity certification;
- installation of firewalls in front of important network entities (such as base station controllers) or interconnecting nodes;
- creation of IPsec tunnels on the route from base stations to a controller.

Bandwidth available in PSN is normally shared with unknown traffic. Besides, complex and non-homogeneous transport networks working under variable load conditions result in bandwidth fluctuations. Thus, the bandwidth required to avoid overload in the whole transmission path between the base station and the base station controller must be protected by a congestion control mechanism.

GERAN networks are very different from each other and therefore no single recommendation for the migration from TDM-based to IP-based transport exists. Quite the contrary, several possible migration paths are available. To choose the best one, the requirements of each operator must be elaborated separately and a final decision depends on many incomparable factors including, but not limited to, customer strategy and preferences, installed base in terms of HW platform, network configuration and traffic load, price scheme and subscribers' profile. Nevertheless availability of reliable PS networks is always the supreme precondition for starting the migration, so it becomes a decisive factor which definitely affects the pace of the transport evolution and is taken into account during the elaboration of the network evolution steps and timeframe.

12.8.3   GERAN Transport Evolution for Different Network Scenarios

Packet Abis provides significant savings in terms of a required transmission bandwidth. As already explained, bandwidth savings are possible owing to decomposition of a PCM time structure and the introduction of optimization mechanisms aimed at transmitting solely the informative bits. As a result, in the E1 environment, even up to 24 TRXs can be served by a single line with Packet Abis over TDM depending on a traffic load and user profile. Although the achievable bandwidth savings definitely vary with different traffic volumes to be carried over the Abis interface, and thus the number of TRXs that can be handled by an individual TDM line needs to be evaluated case-by-case, quite a few overall conclusions can be drawn.

For the Packet Abis over TDM, the crucial base station configurations are those where the entire TDM line can be off-loaded just due to introduction of Packet Abis. In such a case the achievable savings can be employed twofold: either by allowing more traffic to be handled by the base station with the same transmission capacity in terms of TDM lines or by moving the redundant TDM lines to other base station sites where shortage of transmission capacity is experienced. Both ways are helpful to improve cost efficiency of a transport network even before the migration to PSN.

Further cost optimization is possible after a replacement of TDM lines with PSN even though Packet Abis over PSN consumes more bandwidth than Packet Abis over TDM due to a bigger transport overhead. Nevertheless, better cost efficiency is achievable after the migration to the IP transport because costs per bit are significantly lower in PSN than those in TDMlines. The related savings, compared to the same bandwidths, can even exceed 50%, depending on the network configuration and price scheme.

Figure 12.34 summarizes typical migration paths in GERAN networks. The traditional TDM-based transport can evolve either directly to a PSN-based one or gradually in intermediate steps where a Packet Abis over TDMis in use to protect installed base and optimize bandwidth consumption on Abis. As mentioned in the previous section, the evolution path depends on many factors and must be elaborated for each individual customer case.

IP-based transport combines bandwidth savings due to explicit optimization done prior to transmission and low cost transport media. Therefore it is the most cost effective one. Besides, in addition to Packet Abis, a similar concept of transport is already applicable to GERAN interfaces towards core network (A over IP, Gb over IP) to enable seamless IP transport throughout the entire GERAN network. Finally, further synergies and in consequence additional cost savings are possible when the same transport network is shared by various technologies, for example, by co-located 2G and 3G sites, which is also possible with a common IP-based transport. To satisfy market demands that expect a consistent plan of GERAN transport evolution, future proof products are now available, together with various

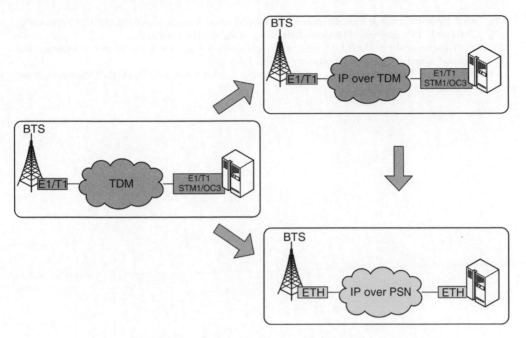

Figure 12.34 Different migration paths for GERAN transport evolution.

migration paths suitable for any network configuration, traffic load and different customer preferences.

References

[1] J. Wigard, P. Mogensen, P.H. Michaelsen, J. Melero and T. Halonen, "Comparison of Networks with Different Frequency Reuses, DTX, and Power Control Using the Effective Frequency Load Approach", in *Vehicular Technology Conference Proceedings, 2000. VTC 2000*, Spring Tokyo, 2000. IEEE 51st, 15–18 May 2000, Vol. 2, pp. 835–839.

[2] K. Ivanov et al., "Frequency Hopping Spectral Capacity Enhancement of Cellular Networks", in *Proc. of IEEE 4th International Symposium on Spread Spectrum Techniques & Applications*, 1996.

[3] J. Wigaard et al., "Capacity of a GSM network with Fractional Loading and Random Frequency Hopping", in *Proc. of IEEE Symposium on PIMRC*, 1996.

[4] K. Ivanov et al., "Spectral Capacity of Frequency Hopping GSM", in Z. Zvonar (ed.) *GSM Evolution Toward 3rd Generation Systems*, New York: Kluwer, 1999.

[5] U. Rehfuess and K. Ivanov, "Comparing Frequency Planning against 1 × 3 and 1 × 1 Reuse in Real Frequency Hopping Networks", in *Proceedings of IEEE VTC*, Fall 1999, pp. 1845–1849.

[6] T. Nielsena and J. Wigaard, *Performance Enhancements in a Frequency Hopping GSM Network*, Dordrecht: Kluwer Academic Publishers, 2000.

[7] G. Heine, *GSM-Signalisierung verstehen und praktisch anwenden*, Franzis Verlag, 1998.

[8] C.F. Ball, K. Ivanov, H. Winkler, M. Westall and E. Craney, "Performance Analysis of a GERAN Switched Beam System by Simulations and Measurements", in *Proc. IEEE VTC*, Spring, 2004.

[9] K. Ivanov, C.F. Ball, R. Müllner and H. Winkler, "Smart Interference Reduction Dynamic MAIO Allocation Strategy for GERAN Networks", in *Proc. IEEE VTC*, Fall, LA, 2004.

[10] T. Halonen, J. Romero and J. Melero, *GSM, GPRS and EDGE Performance*, Chichester: John Wiley & Sons, 2002.

[11]   3GPP TS 44.018, "Mobile Radio Interface Layer 3 Specification; Radio Resource Control (RRC) Protocol".

[12]   GP-041044, "Enhancements of Handover-Related Signalling in AMR Networks".

[13]   ITU-T Recommendation G.8261, "Timing and Synchronization Aspects in Packet Networks: Ethernet Over Transport Aspects – Quality and Availability Targets", May 2006.

[14]   IEEE 1588v2, "Standard for a Precision Clock Synchronization Protocol for Networked Measurement and Control Systems", July 2008

13

Green GSM

Environmentally Friendly Solutions

Sebastian Lasek, Krystian Krysmalski, Dariusz Tomeczko, Andrzej
Maciolek, Grzegorz Lehmann, Piotr Grzybowski, Alessandra Celin
and Cristina Gangai

13.1 Introduction

Climate change due to global warming is one of the most challenging global problems today
and for the next decades. The root cause of global warming is the greenhouse effect induced
by massive emissions of carbon dioxide (CO_2) and other greenhouse gases [1], especially in
highly industrialized and fast developing countries. Although the telecommunications sector
contributes to only 0.5% of the total CO_2 emissions [2], it is crucial that it minimizes its
ecological footprint as well.

The energy costs are currently one of the biggest components of the operator's OPEX[1]
and may become even more significant in the coming years if the energy prices increase.
So by applying measures to reduce energy consumption such as efficient network planning
or solutions to optimize the GSM network energy performance, operators may lower their
operational expenditures while limiting the negative impact upon our natural environment.

Among solutions to improve the energy efficiency, the main focus is placed on features
reducing the power consumption of base station sites, which amounts to about 90% of the total
energy consumed by a mobile network infrastructure. The remaining 10% are dedicated to
power other network elements such as controllers, OSS,[2] core network, etc. About 70% of the
power consumed by a base station site is directly utilized to form communication signals to
mobile terminals and controllers. The remaining 30% are expended mostly by air-conditioning
systems.

---

[1] OPEX: Operational expenditures.
[2] OSS: Operational Support System.

GSM/EDGE: Evolution and Performance   Edited by Mikko Säily, Guillaume Sébire and Eddie Riddington
© 2011 John Wiley & Sons, Ltd

In this chapter some of the solutions to improve energy efficiency are presented. Section 13.2 provides an overview of an energy optimized network design. Sections 13.3 and 13.4 focus on the functionalities developed to enhance network coverage and capacity with the help of which the base station footprint can be minimized and consequently substantial power savings can be achieved. Section 13.5 addresses dedicated "green" software features to optimize the power transmitted over the air interface and simultaneously improve the environmental performance of GSM networks. Section 13.6 highlights hardware solutions lowering power consumption of the signal processing units and the air-conditioning systems. In Section 13.7, renewable energy sources as alternative ways to power base station sites are described. Finally, Section 13.8 presents the concepts of the energy-aware mechanisms applied to controllers and transcoders.

13.2   Energy Optimized Network Design

When deploying an environmentally friendly network, the energy efficiency should be considered in several areas. One of the key enablers of highly energy-efficient mobile networks is the optimum planning and design aiming at building the network with a minimum number of hardware elements such as BTSs and TRXs. This can be achieved by applying various advanced coverage and capacity enhancements.

Coverage extension beyond the one offered by a commonly used two-branch (2RX) receive diversity in uplink is possible by means of advanced techniques to improve the radio link performance. Moreover, specific BTS site architectures have been developed to increase the range of the cells. For instance, by installing feederless sites, in which a base station radio unit is located close to an antenna system, a feeder cable is eliminated and with it the related feeder losses. As a result of coverage improvements, the number of radio sites can be potentially reduced without jeopardizing the overall perceived radio quality. Owing to the decreased number of sites in the network, the total power consumption can thus be minimized, as illustrated in Figure 13.1.

Furthermore, the growth of mobile traffic introduces challenges to operators on how to boost the network capacity while sustaining sufficient quality and keeping CAPEX[3]/OPEX down. Therefore, features to increase the spectral and hardware efficiencies have become highly desirable.

By improving the network spectral efficiency, it is possible to boost the network capacity without quality degradation. In some scenarios, higher spectral efficiency may allow the operators to improve the capacity by extending cell configurations with additional TRXs. This, in turn, may avoid the deployment of extra capacity sites that would increase the energy consumption in the network. By increasing the hardware efficiency, it is possible either to provide the same capacity with a reduced amount of resources or to serve higher traffic without configuration extensions. The resultant savings in the number of actively used radio time slots or even in the number of deployed TRXs, may translate directly into the reduction of energy consumption. A potential package of "green" features could include: Power Control,

---

[3] CAPEX: Capital expenditures.

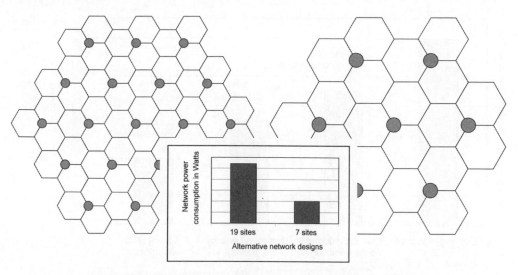

Figure 13.1   Site count reduction by energy-aware planning and design.

DTX,[4] AMR FR and HR, SAIC, OSC and DFCA.[5] In the following sections, the selected coverage and capacity boosters and their influence on the energy savings are discussed in more detail.

Although particular features aiming to enhance either coverage or capacity may bring significant energy savings, further savings can be achieved by applying dedicated "green" software solutions to optimize the power transmitted on the radio interface. This optimization may be realized by intelligent (radio) resource allocation, by extending the energy saving mode on BCCH TRX or by putting network elements onto standby mode when not in use. Energy-aware software solutions are presented in Section 13.5.

Another way to slash the power consumption is to use the latest most energy-efficient base station technology which can yield savings up to 35%.

Air-conditioning units are nowadays the most common solution for cooling telecom shelters that contain radio equipment. They consume approximately 30% of the total BTS site energy. Introducing instead a new cooling method, that is air-flow cooling may bring additional energy savings.

In addition, renewable energy sources are becoming viable alternatives to power base station sites. The sun and wind are the "greenest" solutions as the energy is harnessed without any impact on the natural environment. For example, while today's primary choice to power remote sites that are out of range of an electricity grid is a diesel generator, using solar power is a much "greener" alternative. And although the initial investments for a solar power generator are much higher than that of a diesel generator, it can prove to be a much cheaper option over the years since regular on-site maintenance including refueling can be avoided.

---

[4] DTX: Discontinuous transmissions.
[5] DFCA: Dynamic Frequency Channel Allocation.

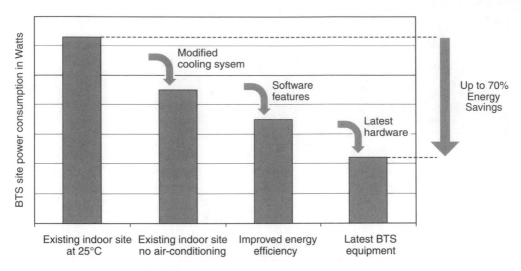

Figure 13.2   Evolution of the BTS site energy efficiency.

Altogether, an energy optimized network design can yield substantial energy savings. With the solutions outlined above, up to 70% of a BTS site power consumption can be saved, as illustrated in Figure 13.2.

13.3   Coverage Improvement Techniques

As introduced earlier, techniques to improve the radio coverage can contribute to the reduction of the number of radio equipments to cover a given region, and in turn reduce the energy consumption of the corresponding network. It is important to note, however, that a balance between the uplink[6] and downlink[7] ranges and the gains thereof is necessary in order to provide an overall radio coverage gain. Indeed, the radio coverage is downlink-limited when the uplink coverage is better than the downlink coverage, and uplink-limited otherwise.

13.3.1   Uplink Techniques

13.3.1.1   Antenna Diversity – 2RX, 4RX

By using multiple receiving antennas, more signal energy can be collected and therefore signal-to-noise ratio (SNR) can be improved when the signals from the antennas are coherently combined. Moreover, since these radio signals are independently faded, it is also possible to achieve diversity gain in the environment dominated by multipath propagation. The efficiency of antenna diversity, called diversity gain, depends on the grade of de-correlation between the diversity receiving paths.

---

[6] Mobile to network link.
[7] Network to mobile link.

This gain can be achieved by deploying antennas with orthogonal polarizations[8] (X-pol) or by ensuring sufficient horizontal spatial separation between vertically polarized[9] (V-pol) antennas. For example, for a voice service, it can be assumed that a two-branch (2RX) receive diversity provides from 4 to 5 dB gain in the link budget. Four-branch (4RX) may be realized by means of two cross-polarized (X-pol) antennas, however, space separation is required between them. For the most space-critical cases, it is possible to use a so-called XX-pol antenna which is composed of 2 separate X-pol antenna elements embedded in a single antenna panel. The 4RX diversity improves sensitivity performance by an average 2 to 3 dB on top of a standard 2RX diversity technique.

13.3.1.2   Mast Head Amplifiers – MHA

An additional way to improve uplink performance is to place a low noise amplifier between the antenna and the feeder cable. This solution is known as the mast head amplifiers (MHA). It aims at compensating the feeder loss in the power budget in the uplink direction and reduces the composite noise figure of the base station receiver system. Furthermore, if the noise figure of MHA is better than the one of the base station receiver, MHA may further improve the sensitivity performance.

13.3.2   Downlink Techniques

In scenarios where the maximum cell range is limited by the performance of the downlink direction, the cell coverage may be improved by means of for example the downlink transmit diversity techniques. Among others, the delay diversity (DD) and antenna hopping (AH) methods have become the most popular ones since they do not require any changes in mobile stations receiver and hence provide full gains for all terminal types.

13.3.2.1   Delay Diversity – DD

The delay diversity technique refers to a transmission scheme in which delayed copies of the same modulated signal are transmitted over multiple antennas. The antenna signals are de-correlated by ensuring sufficient antenna spacing or by applying cross-polarized antennas. On the MS side, the receiver equalizer is able to separate the signals thanks to the applied time delay (up to several symbols) and in turn to coherently combine them. On the transmitter side, a delay is first induced in the digital domain after which the signals are amplified separately in the analogue domain. When two antennas are used, this gives 3 dB extra transmission power that is double transmission power, to which an additional 2 to 3 dB diversity gain can be assumed in link budget calculations.

Moreover, the performance of the delay diversity technique can be further enhanced by rotating the phase of the signal transmitted by the second antenna. The idea behind the

---

[8] Cross-polarized antenna contains two polarization planes ($+45°$, $-45°$) in a single unit. It achieves diversity in the same antenna because the main and diversity dipoles are separated by orthogonal polarization planes.

[9] Antenna that produces an electrical wave vector in a vertical plane.

phase hopping is to prevent continuous, destructive summing of the received signals (e.g. when the phases of the received signals are in opposition). Since the diversity gain of phase hopping comes through channel coding and interleaving, similar to that of terminal movement and frequency hopping, this technique is therefore most beneficial in case of slow-moving terminals and non-hopping networks.

13.3.2.2 Antenna Hopping

Antenna hopping refers to a scheme in which the transmitter alternately switches between the two antennas. The idea is to de-correlate the fading profile between bursts, and reach an improved BLER/FER performance through interleaving and channel coding. This technique brings the highest benefits in non-frequency-hopping scenarios or in cases when frequency hopping is not efficient because of for example high correlation between the frequencies.

13.3.2.3 Feederless Sites – FL

Certain undesirable attenuation in the antenna system of a typical BTS site is caused by a feeder cable between the antenna and the transceiver module. In a feederless site, as depicted in Figure 13.3, the transceiver modules are located as close as possible to the antenna. In addition, either the whole base station hardware is located on top of the tower or the baseband system modules are connected by means of fiber optic cables and optical converters.

Hence, in both downlink and uplink directions, the signal attenuation due to the feeder cable is eliminated or greatly reduced, which results in link budget improvements.

The combinations of the afore-mentioned uplink and downlink features might be considered when planning a well-balanced system from the coverage point of view, as illustrated in Figure 13.4.

13.3.3 Energy Savings Due to Coverage Enhancements

Coverage enhancements solutions presented in Sections 13.3.1 and 13.3.2 can lead to significant site count reduction without affecting the quality of the network. However, this gain

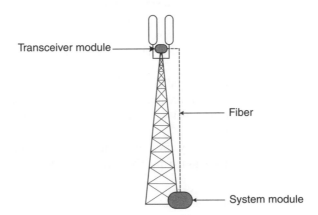

Figure 13.3   Tower installation of a feederless site.

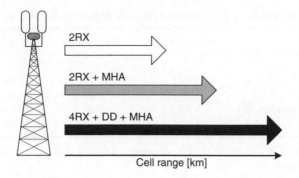

Figure 13.4 Coverage improvement solutions.

cannot be directly transformed into energy savings at the network level since the deployment of coverage boosters at a BTS site may increase its power consumption.

In Table 13.1, the performance of various functionalities that improve the cell coverage is compared against the reference 2RX diversity technique. For each scenario the number of BTS sites required to cover a rural area of 10000 sq.km with 90% probability is estimated. Three-sector sites with 2 TRXs per cell and 30-meter-high antennas with 18 dBi gain are assumed in the link budget calculations. The typical output power and sensitivity values for the GSM 900 MHz band and state-of-the-art BTS and MS equipment are used.

Although the MHA deployment on top of two-branch receive diversity increases the cell range significantly, the scenario remains still uplink-limited – as in the reference case. The application of the 4RX solution further extends the cell range but becomes downlink-limited. The additional application of DD allows a site-to-site distance increase by approximately 43% compared to the reference scenario.

The increase of the site area allows a reduction of the site count, the level of which for different coverage boosters is presented in Table 13.2, together with the influence on the power savings for the exemplary BTS power consumption figures.

Tables 13.3 and 13.4 depict the influence of the feederless site deployment on the cell coverage and power savings respectively. Downlink coverage improvement solutions are not applicable here since the scenario remains uplink-limited even for the 4RX diversity case. With a feederless site with four-branch receive diversity technique, a site count reduction of 48% can be translated into 41% of power savings.

Table 13.1 Coverage improvements

Coverage scenario	Reference 2RX	2RX + MHA	4RX + MHA	4RX + DD + MHA
Cell range (km)	11.1	13.5	14.2	15.9
Limitation	Uplink	Uplink	Downlink	Uplink
Site area (sq. km)	239	353	392	490
Site-to-site distance (km)	16.6	20.2	21.3	23.8
Sites per 10000 sq.km (rounded)	42	29	26	21

Table 13.2   Power savings due to deployment of coverage improvement solutions

Coverage scenario	Reference 2RX	2RX + MHA	4RX + MHA	4RX + DD
+MHA				
Site count	42	29	26	21
Site count reduction	–	31%	38%	50%
Power consumption per BTS (W)	940	940	1060	1610
Power savings[1]	–	31%	30%	14%

Note: [1]Power consumption figures of auxiliary equipment installed on a site, e.g. air conditioning systems, are not included.

13.4   Capacity Improvement Techniques

13.4.1   AMR, SAIC and DFCA as Green Features

13.4.1.1   AMR

AMR is an efficient quality and capacity improvement functionality that relies on improved robustness of the AMR codec modes against errors in poor radio conditions. With AMR, the need for additional base station sites due to capacity reasons can be significantly reduced as more TRXs per site can be deployed. This is because of the higher spectral efficiency it offers, that is more subscribers per MHz can be supported with AMR than without.

AMR half-rate (AMR HR) can carry the same traffic with less radio resources as two users can be served by a single radio time slot. These inherent properties of AMR can be utilized to slash the energy consumption in GSM networks.

In the case of AMRHR codec modes, although the same speech codecs as in full-rate mode are defined, better radio conditions (higher C/I values) are required for the same frame error rate (FER) performance. Therefore, half-rate utilization and consequently hardware efficiency depend strongly on the C/I distribution in the network. For example, higher HR utilization can be achieved at lower load operation points as better C/I distribution enables higher HR usage. On the other hand, in high network load conditions and worse radio interface quality, AMR HR penetration must be decreased. In that case a higher utilization of AMR FR will guarantee sufficient speech quality owing to very robust codecs such as 5.9 or 4.75 kbit/s.

The introduction of AMR codecs which can tolerate lower C/I values provides additional means to reduce the power consumed by the BTSs. By defining tighter power control thresholds for AMR calls, the transmission power used in uplink and in downlink may be lowered

Table 13.3   Coverage improvements with a feederless site

Coverage scenario	Reference 2RX	2RX + FL	4RX + FL
Cell range (km)	11.1	13.0	15.4
Limitation	Uplink	Uplink	Uplink
Site area (sq. km)	239	331	459
Site-to-site distance (km)	16.6	19.5	23.0
Sites per 10000 sq.km (rounded)	42	31	22

Table 13.4 Power savings with a feederless site

Coverage scenario	Reference 2RX	2RX + FL	4RX + FL
Site count	42	31	22
Site count reduction	–	26%	48%
Power consumption per BTS (W)	940	940	1060
Power savings	–	26%	41%

efficiently, and energy savings ensued. The impact of AMR on BTS transmission power has been verified by means of system level simulations (see Figure 13.5) performed on top of a cluster comprising 122 BTSs with maximum output power varying between 37 and 42 dBm. A dynamic range of a downlink power control equal to 30 dB was used.

As Figure 13.5 depicts, the introduction of the AMR codecs combined with AMR optimized power control settings reduces the usage of the high transmit power levels. The application of tighter power control thresholds for AMR mobiles would help reduce the BTS transmit power even more. However, further energy savings would be realized at the cost of a compromised speech quality.

The average BTS power reduction due to AMR optimized power control settings is shown in Figure 13.6 (a). It can be seen that the reduction is equal to approximately 2.5 dB for a conservative power control strategy while for an aggressive one it exceeds 3 dB. The transmit power reduction can be transformed into a lowered BTS energy consumption. As shown in Figure 13.6 (b), the energy savings are even up to 7% per BTS for aggressive power control settings. The aggressive power control strategy means that the PC thresholds are set to less stringent values, hence effectively reducing the corresponding transmission power.

13.4.1.2 Dynamic Frequency Channel Allocation (DFCA)

DFCA is another functionality that can help to reduce the energy consumption in the radio access network. Through an intelligent management of radio resources DFCA improves the

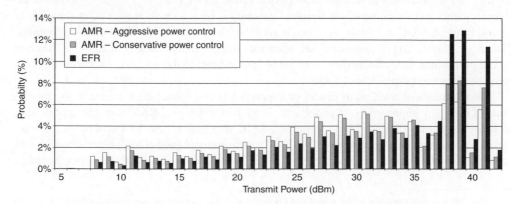

Figure 13.5 Transmit power distribution with AMR and EFR codecs.

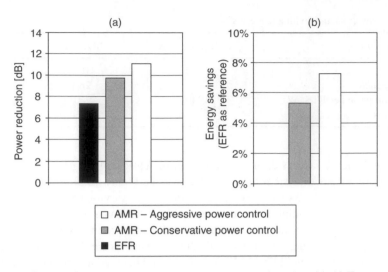

Figure 13.6   Power reduction (a) and energy savings (b) with AMR.

overall C/I distribution in the network. This improvement can be utilized to increase the network capacity without building new sites. In addition, DFCA allows very high HR use without compromising the speech quality. A higher HR penetration leads to a higher BTS hardware utilization which in turn requires a lower number of TRXs to support the same traffic load. And thanks to lower TRXs configurations, the combining losses[10] might be minimized as well. Therefore, the number of sites due to enhanced cell ranges can be reduced. A detailed description of the DFCA functionality can be found in Chapter 9.

13.4.1.3   Single Antenna Interference Cancellation (SAIC)

SAIC is a generic term for single antenna receiver techniques which attempt to cancel or suppress interference through signal processing. It is mainly intended for use in the downlink direction, where terminal space and aesthetics typically preclude the use of multiple antennas. SAIC was standardized in Release 6 as DARP Phase 1 (see Chapter 1), that is performance requirements and signaling of the terminal capability were defined. However, at some operators' request, SAIC terminals were made available that were non-DARP Phase 1 terminals. The first commercial pre-release 6 SAIC terminals were available already in 2004 but they had some limitations, for example they were not able to indicate to the network that they supported advanced receiver capabilities [2, 4].

The main idea of SAIC is to cancel or suppress the co-channel interference at the mobile side. Owing to improved co-channel link level performance a SAIC-capable mobile can tolerate higher interference level than a non-SAIC terminal. Figure 13.7 compares the link-level performance of an exemplary SAIC and a conventional (non-SAIC) receivers. Channel conditions in these simulations are typical urban 3 km/h (TU3) and ideal frequency hopping.

---

[10] Combining losses stands for all the attenuations between a TRX output and an antenna feeder input, introduced by units separating the TX and RX paths and coupling several TRX into a single antenna.

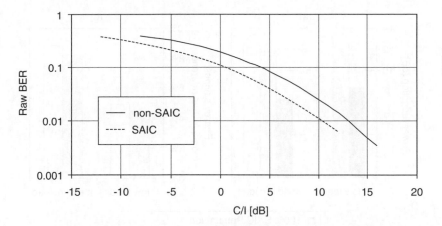

Figure 13.7   Link-level performance, SAIC vs. non-SAIC receiver.

As can be seen (and expected), the SAIC receiver has better Raw BER[11] performance than a non-SAIC receiver. Consequently better received signal quality values (RXQUAL[12] [5]) can be reported by a SAIC mobile station in certain radio conditions. Since RXQUAL is one of the inputs to the power control algorithm, a lower transmission power in downlink direction in turn can be expected for SAIC mobiles.

A BTS serving SAIC mobiles can thus transmit with lower power. This in turn reduces the overall level of interference which improves the performance of the conventional non-SAIC mobiles as well.

Moreover if the SAIC terminal penetration increases, the downlink power control algorithm is able to reduce the BTS transmit power more efficiently. This can be seen in Figure 13.8

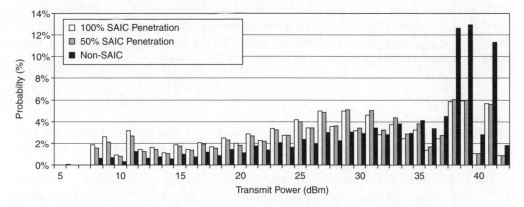

Figure 13.8   Tx power distribution for different SAIC terminal penetration.

---

[11] Bit error rate before channel decoding.

[12] A reported RXQUAL value indicates a Raw BER range experienced over the measurement period. The lower the RXQUAL value, the lower the bit error rates, that is the better the received signal quality.

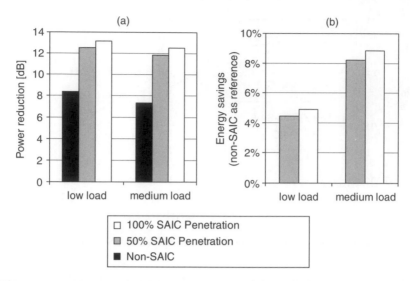

Figure 13.9   Power reduction and energy savings with SAIC: (a) power reduction; (b) energy savings.

which shows the distribution of BTS transmitted carrier power for different SAIC terminal penetration rates. One can note that the percentage of radio bursts transmitted with high power drops significantly once SAIC capable mobiles are present in the network. At the same time the share of bursts transmitted with reduced power increases compared to the reference case with non-SAIC terminals only.

The average BTS power reduction resulting from SAIC introduction is shown in Figure 13.9 (a). It can be noticed that for 50% penetration of SAIC capable terminals the average reduction was approximately 4 dB higher compared to the non-SAIC scenario. With 100% of SAIC terminals, the reduction was slightly higher, that is approximately 4.5 dB compared to the non-SAIC scenario. There is a clear correlation between the transmit power reduction and the BTS energy consumption. As shown in Figure 13.9 (b), the energy savings coming from SAIC technique reach up to 9% per BTS for a pure SAIC scenario.

13.4.2   OSC

Orthogonal sub-channels (OSC) is a voice capacity feature that aims to improve the cost efficiency of GSM voice services by allowing two mobile stations to be multiplexed simultaneously on the same radio channel, each occupying a sub-channel.

The technical method of doubling the voice channel capacity in downlink is based on a QPSK-like modulation, where each of the sub-channels is mapped so that it can be received as a (legacy) GMSK signal. In the uplink direction the traditional GMSK modulation is applied in combination with 2×2 multi-user MIMO[13] technique. Both orthogonal sub-channels are simultaneously received by the BTS that needs to employ, for example diversity and interference cancellation means separating the two orthogonal sub-channels. Different

---

[13] MIMO: Multiple input multiple output.

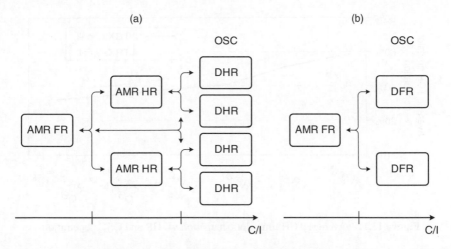

Figure 13.10 Double half-rate and double full-rate modes.

training sequence codes are assigned to terminals simultaneously sharing the same channel in order to optimize the link performance, in both uplink and downlink directions. With these techniques combined with AMR half-rate mode, up to four legacy SAIC handsets can be served within a single radio time slot in a so-called Double Half-Rate (DHR) mode (see Figure 13.10). The OSC concept can also be extended to AMR full-rate mode offering the capability of multiplexing two full-rate connections on the same radio time slot in Double Full Rate (DFR) mode. Note that in this sub-section the focus is on DHR mode only.

OSC enables up to four users to be allocated on the same radio time slot which significantly increases the BTS hardware efficiency. This means that in contrast to a non-OSC scenario either less radio resources are needed to carry the same CS traffic or a higher traffic volume can be served with the same number of TRXs. The BTS hardware efficiency gain greatly depends on DHR mode utilization itself, the penetration of SAIC-capable mobiles, the network load conditions, the interference distribution, etc. Additionally, the path-loss criterion plays an important role in the multiplexing procedure as only a limited difference between link budgets of paired users is allowed.

Figure 13.11 shows the average TCH time slot occupation as a function of different AMR FR/HR and OSC penetrations. In all the scenarios for the given TRX configuration the same offered traffic was used. It was calculated for 100% AMR, the FR case was at 2% of blocking probability. As can be observed, AMR HR significantly decreases the average TCH time slot occupation while OSC may reduce this KPI[14] further by around 20–25%.

The TCH time slot occupation has a substantial impact on the amount of energy expended per BTS hardware. The major portion of the energy is consumed by TRXs, especially by power amplifiers during on-transmission periods. Hence higher energy savings can be achieved with OSC as the number of active time slots for a given traffic load is lower than in non-OSC scenarios. The potential savings in the BTS power consumption are depicted in Figure 13.12. The savings obtained in the scenario with high HR utilization can be extended by an additional

---

[14] KPI: Key performance indicator.

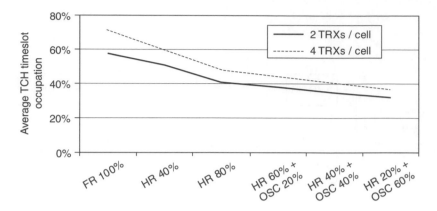

Figure 13.11   Average TCH time slot occupation vs. HR and OSC penetrations.

approximately 30% with the introduction of OSC. It is worth noting that with sufficient OSC penetration, it is possible to reduce the number of deployed TRXs without compromising the network grade of service. Thus the afore-mentioned energy savings could be boosted even further.

13.4.3   Power Control on Non-BCCH Transceivers

The aim of power control (PC) is to adjust the level of the power transmitted by a BTS or by a mobile so that an adequate transmission quality is maintained on the radio communication path in both downlink and uplink directions. This results in a decreased interference level in the network. However, another positive effect is a reduced power consumption at the BTS and MS sides.

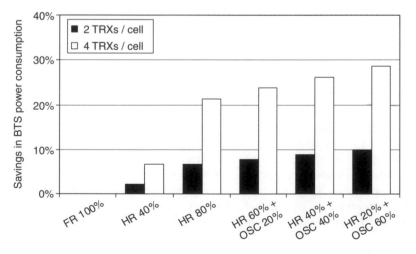

Figure 13.12   BTS power consumption savings vs. HR and OSC penetration.

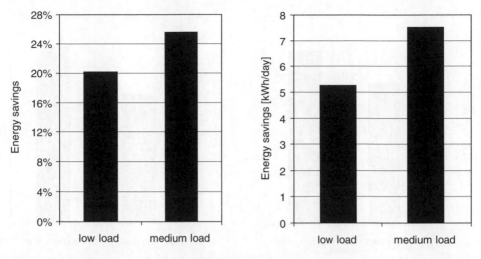

Figure 13.13   Energy savings from power control referenced to the case with no power control.

The achievable savings of the PC algorithm on the BTS power consumption were verified by network level simulations executed on top of the same scenario as described in Section 13.5.1. Figure 13.13 depicts the absolute (in kWh/day) and relative savings as a result of PC activation for two different network load conditions. As proved by simulations, PC reduces the BTS power consumption to a large extent. Up to 8 kWh/day can be saved, which corresponds to 25% cuts in energy expanded by the BTS.

13.4.4   Discontinuous Transmission

Other effective means of decreasing the BTS power consumption are to activate the discontinuous transmission (DTX) functionality. DTX avoids transmitting when a silent period is detected during a speech connection. In a typical speech call, the involved parties are active alternately, hence leaving periods of silence. A subscriber talks for on average 50–60% of the conversation time. This, however, depends on the language family and speaker's temperament. As a result, when a subscriber does not speak, the transmitter can be switched off. Thus, not only can the interference level in a network be decreased (which is the main motivation to introduce DTX), but the power consumption of the BTS and the MS could be decreased as well.

The savings in the BTS power consumption due to the application of the DTX feature were verified by means of network level simulations performed on the same scenario as described in Section 13.5.1. Figure 13.14 depicts the absolute (in kWh/day) and relative savings from DTX for two different network load conditions. As shown, the DTX can save up to 3.5 kWh/day per base station. This is approximately 12% of the power consumed by the BTS in a non-DTX case.

13.4.5   Energy Savings Due to Capacity Enhancements

As presented in the previous sections, the features intended to improve hardware and spectral efficiency may lead to significant reductions of the energy consumption in GSM networks.

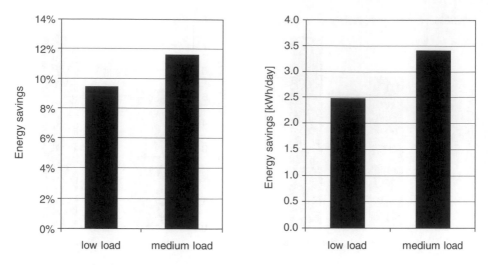

Figure 13.14   Energy savings from DTX referenced to the case with DTX off.

In Table 13.5, exemplary combinations of particular features are compared in the context of power consumption savings resulting from possible reductions in the site count and cell configurations. As the reference, the EFR-only network scenario with a basic set of standard GSM functionalities such as radio frequency hopping (FH), Power Control (PC) and Discontinuous Transmission (DTX) was considered. Assuming a maximum effective frequency load (EFL) of 9–12% in the hopping layer for a well-optimized network scenario with 5-MHz bandwidth deployment, approximately 16.5 Erlangs can be served by every single cell without compromising the network quality in terms of speech quality (reflected by bad quality samples (BQS)) as well as dropped call rate (DCR). This traffic has to be carried on 4 TRXs in order to keep the blocking probability below 2%.

The introduction of AMR FR, owing to its increased robustness against errors in poor radio conditions, achieves higher EFL without violating the soft blocking limit (a maximum EFL for AMR FR of 20% has been assumed). This, in turn, makes it possible to increase the site capacity by deploying additional TRXs and consequently to reduce the number of sites needed to provide a certain guarantee of service. In the analyzed example, the number of sites

Table 13.5   BTS site count and TRX configuration with capacity improvement solutions

Capacity scenario	Reference EFR, 1/1 FH	AMR FR	AMR HR 80% + DFCA	AMR HR 40% + OSC 40% + DFCA
Number of TRXs/cell	4	6	4	3
Number of TRXs/network	1 200	972	648	486
Number of Sites	100	54	54	54
Erlangs/cell	16.5	31.0	31.0	31.0
Erlangs/network	4950	4950	4950	4950

Table 13.6 Energy savings with capacity improvement solutions

Capacity scenario	Reference EFR, 1/1 FH	AMR FR	AMR HR 80% + DFCA	AMR HR 40% + OSC 40% + DFCA
Site count	100	54	54	54
Site count reduction (%)	–	46%	46%	46%
Power consumption per BTS (W)	1 500	2 400	1 600	1 350
Power consumption per network (W)	150 000	129 600	86 400	72 900
Power savings (%)	–	14%	42%	51%

can be decreased by 46% which corresponds to a 14% lower power consumption (see Table 13.6). The activation of AMR half-rate mode enables a considerable reduction in the number of TRXs needed to carry a given traffic without any hard blocking deterioration. This yields further energy savings. Forcing AMR HR utilization (80%) with the help of the DFCA feature yielded about 40% energy savings. Further cuts in power consumption were obtained thanks to the introduction of the OSC functionality. With 40% OSC penetration, the energy savings reached 51%.

The above figures expressed as the ratio between the consumed energy and the traffic carried by the BTS are presented in Figure 13.15.

13.5 Energy Savings through Software Solutions

13.5.1 Energy-Aware Resource Allocation

The aim of a radio resource allocation algorithm in the circuit-switched domain is to search and allocate radio channels for requests coming to the network during for example a call establishment or a handover procedure. The radio channel selection algorithm can be based on criteria such as cell load conditions, terminal and cell capabilities, quality of the available radio time slots or preferred traffic allocation strategy, that is on a non-BCCH layer or on a BCCH layer. A typical approach is to prioritize the non-BCCH transceivers for voice call while the

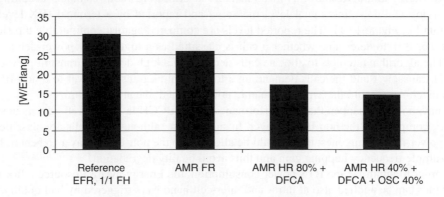

Figure 13.15 BTS energy consumption per Erlang.

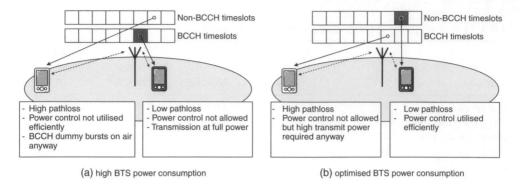

(a) high BTS power consumption   (b) optimised BTS power consumption

Figure 13.16 Typical (a) vs. energy-aware (b) resource allocation.

BCCH transceiver is predominately reserved for packet-switched connections. This is mainly because of a better quality offered on the BCCH layer resulting from a looser frequency reuse pattern. With this approach a voice call is allocated to the BCCH transceiver only if the TCH ones are fully occupied. However, the opposite strategy can be justified as well, for example in narrowband deployments where more aggressive BCCH frequency reuse patterns are applied, leading to poorer quality experienced on this layer.

The above-mentioned allocation rules are not optimal in the context of BTS power energy consumption. One of the reasons is that calls being able to operate with low downlink transmit power may be allocated to the BCCH transceiver on which power control is not allowed and therefore full power is constantly used as shown in Figure 13.16 (a). If the calls were allocated to a non-BCCH transceiver, the transmit power could be reduced by means of the power control procedure, consequently leading to lower BTS power consumption. On the other hand, the allocation to the non-BCCH transceiver of calls that require high downlink transmit power significantly limits the usage of the power control algorithm. Instead, by assigning such calls to the BCCH transceiver on which the maximum transmit power is used regardless of whether time slots are occupied or not, the BTS power consumption could be optimized. Through allocation of high power demanding calls to the BCCH transceiver, the unused time slots on the non-BCCH transceivers could be assigned to calls that can utilize power control more efficiently as presented in Figure 13.16 (b).

The Energy-aware Resource Allocation aims at allocating radio channels according to the serving cell downlink signal level measured and reported by a terminal (RX level) as illustrated in Figure 13.17. The reported levels are compared against configurable thresholds in the network to determine whether a call is classified as a high or low power-demanding one. The algorithm attempts to allocate calls demanding a high downlink transmit power to a BCCH transceiver and the calls demanding a low downlink transmit power to a non-BCCH transceiver. As a result, the average dynamic power reduction can be maximized.

It may, however, happen that the performance of the BCCH layer at a cell edge is unsatisfactory compared to the non-BCCH layer. In such a case, although the calls are classified as high power-demanding ones, they should be allocated to the non-BCCH layer to benefit from for example frequency hopping gain and thus avoid quality degradation.

In order to minimize the BTS power consumption, the Energy-aware Resource Allocation principles can be applied also at intra- and inter-cell handovers triggered by bad quality, low signal level, power budget, etc.

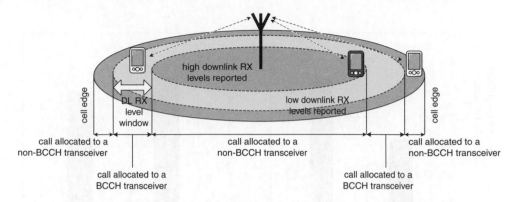

Figure 13.17   Energy-aware resource allocation based on downlink RX level.

The performance of the Energy-aware Resource Allocation has been evaluated by system level simulations run on top of a real network scenario with 3-sectorized BTS sites, the configuration of which was 4 TRXs per cell that is 1 BCCH and 3 non-BCCH TRXs. The standard channel allocation priority strategy, that is non-BCCH layer preferred, was used as the reference for the Energy-aware Resource Allocation. The threshold level that determines whether a call is classified as a high or a low power demanding one was set in correlation with the BCCH downlink reception level distribution experienced in the simulated network. The power control window was adjusted to ensure that the calls allocated to the non-BCCH transceivers were immediately subject to the power reduction as shown in Figure 13.18. The simulations were performed for different network load scenarios.

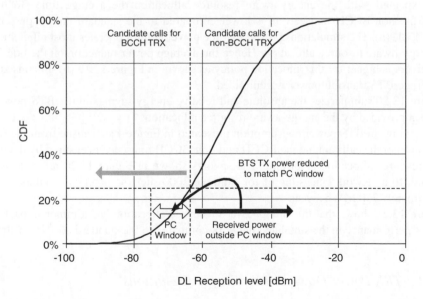

Figure 13.18   BCCH RX level distribution in the simulated network and the corresponding parameter settings.

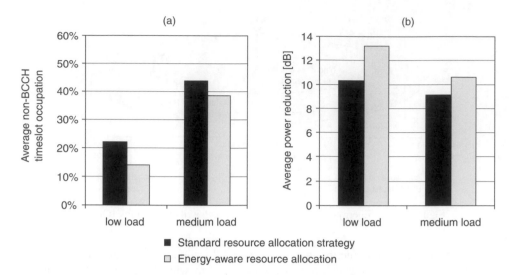

Figure 13.19   Average time slot occupation and average power reduction levels on the non-BCCH transceivers for the simulated resource allocation strategies.

The potential savings in the BTS power consumption thanks to energy-aware resource allocation were benchmarked against the standard resource allocation strategy by evaluating the average time slot occupation and the average power reduction on the non-BCCH transceivers. Moreover, the network performance was monitored by evaluating the overall bad quality samples in uplink and downlink with a frame erasure rate (FER) greater than 2.1%.

As expected, with the energy-aware resource allocation the average time slot utilization of the non-BCCH transceivers is lower compared to the standard one as presented in Figure 13.19(a). The simulations proved that the power control operates most efficiently with the energy-aware resource allocation. Hence the average power reduction on the non-BCCH transceivers is higher for this allocation policy as shown in Figure 13.19 (b). In consequence, the average BTS transmit power is minimized.

Figure 13.20 summarizes the absolute and relative energy savings in the BTS power consumption provided by the energy-aware resource allocation.

The savings in BTS power consumption presented in Figure 13.20 ensue from the optimal distribution of the calls among the BCCH and non-BCCH transceivers ensured by the energy-aware resource allocation algorithm. The savings shown in Figure 13.20 refers to a single BTS site. In a network consisting of for example 2000 sites the potential savings would be approximately 730 MWh a year.

Figure 13.21 shows that the energy-aware resource allocation has a minor impact on the network performance in the simulated network compared to the standard allocation strategy.

13.5.2   TRX Power Down in Low Traffic Conditions

It is a well-known fact that the network resources have to be dimensioned according to the busy hour traffic. This means that on the radio interface the number of transceivers must be

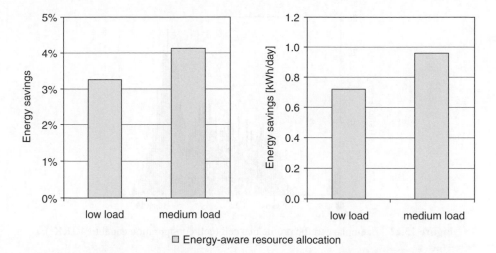

Figure 13.20   Absolute and relative energy savings referenced to the standard allocation strategy.

sufficient to handle the peak traffic demand in the cell according to required grade of service and/or other quality of service requirements. The network load varies a lot in a day and over the days of the week. In addition, the traffic during off-peak hours might be significantly lower than during the busy hour. Consequently a lot of hardware resources are simply unused during low traffic periods and part of them can be switched off with no impact on network accessibility.

Figure 13.22 illustrates the hourly profile of the served traffic over a day derived from a 4-TRX cell. As can be seen, there is a huge variation between the served amount of traffic during off-peak hours (between 11 pm and 8 am) and during busy hours (between 7 pm and 10 pm).

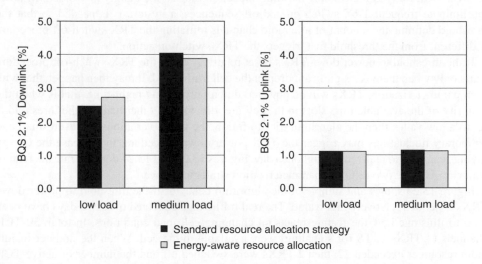

Figure 13.21   Bad quality samples in uplink and downlink for the simulated resource allocation strategies.

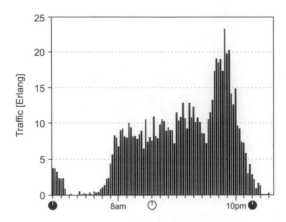

Figure 13.22 Exemplary traffic profile in a cell (cell configuration equal to 4 TRXs).

It can be noted that in off-peak hours there is no need to keep all four TRXs active as the traffic is low enough to be served even by a single TRX without compromising on network accessibility. By exploiting the traffic variations in the network, it is possible to achieve significant power consumption savings.

The TRX power down solution monitors the number of unused time slots in the cell. When that number exceeds a certain threshold, some TRXs are automatically disabled so that only those necessary for managing the current traffic load are supplied. However, the decision to disable TRXs is taken only if the number of free channels remains high enough for the specified supervision period.

As soon as the traffic in the cell starts to increase and the number of unused time slots falls below a predefined threshold, the powered-down TRXs are immediately brought back into service. With this mechanism any harmful impact on the service quality is avoided. In order to eliminate frequent TRX switch on and off sequences, a hysteresis is needed so that the threshold defining the amount of idle radio channels initiating the TRX switch-off procedure is different from the threshold that triggers the TRX switch-on action.

If the threshold to power down TRXs is set relatively high, the TRX switch-off procedure starts only when there is a significant drop in the cell traffic load. It may then happen that with such parameterization, TRXs will be powered down only during night-time, that is when the majority of the available time slots in the cell are idle. On the other hand, if the threshold is set to a low value, then the algorithm tries to follow the traffic variations and even small and temporary fluctuations may trigger the TRX power down procedure. In this case the energy savings can be also realized throughout a day. Figures 13.23 and 13.24 demonstrate the feature behavior with the two different parameterization strategies above.

Figure 13.23 shows an example of the algorithm behavior if the threshold to power down TRXs is set to a relatively high value. The real traffic data collected over two days have been used to illustrate how the feature reacts on changing cell load conditions. In total, 30 TCH channels (4 TRXs, 2 TS for signaling) were available in the cell. When the number of idle radio resources exceeded 22, then 2 TRXs were switched off and the number of active TCH channels was reduced to 14. It can be observed that TRXs were mainly deactivated at night. The reason is that the traffic during the day rarely dropped below the level which initiates

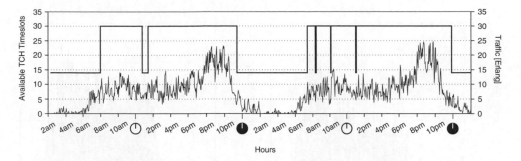

Figure 13.23 Example of TRX automatic power down behavior – number of active TCH channels in a cell (case 1).

TRX power down procedure. On the contrary, Figure 13.24 presents the algorithm behavior in case the threshold to power down TRXs was set relatively low, that is when just 17 idle radio resources triggered TRX to shut down. It can be seen that TRXs in that scenario are switched on and off more frequently than in the previous case. This means that energy could be saved during daytime as well.

The cell traffic pattern is another factor that has an impact on the level of reduction of the energy consumed by the BTS. If the traffic load in the cell during busy hours is significantly higher than in the remaining hours (see Figure 13.25 (a)), TRXs may be switched off earlier and remain in standby mode for a longer time. This will account for higher energy savings. If the cell has a more flat traffic distribution (see Figure 13.25 (b)), then energy saving will be lower as more TRXs are required to serve traffic in off-peak hours.

Besides traffic profile, the amount of traffic in the cell can also strongly influence the efficiency of the feature. The higher the traffic volume carried by the cell during busy hours, the more the number of TRXs that need to be installed to meet the required grade of service. However, in such cells, especially at night, the traffic often drops to a very low level so that the difference between the maximum and minimum loads is relatively high. As a result, in such cells, potentially more TRXs may be switched off and eventually a higher portion of the energy may be saved.

Certain priorities are considered by the TRX switch-off procedure so that some TRXs are preferred for powering down earlier than other. For instance, TRXs with a lower number of

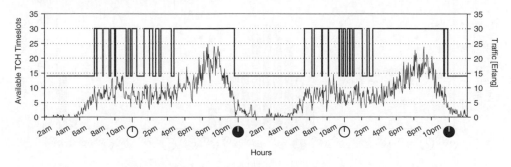

Figure 13.24 Example of TRX automatic power down behavior – number of active TCH channels in a cell (case 2).

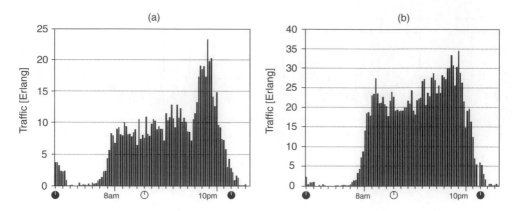

Figure 13.25 Exemplary traffic profiles, (a) distinctive busy hour; (b) flat profile.

traffic channels configured (half-rate, dual-rate, full-rate) are the preferred ones for powering down. Additionally, some TRXs cannot be switched off at all as it would prevent normal BTS site operation, for example the BCCH TRX cannot be powered down. Moreover, the TRX management algorithm must not disturb the already established connections. Before being switched off, a TRX is cleared from the traffic by standard intra-cell handover procedure.

The amount of energy that can be saved by deactivating TRXs during low traffic periods depends as well on the type of hardware installed on a BTS site and on the methods that are applied to power down a TRX. In some solutions a whole TRX is switched off, that is baseband processing together with the RF part. Another method is to switch off the power amplifier (PA) unit only. Depending on how TRXs are powered down, up to 50 W per TRX might be saved. For example, in the case of 3-sector BTS site, each sector with 4 TRXs, the energy savings can reach approximately 1 kWh per day provided that 2 TRXs per cell are switched off for 8 hours.

The time-domain behavior of the TRX Power Down functionality activated in a single cell with 6 TRXs is presented in Figure 13.26. The number of occupied time slots reflects the utilization of these TRXs and follows the CS traffic variation in the network reaching its minimum level with only a few active channels during night-time period. The corresponding figures that represent the number of switched off TRXs are given on hourly basis. The achieved

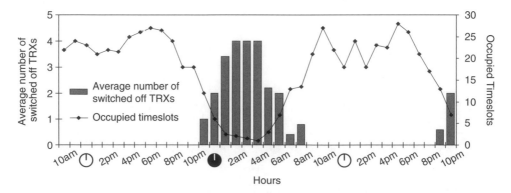

Figure 13.26 Example of TRX power down time.

mean electricity savings due to putting the TRXs in standby mode were equal to approximately 1 kWh/day per cell during extensive observation time. This proves that a considerable gain from the energy consumption viewpoint can be expected at network level thanks to deploying TRX Power Down enhancement.

13.5.3   Energy-Saving Mode on BCCH Transceiver

Additional energy savings might be obtained by activating a so-called energy saving mode for a BCCH transceiver. With this software-based solution, an overall power consumption is reduced by transmitting dummy bursts on idle channels with 2 dB lower power level compared to the maximum power of BCCH transceiver. The highest energy savings achievable by the power reduction on dummy bursts are possible at low traffic periods for example at night when there is no traffic served on the BCCH transceiver. Further potential energy savings come from dynamic power control applied on active dedicated circuit switched channels (TCH/SDCCH) of the BCCH transceiver. With the dynamic power control on busy traffic channels of the BCCH transceiver, the reduction of power consumption is possible at high traffic periods as well. The downlink power control on the BCCH transceiver does not differ from a standard power control utilized on non-BCCH transceivers apart from the fact that the maximum dynamic power reduction is limited from 30 dB to only 2 dB. As in case of the standard power control in downlink, the received signal quality (RXQUAL) and the signal level (RXLEV) of a serving cell measured and reported by a mobile station are used as inputs to the power control algorithm. If the averaged signal level and quality experienced by a call handled on the BCCH transceiver indicate good radio conditions, then a 2 dB power reduction is used. Otherwise, transmission is done at full power. The broadcast and common control channels (BCCH/CCCH) that are carried on time slot 0 are always transmitted at full power. This ensures that the cell range from system access viewpoint is not affected.

In the case of packet switched connections served on BCCH time slots, the dynamic power reduction is not applied. This prevents decreasing the user throughput due to downgrading of the coding schemes that could result from the reduction in the transceiver output power. Furthermore, this restriction guarantees that the cell range and data rate distribution for the GPRS/EDGE services are not affected by this feature. Figure 13.27 illustrates the concept of energy-saving mode for the BCCH TRX.

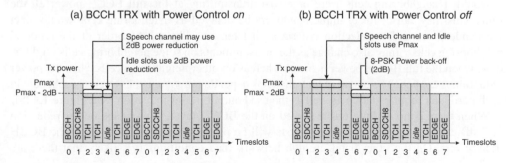

Figure 13.27   Concept of the energy saving mode for the BCCH TRX (a) vs. BCCH TRX operating at full power (b).

Table 13.7 Savings in BTS power consumption due to application
of the power control on the BCCH transceiver

TSLs/BCCH TRX with	Exemplary power savings	
2 dB power reduction	(W)	(%)
1	9	2
2	18	4
3	27	6
4	35	7
5	44	9
6	53	11

The potential gains from that functionality depend on how many time slots of the BCCH a transceiver can operate with a reduced output power. The number of time slots of the BCCH TRX that are subject to 2 dB power reduction will depend on factors such as the traffic load in the cell, distribution of the users, traffic management policy, for example BCCH preferred vs. non-BCCH preferred time slot allocation for CS/PS services, and quality of the BCCH layer. Table 13.7 summarizes the exemplary energy savings per 3-sector BTS site with 1 TRX configuration as a function of the number of time slots operating with 2 dB power reduction. The percentage values in Table 13.7 indicate the energy savings compared to a case when all the time slots of the BCCH transceiver are transmitted with full output power.

The BCCH carriers are continuously transmitted in the downlink direction so that the minimum reuse for the BCCH layer is clearly limited by the performance of that link. In this perspective, the energy-saving mode on the BCCH TRX may provide another benefit, that is the reduction of the interference level that is caused by BCCH transmissions. While this may lead to an improved network quality, the gain strongly depends on the average power reduction applied on the BCCH layer.

If an operator decides to enable the energy-saving mode on the BCCH TRX in certain cells, the cell selection, cell reselection and handover procedures involving these cells might be slightly affected. This is due to the fact that the signal level measured by the mobile at some instances in time will be affected by the possibly lower output power level of the BCCH carriers. When the power reduction is applied, a terminal reports lower signal levels than it would if neighboring cells were otherwise transmitting at the full BCCH power all the time. This may cause, for example handovers be delayed as neighboring cells would meet the handover candidate selection criteria a bit later. In addition, the order of the reported neighboring cells[15] may also change as the measurements on some neighboring cells might be done when the full BCCH power is in use whereas on other neighboring cells when the power reduction is applied. The extent of that impact is dependent upon the measurement schedule in each particular mobile station as well as the used output power reduction (See Figure 13.28).

When there is no or little traffic served on the BCCH layer, the average power reduction of 2 dB can be expected as dummy bursts will be predominantly transmitted by the BCCH transceivers. On the other hand, in high network load conditions it is likely that there are

---

[15] A mobile reports neighboring cell measurements in decreasing order of strength, the strongest cell first.

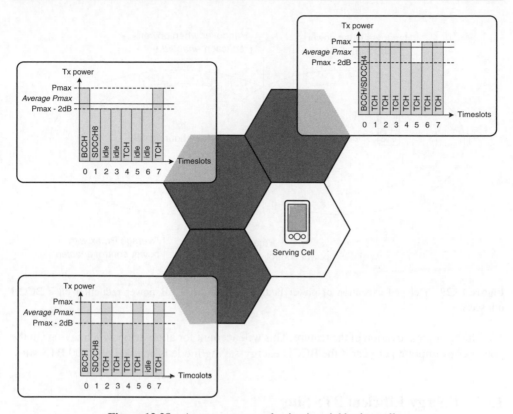

Figure 13.28   Average power reduction in neighboring cells.

hardly any dummy bursts transmitted and on the dedicated SDCCH/TCH channels the power reduction may not be applied due to quality reasons or insufficient signal level. In such a case, the expected average deviation from the maximum BCCH signal level is approximately 1 dB.

The afore-mentioned side effects of energy-saving mode on BCCH TRX can be minimized simply by tuning the corresponding network parameters. However, the adjustment of the parameters might be needed only in specific cases when for example handover performance degradation would be noticed. It is worth noting that experience from EDGE deployment showed no need for global parameter adjustments due to 8-PSK back-off and reduced BCCH power in scenarios with EDGE services deployed on the BCCH TRX.

Figure 13.29 presents the potential impact of the BCCH TRX power reduction on the handover procedure. The handover trigger point can be shifted as the result of a lower transmitted power in the neighboring cell, for example execution of the power budget handover might be delayed.

Assuming that there are, during one day on the BCCH transceiver, on average 3 time slots idle and that 2 dB power reduction is applied to 50% of the active BCCH time slots, energy savings of approximately 1 kWh/day per BTS site can be achieved. The reductions in the BTS power consumption achievable through BCCH energy saving mode were demonstrated in a field trial. Figure 13.30 shows the BTS power consumption for periods with and without the feature activated. It can be observed that the average BTS power consumption drops by around

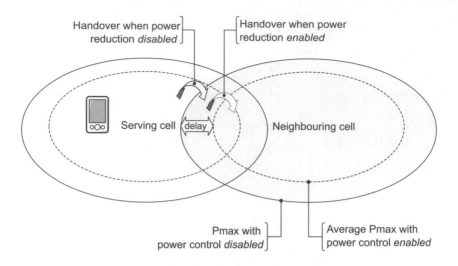

Figure 13.29 Delayed execution of power budget HO due to 2 dB power reduction on a BCCH transceiver.

1 kWh/day after activation of the feature. This will account for almost 365 MWh savings in the power consumption per year if the BCCH energy-saving mode is enabled on 1000 BTS sites.

13.6 Energy-Efficient BTS Site

The hardware and cooling system of a base station's hardware can be identified as the main contributors to the energy consumption of a BTS site. Although the cooling system is only part of the auxiliary equipment installed on the BTS site, its power consumption reaches 30% of the total BTS site power consumption. The remaining 70% is utilized by the base station hardware itself (see Figure 13.31). Of this, the largest portion, up to 80%, is consumed by the radio part and only 20% is needed for the baseband processing, transmission equipment, etc. The energy split within the base station hardware depends on the number of installed TRXs,

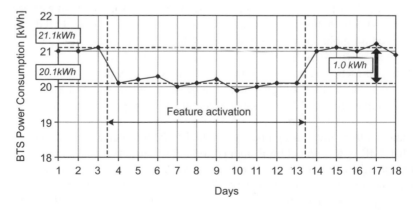

Figure 13.30 BTS power consumption with energy-saving mode for BCCH TRX feature on/off.

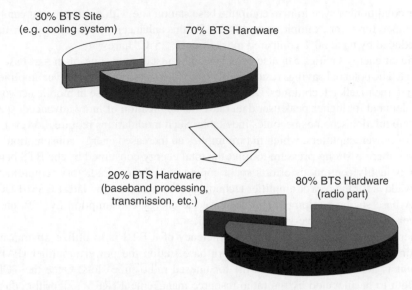

Figure 13.31 BTS site power consumption.

the utilization of radio time slots and the transmission power. These factors directly determine the amount of energy spent to supply the power amplifier stage of each TRX. Since the power amplifier is the largest energy-consuming unit of the radio part, its demand drives the share of the radio part in the overall base station power consumption.

A lower BTS power consumption can be achieved by elimination of the power-hungry air conditioning that is normally used as the cooling system. Air conditioning equipment aims at maintaining a desired temperature, usually 25°C, which is the maximum that the BTS HW can tolerate in a long time period. Significant reduction in power consumption can be obtained by replacing the air conditioning unit by a fresh-air cooling system with fans.

In a standard system, the air circulates inside the BTS cabinet through the cooler. Without the air-conditioner the cold air intake from outside and the hot air from inside are circulated by means of fans as presented in Figure 13.32. Another method to reduce the power consumption

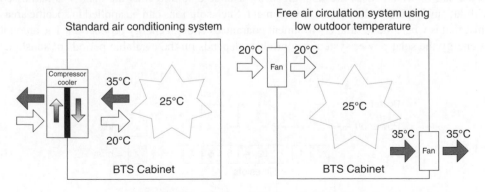

Figure 13.32 Energy-efficient cooling system.

of the air conditioning system is to equip the base station site with hardware that can tolerate higher temperatures, for example up to 40°C. With this enhancement the energy consumption can be reduced by up to 30% compared to traditional "25°C" hardware.

Significant energy savings can also be achieved by installing a state-of-the-art base station equipment. The potential savings result mainly from the use of improved power amplifier technology and linearization techniques. Due to continuous traffic increase in mobile networks and growing demand for higher peak user data rates, the application of more advanced, spectrally efficient modulation schemes becomes inevitable. Such modulations require, however, a high linearity of power amplifiers, which in turn implies an increased energy consumption. On the other hand, there is strong pressure to limit the total energy consumed by the BTS hardware. These facts justify continuous efforts spent coping with the contradictory requirements defined nowadays for the power amplifier technology. For example, the latest GSM/EDGE and WCDMA base station equipment can decrease the energy consumption by 35% and 75%, respectively, compared to five-year older equipment.

Another technique to boost the energy efficiency of a BTS is to utilize off-transmission periods on idle TCH time slots. In a modern base station the power amplifier (PA) of the carrier unit is automatically switched off for unused radio time slots. Once the TCH time slot is about to be allocated by the radio resource management (RRM) algorithm due to for example call set-up or handover, the PA is automatically activated. The corresponding power savings are of course a function of the radio time slot occupation. The highest savings can be obtained in low traffic periods when there are more radio time slots that do not carry any circuit-switched speech or user data. Depending on the hardware and traffic characteristics, the reductions in power consumption from the above mechanism may vary between 20 W to 40 W per TRX. Figure 13.33 illustrates the basic concept of efficient power amplifier management.

If the considered improvements are deployed together, the energy savings on a BTS site may reach even 50–60%. Thus energy-aware hardware solutions are powerful means operators have at hand to satisfy "green objectives" and improve the site energy efficiency while reducing their energy bill.

13.7 Renewable Energy Sources

Further energy savings might be achieved by utilization of renewable energy sources, such as the sun or wind. These are the "greenest" available solutions as the energy is harnessed without any impact on the natural environment. Such solutions can be applied in specific areas only, that is where the natural environment guarantees enough energy production. The amount of energy the solar power system can produce depends on the available periods of sunshine.

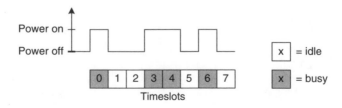

Figure 13.33   Efficient power amplifier management.

These vary depending on the geographical location, the season, the local weather and the time of the day. In Europe, the average solar radiation is relatively low and it is strongly influenced by seasons. In the Sahel region, for instance, the solar radiation is significantly higher and evenly distributed throughout the year. Depending on the latitude the sun strikes the surface at different angles ranging from 0° (just above the horizon) to 90° (directly overhead at the equator). Zones with the highest sun radiation are located mainly in Africa, Australia and South America.

The efficiency of the solar power system depends also on the number of solar panels it is composed of. Nevertheless, in areas with low sun radiation, even large solar panel constructions do not guarantee the required amount of power. In addition, the additional space needed to install them might not be available. Hence each power system has to be designed individually.

The number of solar panels has to be calculated assuming the expected power consumption of the BTS configuration, which is installed on-site. Then, the sunshine periods have to be analyzed taking into account average conditions as well as the worst case scenario sunshine point of view. Finally, solar panels have to be correctly installed to generate the highest output, typically on the North–South axis and tilted as a function of latitude.

The solar power system is usually equipped with the following main elements: solar panels (energy source), batteries (energy storage) and a system controller, which is responsible for the correct charging of batteries and optimal sharing of the load between the solar panels and the batteries (see Figure 13.34).

Solar systems operate on a fixed 24-hour cycle: a peak sunny phase, a partly sunny phase and a night phase. The peak phase covers full-load capacity and in this phase batteries are also charged. The partly sunny phase uses 100% of the capacity with surplus energy supplied by the batteries. While at the night phase, 100% of the load is drawn from the batteries.

Solar power systems can also be supplemented with a wind turbine. Such a solution guarantees higher efficiency of the system as the energy is harnessed from more than one source. The wind turbine can be an option especially in coastal and mountain areas and generally in locations of a particularly high elevation.

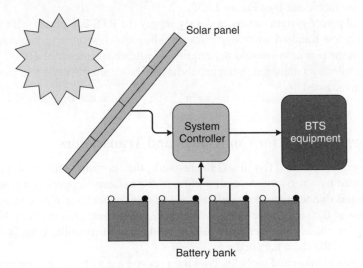

Figure 13.34   Solar power system.

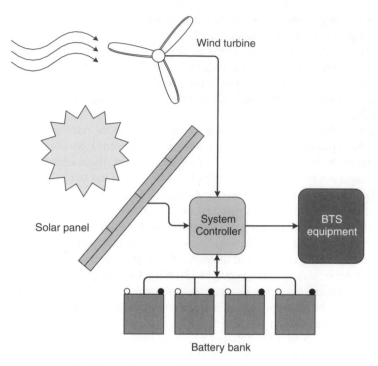

Figure 13.35   Solar/wind power system.

When a wind turbine is installed, the system controller decides which source of energy (sun or wind component) should be used to produce the energy. Of course, both sources can be utilized at the same time depending on the weather conditions. Batteries are employed when both sources are inefficient (see Figure 13.35).

A solar/wind power system can continuously supply the BTS equipment with stable power ranging from a few hundred watts up to several kilowatts. Importantly, the solar and wind components of the power system do not require significant maintenance effort. They can be remotely controlled and managed. Solar panels have to be cleaned every 4 to 6 months so their efficiency is maintained.

13.8   Energy Savings for Controllers and Transcoders

Given the large number of BTSs in a GSM network, they constitute the main portion of the energy consumed by a network. A small portion is due to Core Network equipment and only a minor portion is due to controllers and transcoders (plus an additional percentage due to the cooling process of the cabinets or rooms where these equipments are installed). Nevertheless, energy savings in these equipments are also possible and worthwhile, even if on a smaller scale compared to the radio network.

In the European Union and globally, growing pressure exists for higher energy efficiency. For example the International Energy Agency (IEA) urges new policies to reduce the energy consumption of Information Technology and Communication (ITC) equipments without referring to specific products. Moreover, European projects, such as OPERA-net (Optimizing Power Efficiency in Mobile Radio Networks), address the complete end-to-end cellular systems.

Also mobile operators are showing a growing interest in the energy consumption figures of the whole radio access network, not only of the BTSs. Up to now, the optimization of controllers and transcoders focused on features aimed at reducing costs and increasing equipment capacity, with energy savings arising as a side effect of the parallel evolution of the component efficiency.

A more conscious step on the energy-saving path would be the use of new or existing ad-hoc mechanisms to reduce the energy consumption of equipments. A possibility could be to redesign controllers and transcoders in order to be able to work at higher room temperature, thereby reducing the energy required by the cooling process. Another possibility could instead be to "modulate" the hardware consumption with the shape of the managed traffic. With this approach, it would be possible to exploit the mechanisms available in components such as for example a processor's in-built technology that dynamically adapts processor speed according to the computing load, or the execution of "IDLE" or "RESET" commands in DSPs (Digital Signal Processors) when there is no traffic to process.

Another approach instead would be to power on and off cards via software, depending on the handled traffic (load-based cards power on/off). As seen earlier in Section 13.5.2, traffic changes with time. And usually voice and data traffic inactivity periods are different. In general, it may be assumed that voice traffic is mainly present during the day while data traffic is mainly concentrated in the late afternoon and evening hours (see Figures 13.36 and 13.37).

At present in both controllers and transcoders, all HW resources are active despite traffic variations. As a result, their energy consumption remains almost constant during the whole day and night.

The main idea of the load-based power on/off of cards approach is to exploit traffic load variations by monitoring traffic requirements and powering only cards which are necessary to manage the current traffic load. In practice, this means that the current traffic is clustered on a subset of the available cards while the unused cards are powered off, thus saving energy. As traffic increases, the required cards are powered on again.

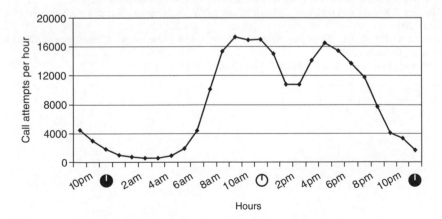

Figure 13.36   Exemplary distribution of call attempts in BSC, urban area.

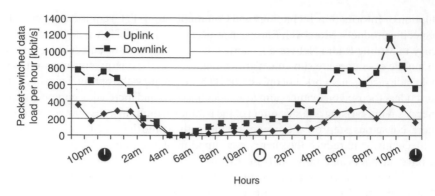

Figure 13.37   Exemplary distribution of data (packet-switched) load per BSC, urban area.

In the case of controllers, such a mechanism may be applied to packet-switched and circuit-switched services related cards. The maximum gain is obtained if the controller architecture allows powering them off independently, thus fully exploiting the different inactivity periods. In the case of transcoders, the above mechanism may be applied to speech transcoding cards by following voice traffic variations.

Preliminary energy-saving evaluations performed both on processor's speed adaptation methods (based on load) and IDLE/RESET commands-based approaches have revealed minor gains. However, a better performance has been observed in the case of load-based cards power on/off mechanism. For example, considering the traffic reported in Figure 13.36 and Figure 13.37 where both PS and CS traffic are characterized by about 6 hours of inactivity, 9% of energy saving is obtained independently of the configurations.

References

[1] WWF, "What Are the Causes of Climate Change?", http://www.panda.org/about_our_earth/aboutcc/cause/.
[2] "Environmentally Sustainable Business Is an Opportunity for Communication Service Providers", http://www.nokiasiemensnetworks.com/sites/default/files/document/Sustainability.pdf.
[3] T. Halonen, J. Romero and J. Melero, *GSM, GPRS and EDGE Performance*, 2nd edition, Chichester: John Wiley & Sons, 2003.
[4] 3GPP TR 45.903 v8.0.0, "Feasibility Study on Single Antenna Interference Cancellation (SAIC) for GSM networks", December 2008.
[5] 3GPP TS 45.008 v8.4.0, "Radio Subsystem Link Control", September 2009.

Part III

Extending the GSM Paradigm

14

GSM in Multimode Networks

Jürgen Hofmann

14.1 Introduction

This chapter deals with standardization and deployment aspects of GSM Multi-carrier Base Transceiver Stations (GSM MCBTS) in multimode networks.

The performance of GSM MCBTS has been standardized for a GSM/EDGE stand alone network in 3GPP Release 8, see Chapter 3. As an extension of the multi-carrier concept, the standardization of base stations in Multi-Mode networks is described hereafter referring to the Release 9 feature related to the specification of the multi-standard multicarrier radio base station (MSR) including a brief overview of the status in standardization and regulation. Finally, the deployment of such a multi-carrier BTS supporting multiple 3GPP radio access standards including impacts on radio resource management is shown. The introduction of base stations based on the multi-carrier architecture provides the platform for operating multiple 3GPP radio access technologies simultaneously.

14.2 Standardization of MSR Base Station for Multimode Networks

A consequential extension of the multi-carrier GSM architecture, where a single wideband transmitter and receiver is used for transmitting and receiving multiple GSM carriers, is the establishment of a multi-standard architecture for a base station, that is to enable a base station operating simultaneously different radio access technologies. In this respect 3GPP opened the work item "RF requirements for Multi-carrier and Multi-RAT BS (MSR)" [1], in September 2008. It aims at specifying the RF performance requirements of an MSR base station equipment supporting multiple 3GPP radio access technologies (RAT) without touching the baseband receiver performance specification. The technical progress of this work is reflected in 3GPP TR 37.900 [3].

GSM/EDGE: Evolution and Performance   Edited by Mikko Säily, Guillaume Sébire and Eddie Riddington
© 2011 John Wiley & Sons, Ltd

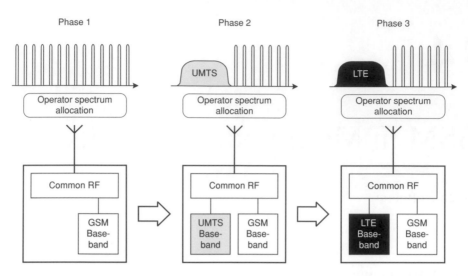

Figure 14.1 Exemplary migration path for an operator with licenses for GSM/EDGE, UTRA FDD and E-UTRA FDD.

14.2.1 Motivation

The motivation to deploy multi-standard base stations is the fact that many mobile operators to-day operate GSM (i.e. GSM/EDGE) and UMTS (i.e. UTRAN networks using WCDMA/HSPA access) radio networks in parallel and may additionally deploy LTE (i.e. E-UTRAN) radio networks in the coming years. MSR supports such migration path to allow higher bit rates and larger RF bandwidths with minimum disruption to GSM/EDGE, that is the coverage network which across the globe today accounts for over 4 billion subscribers (and growing fast). A realistic exemplary migration in three phases is shown in Figure 14.1 for an operator with licenses for GSM/EDGE, UMTS (i.e. UTRAN networks using WCDMA/HSPA access), and LTE (i.e. E-UTRAN networks) substituting a part of the GSM spectrum with WCDMA/HSPA spectrum in Phase 2 and later on replacing the UMTS spectrum with LTE in Phase 3.

The start of Phase 3 largely depends on the time of availability of E-UTRAN and the user experience it provides (i.e. vs HSPA), as well as on the population of GSM-only mobiles in the field at that time (including roaming users). Eventually all three RATs may be operated by a MSR base station in the transition from Phase 2 to Phase 3.

Note that other migration paths exist such as a direct transition from Phase 1 to Phase 3 (i.e. UMTS skipped altogether), or from UTRAN to E-UTRAN (e.g. in Japan where GSM is not deployed). The MSR standard is to cater for *all* migration paths, but from a GSM/EDGE perspective, only the ones depicted above, that is from Phase 1 via Phase 2 to Phase 3, or from Phase 1 directly to Phase 3, are relevant.

14.2.2 Deployment Scenarios

When standardizing the radio performance requirements for a MSR base station, the deployment scenarios need to be first investigated. 3GPP RAN WG 4 has taken the approach of defining frequency bands for MSR and classifying them into three band categories:

- *Band Category 1 (BC1):* MSR is defined into frequency bands that simultaneously support two RATs, namely UTRAN FDD and E-UTRAN FDD.
- *Band Category 2 (BC2):* MSR is defined in frequency bands that simultaneously support three RATs, namely GSM/EDGE, UTRAN FDD and E-UTRAN FDD.
- *Band Category 3 (BC3):* MSR is defined in frequency bands that simultaneously support two RATs, namely UTRAN TDD and E-UTRAN TDD.

It has to be noted that no mixture of FDD and TDD RATs is foreseen for MSR. Table 14.1 depicts the assignment of BC1 and BC2 band categories to paired frequency bands using FDD.

The support of GSM/EDGE is foreseen merely for Band Category 2 for GSM 850, E-GSM 900, DCS 1800 and PCS 1900 frequency bands.

14.2.3   Standardization in 3GPP GERAN

The standardization work for MSR is reflected in the following documents:

- An overview of the specification impacts due to the standardization of the MSR base stations is provided in the Work Item Technical Report 3GPP TR 37.900.
- The core specification for MSR radio performance is included in 3GPP TS 37.104 detailing the TX and RX performance requirements for the specified three band categories, both in

Table 14.1   Paired frequency bands for MSR operating 3GPP RATs, i.e. GSM/EDGE, UTRA FDD and E–UTRA FDD

E–UTRA Band	UTRA Band	GSM Band	Uplink (UL) BS receive UE transmit	Downlink (DL) BS transmit UE receive	Band category
1	I	–	1920–1980 MHz	2110–2170 MHz	1
2	II	PCS 1900	1850–1910 MHz	1930–1990 MHz	2
3	III	DCS 1800	1710–1785 MHz	1805–1880 MHz	2
4	IV	–	1710–1755 MHz	2110–2155 MHz	1
5	V	GSM 850	824–849 MHz	869–894 MHz	2
6	VI	–	830–840 MHz	875–885 MHz	1
7	VII	–	2500–2570 MHz	2620–2690 MHz	1
8	VIII	E–GSM	880–915 MHz	925–960 MHz	2
9	IX	–	1749.9–1784.9 MHz	1844.9–1879.9 MHz	1
10	X	–	1710–1770 MHz	2110–2170 MHz	1
11	XI	–	1427.9–1452.9 MHz	1475.9–1500.9 MHz	1
12	XII	–	698–716 MHz	728–746 MHz	1
13	XIII	–	777–787 MHz	746–756 MHz	1
14	XIV	–	788–798 MHz	758–768 MHz	1
...					
17	–	–	704–716 MHz	734–746 MHz	1*
18	–	–	815–830 MHz	860–875 MHz	1*
19	XIX	–	830–845 MHz	875–890 MHz	1

Note: * The band is for E–UTRA only.

mixed RAT mode, that is MSR operates different RATs simultaneously and in single RAT mode, that is MSR operates only one RAT.
• The MSR conformance testing is specified in 3GPP TS 37.141, depicting the selected test scenarios for the specified three band categories BC1, BC2 and BC3.

3GPP GERAN is contributing to the work led by 3GPP RAN on MSR standardization for equipment foreseen to support GSM/EDGE together with other 3GPP RATs dealt with in BC 2 category. The following aspects are treated in this context:

14.2.3.1   Manufacturer's Declaration

The manufacturer's declaration includes the declaration of the manufacturer related to the total output power of the MSR BS, the maximum power per supported RAT and the maximum carrier power as well as the maximum supported carrier number in case of a multi-carrier BS for a specific RAT. To allow a more flexible power distribution in case of a multi-carrier BS for a specific RAT, the specification of a maximum power difference between multiple carriers within one RAT is foreseen. According to TR 37.900, the manufacturer of a MSR base station should declare the following parameters according to the supported RATs:
   General parameters:

• supported band(s);
• the maximum RF bandwidth supported by a MSR BS within an operating band;
• the rated total output power as a sum over all RATs;
• maximum supported power difference between carriers;

   RAT specific parameters:

• whether the MSR BS supports GSM/UTRA or E-UTRA carriers;
• the maximum number of supported GSM/UTRA/E-UTRA carriers;
• the maximum RF bandwidth supported by the MSR BS when configured with GSM carriers only/when configured with UTRA carriers only/when configured with E-UTRA carriers only;
• which of the E-UTRA channel bandwidths specified in TS36.104 clause 5.6 are supported;
• the rated output power for GSM/UTRA/E-UTRA as a sum of all GSM carriers/all UTRA carriers/all E-UTRA carriers;
• the rated output power per GSM carrier/per UTRA carrier/per E-UTRA carrier.

14.2.3.2   MSR Single RAT Performance

Based on the work task description in GERAN [2] to adapt existing GERAN single RAT requirements to an MSR specification aiming at minimizing changes to these requirements, the MSR base station is required to exhibit almost the same performance as a GSM MCBTS (either class 1 or class 2) if being operated in single RAT mode, that is if only GSM/EDGE is operated by the MSR BS. The same is valid, if only UMTS or LTE is operated by the MSR BS, where the existing requirements for the corresponding RAT apply.

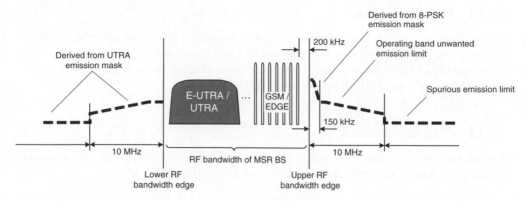

Figure 14.2 Generic UEM mask principle for MSR BS in Band Category 2 with GSM adjacent to the RF bandwidth edge.

14.2.3.3   TX Requirements

Transmitter requirements comprise aspects like modulation quality, frequency error, transmitter intermodulation and unwanted emissions. Unwanted emissions can be distinguished into operating band unwanted emissions which apply within the downlink operating band and in addition the frequency ranges 10 MHz above and 10 MHz below this band and into transmitter spurious emissions, defined further apart from the used spectrum. For MSR BS transmitting different 3GPP radio access technologies in one contiguous frequency block, that is applying the mixed RAT mode, it is essential to define the level of unwanted emissions. Figure 14.2 depicts an unwanted emission mask (UEM) for Band Category 2.

The unwanted emission is defined outside the operating bandwidth, that is the RF bandwidth of the MSR BS. In this scenario GSM/EDGE is allocated close to the upper RF bandwidth edge, that is with an offset of 200 kHz for its centre frequency, thus requiring that the UEM is based on the 8-PSK spectrum mask as used for EDGE, which is slightly relaxed compared to the GMSK spectrum mask used for GSM. At 150 kHz the UEM needs to meet the UTRA emission mask derived from a carrier power level of 43 dBm, which needs to be met in the given scenario for the frequency range below the allocated frequency block, since UTRA or E-UTRA (5 MHz and higher bandwidth) is operated close to the lower RF bandwidth edge, and both have the same unwanted emission requirements.

Thus the unwanted emissions for MSR comprising the operating band unwanted emissions as well as the transmitter spurious emissions are based on the emission mask of a UTRA carrier with a carrier power level of 43 dBm. In the case of GSM carriers or a narrowband E-UTRA carrier with 1.4 or 3 MHz bandwidth close to the RF bandwidth edge, the mask is relaxed as shown in Figure 14.2, following the 8-PSK spectrum mask for this carrier power level. The operating band unwanted emission mask and the spurious emission mask are hence both based on absolute emission limits in dBm per 30 kHz or per 1 MHz depending on the frequency offset from the RF bandwidth edge. It is noted that this definition of unwanted emission requirements for MSR follows the principle of Block Edge Masks (BEM) as applied in international standards, for example FCC Title 47 requirements in the US and requirements in the "WAPECS" decision in Europe.

14.2.3.4   RX Requirements

RX requirements are defined in terms of receiver sensitivity level, dynamic range, inband selectivity and blocking for a narrowband/wideband interferer, out-of-band blocking, receiver intermodulation for a narrowband/wideband interferer as well as receiver spurious emissions and in-channel selectivity (the latter only for E-UTRAN).

14.2.3.5   EMC Requirements

Electro-magnetic compatibility requirements in terms of conducted emissions and unwanted radiated emissions of the equipment are also being specified for MSR BS.

14.3   Status in Regulatory Bodies

Work on MSR in the regulatory bodies will start once 3GPP has completed the standardization. The author's expectation is that the work will start in the first half of 2010. The Harmonized Standard for MSR in Europe will be created by ETSI TC MSG TFES with the inclusion of coexistence studies in CEPT ECC PT1. Due to the fact that MSR will use the same transmission masks as applied for GSMMCBTS, UTRAN and E-UTRAN for single standard base stations, the coexistence studies will have a limited scope only.

14.4   Use of MSR Base Stations in Multimode Networks

Multi-standard base stations operating different 3GPP radio access technologies are designed to be employed in multimode networks. Typically an operator has a contiguous frequency allocation in a frequency band and aims at serving users connected via GSM/EDGE, WCDMA/HSPA and via LTE. Since the population of multimode terminals is expected to increase steadily in the near future, the share of GSM only terminals will decrease and less capacity is needed for GSM/EDGE traffic. In this context, deployment of multi-standard base stations is advantageous since migration of users connected via state-of-the-art RATs, such as GSM/EDGE and WCDMA/HSPA, to most recent 3GPP RATs, such as HSPA+, Dual Cell HSPA and LTE, is provided on a flexible basis from the start without need for a later major upgrade of the radio access network.

The expected increase of multimode terminals upholding two or more 3GPP RATs also gives support to considerations of establishing a common radio resource management between the different supported RATs. For instance, if the traffic load on GSM/EDGE carriers of such MSR base station is high, the Common Radio Resource Management (CRRM) may hand over some users to UTRA or E-UTRA carriers, actually carrying lower traffic loads. This is not a new mechanism, since common radio resource management between different single RAT base stations was already introduced in 3GPP Release 5 by using the RAN Information Management (RIM) procedure. The major difference here is that the same radio end point is used for the handover and hence the handover can proceed in a faster way due to lower delays for control plane signaling.

One technique being investigated in this context is spectrum sharing. Thus three operational modes can be distinguished:

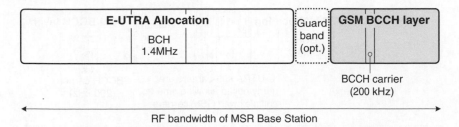

Figure 14.3   No spectrum sharing between different RATs.

1. *No spectrum sharing*. This scenario is depicted in Figure 14.3. The operator deploys MSR base stations serving different RATs, whereby each RAT has a dedicated allocation. For instance, assume that GSM/EDGE and E-UTRA carriers may be operated by a MSR BS. The GSM/EDGE allocation is reduced to one BCCH carrier as only the BCCH layer is deployed. Nevertheless the RF bandwidth of the MSR base station needs to include the complete BCCH layer, which may be as large as 2.4 MHz in case of BCCH reuse 12, the precise allocation being the subject of network planning. For E-UTRA a frequency reuse 1 is assumed. An optional guard band may be used for higher isolation in downlink reception. However, MSR base stations generally avoid such guard bands between different RATs due to implicit coordination between RATs with MSR. If GSM/EDGE traffic needs to be carried, there is only one solution left: the operator shifts dual mode terminals to the RAT with higher throughput performance, that is to E-UTRA, to increase capacity for GSM only terminals. Beyond this, no further flexibility exists.

2. *Static spectrum sharing*. Static spectrum sharing allocates the same resources to different RATs in a pre-assigned manner. For instance, the MSR base station may operate GSM/EDGE and LTE in a specific geographical area. Both spectrums are allocated next to each other and typically no guard band is needed. If LTE traffic decreases or GSM/EDGE traffic increases, some E-UTRA sub-carriers or resource blocks, respectively, will not be needed and hence can be removed from the E-UTRA transmission. Note, only a restricted subset of all E-UTRA sub-carriers may be nullified, while the BCH allocation remains, since neighbor cell measurements of E-UTRAN cells are based on all E-UTRA sub-carriers belonging to the BCH allocation with a 1.4 MHz bandwidth. The resource of the nullified E-UTRA sub-carriers can be reused for carriers operating dedicated GSM/EDGE traffic

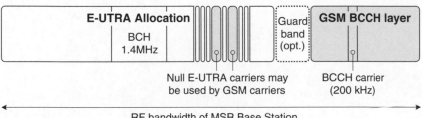

Figure 14.4   Example for static spectrum sharing in the case of simultaneous support of GSM/EDGE and E-UTRA by MSR BS.

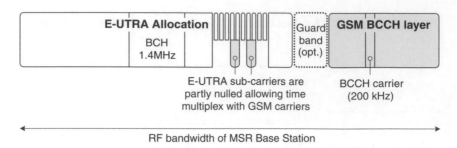

Figure 14.5 Example for dynamic spectrum sharing in case of simultaneous support of GSM/EDGE and E-UTRA by MSR BS.

channels and is beneficial for increasing interference diversity in the case of activated GSM slow frequency hopping. The spectrum sharing is considered static as the configuration only changes slowly over time, for example once per day between day and night operational mode. An exemplary static spectrum sharing operation is shown in Figure 14.4.

3. *Dynamic spectrum sharing*. Dynamic spectrum sharing offers an even increased flexibility for adjusting the traffic loads in different RATs on top of static spectrum sharing. This is achieved, for instance, by applying a time multiplex for the spectrum used by some E-UTRA sub-carriers to be used both for LTE and GSM/EDGE networks. The joint common radio resource management defines the details for the time multiplex and the relevant E-UTRA sub-carriers or E-UTRA resource blocks which are involved. An exemplary dynamic spectrum sharing operation is shown in Figure 14.5.

References

[1] RP-080758, "Work Item Description: RF requirements for Multi-carrier and Multi-RAT BS", TSG RAN#41, September 2008.
[2] GP-081945, "Work Item Description: RF requirements for Multi-carrier and Multi-RAT BS, GERAN Part", TSG GERAN#40, November 2008.
[3] 3GPP TR 37.900 v1.0.0, "RF Requirements for Multi-carrier and Multi-RAT BS (Release 9)", September 2009.

15

Generic Access Network

Extending the GSM Paradigm

Juha Karvinen and Guillaume Sébire

15.1 Introduction

Generic Access Network (GAN), introduced in 3GPP Release 6 [1, 2] may be the furthest expansion of the GSM paradigm as it is not using GSM radio access at all to access GSM services. Instead it relies on any generic IP access network hence its name, GAN.

The core idea of GAN is to securely convey all the GSM-based services such as voice, SMS and data as well as GSM core network signaling (e.g. mobility management) over a generic IP network while also supporting seamless mobility with GERAN and in later releases full mobility with UTRAN as well.

15.2 Drivers for Convergence

The definition of a generic access network was initially motivated by the Unlicensed Mobile Access (UMA) Consortium and later adopted by 3GPP in Release 6 as "GAN". Aiming at offering GSM and GPRS services over an unlicensed radio access such as Wireless LAN (WLAN), the primary drivers behind the UMA/GAN work (hereafter referred to as GAN) were:

- a desire to improve the existing cellular coverage (especially in the US) while faced with the costs and difficulty of new sites acquisition;
- the rapid growth in the penetration of fixed broadband technologies;
- the availability of inexpensive and fast unlicensed radio technologies such as 802.11 (WLAN) and even Bluetooth;
- the threat to incumbent telecoms created by the expansion of (virtually) free internet VoIP calling, in particular in the home environment.

GSM/EDGE: Evolution and Performance   Edited by Mikko Säily, Guillaume Sébire and Eddie Riddington
© 2011 John Wiley & Sons, Ltd

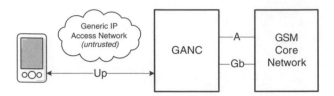

Figure 15.1   Simplified general GAN architecture.

15.2.1   Innovation on Top of Existing GSM Systems

The core innovation of GAN, and a precondition for its success, resides in its ability to reuse the GSM core network capabilities (services and signaling functions) without any changes to existing GSM network elements. That being said, it does not even require any changes to this infrastructure to support active call mobility between these two accesses, GAN and GSM. This is achieved by the camouflage ability of both the GAN access (GAN "cell"), which appears as a GSM cell on the GSM radio interface, and of the GAN controller (GANC), which appears as a BSC from the Core Network viewpoint, see Section 15.3.

15.3   GAN Architecture

15.3.1   General GAN Architecture

The general (simplified) GAN architecture is depicted in Figure 15.1. GAN is characterized by two new main elements: the GAN controller (GANC) and the Up interface (and associated protocols) between the mobile station and the GANC.

The GANC implements the same A and Gb interfaces towards the core network as a GSM BSC does. It is thus seen as a BSC by the core network, as is further explained in Sections 15.3.2, 15.3.3, 15.3.4 and 15.3.5. It interfaces with the mobile station by means of the Up interface, a secure interface which provides IP transport over a (typically untrusted) IP broadband access network such as a fixed line broadband internet accompanied by WLAN or Bluetooth unlicensed radio access. The GANC is therefore responsible, via a security gateway (SEGW), for securely tunneling all traffic between the mobile station and the GANC over the generic IP network while ensuring (seamless) mobility with cellular access (GSM or even UMTS). In the following sub-sections, the required protocol architectures are introduced.

Three types of GANC are defined, each serving a specific (logical) role and helping to distribute the load in the system: the provisioning GANC, the default GANC and the serving GANC:

1. *Provisioning GANC.* Located in the home network of a mobile station, the provisioning GANC is the first GANC contacted by the mobile when initially attempting to gain access to GAN. The contact details of the provisioning GANC reside in the phone itself, or can be derived from information on the SIM card. This GANC supplies the mobile station with the contact details (e.g. IP address) of the default GANC to which it can register.

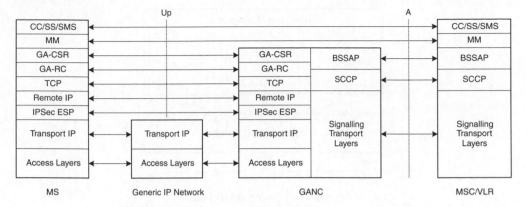

Figure 15.2   GAN CS domain control plane protocol architecture.

2. *Default GANC*. Located in the home network of a mobile station, the default GANC indicated by the provisioning GANC is the one to which the mobile station registers for GAN access, see Section 15.4.3. Upon successful registration, the default GANC may redirect the mobile to a more suitable serving GANC in the same or different network as required or may itself become the serving GANC.
3. *Serving GANC*. Located in the home network of a mobile station or in a visited network, it provides the registered mobile with GAN services. The serving GANC may also redirect the mobile station as necessary to a different serving GANC.

15.3.2   CS Control Plane Protocol Architecture in GAN

As illustrated in Figure 15.2, the GAN CS control plane protocol architecture is characterized by the transparent exchange of GSM mobility management and other core network signaling between the mobile station and the MSC and the reuse as is of the A interface and its protocols. The Up interface provides the actual GAN-related protocols GA-RC[1] and GA-CSR.[2] GA-RC manages the IP connection between the GANC and the MS including the GANC registration procedures. GA-CSR performs an equivalent though lighter functionality to that of GSM-RR; it provides bearer establishment for circuit-switched traffic, mobility between GERAN and GAN such as handover, and other functions such as paging, etc.

All this signaling traffic is then sent through an IPsec[3] tunnel providing a secure link between the MS and the GANC over the IP transport and thus overcoming any potential security issue over the Generic IP network. No assumption is indeed taken as to which access layer is used for the Generic IP network. In practical implementations it is in most cases 802.11 wireless (Wireless LAN). In some implementations Bluetooth radio or even Ethernet wired connection have also been used.

---

[1] GA-RC: Generic Access Resource Control.
[2] GA-CSR: Generic Access Circuit Switched Resources.
[3] IPsec: Internet Protocol security.

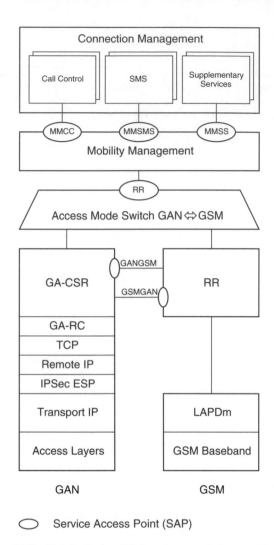

Figure 15.3   Mobile Station CS domain control plane architecture.

One of the key functionalities of GAN is the ability to switch between cellular connectivity and GAN connectivity. In the terminal, this requires a corresponding coordination between GA-CSR and GSM RR protocols for mobility. This capability is well illustrated in Figure 15.3 describing the mobile station's internal architecture (GSM and GAN).

The access mode switch is used to switch between the radio interfaces (GSM and GAN) as a function of their availability and the preferred access mode set in the mobile. There are four different (self-explanatory) access modes: GAN preferred, GAN only, GERAN preferred and GERAN only (see Section 15.4.4). In practical implementations the mode "GAN preferred" is maybe the most used, as it enables GAN to be used whenever the IP connectivity to the GANC is available in a given location.

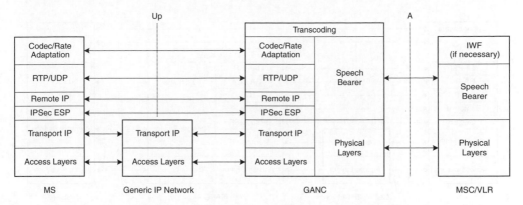

Figure 15.4 GAN CS domain user plane protocol architecture.

15.3.3 CS User Plane Protocol Architecture in GAN

As illustrated in Figure 15.4, the GAN CS user plane protocol architecture is characterized by the reuse as is of the A interface and its protocols and by the use of RTP/UDP to convey user traffic between the mobile station and the GANC. The user traffic can be voice, data or even CS video telephony. CS video can be emulated over GAN in the same way as "CS voice" is conveyed over the GAN IP-based bearer. Figure 15.4 represents the user plane protocol architecture for voice specifically but in principle the same figure also applies to the CS video call case.

The only speech codec mandated in GAN is narrowband AMR (full rate codec). However, the specification also allows use of other codec such as AMR Wide Band (AMR-WB), poised for deployment in GSM networks at the time of writing.

As in the case of the control plane, all user traffic encapsulated within RTP/UDP is then sent through the IPsec tunnel and over the underlying Generic IP network.

15.3.4 PS Control Plane Protocol Architecture in GAN

As illustrated in Figure 15.5, the GAN PS control plane protocol architecture is characterized by the transparent exchange of GPRS LLC and upper layers' protocol information between the mobile station and the SGSN and the reuse as is of the Gb interface and its protocols. The Up interface provides in addition to GA-RC (see Section 15.3.2), GA-PSR.[4] GA-PSR performs an equivalent but narrower functionality to that of the GSM-RLC protocol though adapted to operation over an IP connection.

Similar to the CS side, the ability of terminals to switch between GPRS connectivity and GAN connectivity is illustrated in Figure 15.6, where the necessary coordination between GA-PSR and GPRS (RLC/)MAC protocols for mobility is also represented.

---

[4] GA-PSR: Generic Access Packet Switched Resources.

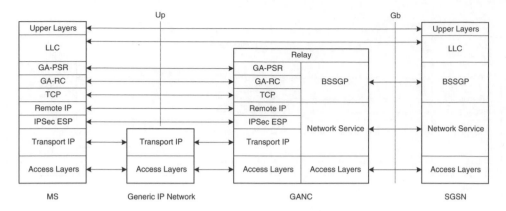

Figure 15.5   GAN PS domain control plane protocol architecture.

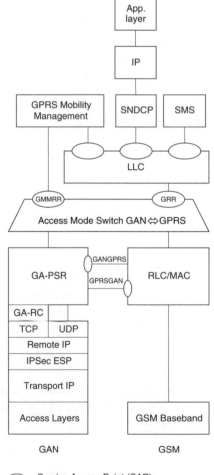

Figure 15.6   Mobile Station PS domain control plane architecture.

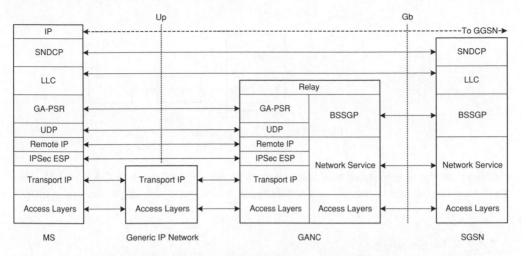

Figure 15.7   GAN PS domain user plane protocol architecture.

15.3.5   PS User Plane Protocol Architecture in GAN

As illustrated in Figure 15.7, the GAN PS user plane protocol architecture is characterized by the transparent exchange of LLC, SNDCP and IP traffic between the mobile station and the core network, and the reuse as is of the Gb interface. The upper IP layer is conveying any user PS data to/from the GGSN in a transparent way compared to normal GPRS usage.

15.4   GAN Management Functionality

15.4.1   Modes of Operation: GAN or Cellular

Knowing the possible states of a GAN-capable mobile station is essential to understand its behavior in different situations. Two modes are defined guiding the operation of the GAN-capable MS:

1. *GERAN(/UTRAN) mode (default)*, when the MS is not in GAN mode and is seeking access to or being served by GERAN (or UTRAN).
2. *GAN mode*, when the MS is served by a GANC.

The MS can be in only one mode of operation at a time. Note, however, that, despite this, a mobile station for example in GERAN(/UTRAN) mode served by a GSM cell, potentially with an ongoing call, can at the same time seek GAN connectivity. "Dual-radio" operation[5] is indeed assumed for GAN as there is no coordination between the 3GPP radio access and the generic access which can use practically any IP access. The wide and growing availability of GSM mobiles supporting multi-radio operation using Bluetooth and possibly Wireless LAN

---

[5] Dual-radio operation: when two radios are simultaneously active for example GSM and WLAN.

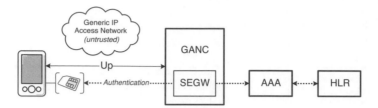

Figure 15.8 GAN authentication.

demonstrate the feasibility of this assumption. The use of dual-radio in GAN is, however, transparent to the user as the services are always provided through a single access at a time.

15.4.2   Authentication

As the MS approaches the GANC over the generic IP access, there is a need to verify and confirm the subscriber's identity before granting it access to the GSM network and its services. The authentication procedure in GAN is based on the same GSM credentials on the SIM card as used in GSM authentication. To this end, the EAP-SIM[6] protocol along with the Internet Key Exchange version 2 (IKEv2) protocol are utilized.[7] On the network side the credentials are authenticated using the security gateway residing in the GANC (GANC-SEGW) interfaced with an AAA[8] -server having access to the subscriber's information in the authentication center or HLR[9] of the network. See Figure 15.8. Upon completion of the authentication procedure, a secure signaling link is achieved between the MS and the GANC after which registration can proceed.

15.4.3   GAN Registration Procedure

Upon successful authentication and establishment of a secure signaling link with GAN, the mobile station (in GERAN/UTRAN mode) then needs to perform registration in order later to be able to access and receive GSM services through the appropriate (serving) GANC. Registration serves indeed two main purposes:

- For the MS to inform the GAN network that it is now cleared (i.e. authenticated) and can be reached in a secured way over GAN.
- For the network to select the appropriate GANC through which GSM services will be accessed by the MS, that is the serving GANC and notify the MS accordingly, so it is able to enter GAN mode ("rove-in").

---

[6] EAP: Extensible Authentication Protocol.
[7] EAP-SIM in IKEv2 is used to authenticate WLAN terminals containing a SIM (and no USIM) [3].
[8] AAA: Authentication, Authorization, Accounting.
[9] HLR: Home Location Register.

Upon successful registration, the mobile station is informed of the parameters that camouflage the GAN access (GAN "cell") as a fake GSM cell in the GSM network, that is an ARFCN[10] and a BSIC.[11] This information can then be used by the GSM network and the mobile station with no impact on GSM signaling for mobility between GSM and GAN. Of course, to ensure the correct operation, these parameters must not conflict with those of existing GSM cells in the network (or at minimum with those of neighboring GSM cells of a GSM cell with which GAN interaction is possible). The simplest alternative is thus to select a unique ARFCN, BSIC pair across the entire network, to ensure the GAN access is always identified in the same way.

15.4.4 Mode Selection in Multimode Terminals

As discussed earlier, there are two main modes of operation for a Generic Access capable terminal: GAN mode and GERAN(/UTRAN) mode. After a normal power up procedure, a Generic Access capable terminal enters GERAN/UTRAN mode, after which it may enter GAN mode as a function of user and/or operator configurable preferences as follows:

- *GAN preferred*: the most used preference mode as the GAN terminal enters GAN mode whenever available (i.e. upon successful registration with the GANC).
- *GAN only*: this may be desired, for example if cellular use is not wanted. In GAN only mode, there is no handover to the cellular network. This implies that the full range of a certain access point, for example WLAN access point is perceived by the user.
- *GERAN/UTRAN only*: with this preference the GAN MS is not attempting to enter GAN mode at all. This may be desired, for example if the user wants to use the CS video call feature over UTRAN and that is not supported over GAN in the terminal. It may also be used as the initial sales package setting in the terminal when the GAN service activation has not been provisioned to it yet.
- *GERAN/UTRAN preferred*: this mode is typically not required by GAN-enabled operators and thus not supported in mobile stations' implementations either. There are very few real-life scenarios where this mode would be beneficial. The most potential use case might be when the cellular network has poor coverage and GAN is then used to fill coverage holes in a kind of "back-up" mode. In practice, however, "GAN preferred" also addresses this situation and does it more elegantly.

15.4.5 PLMN Selection

Principally the PLMN selection is unaffected by the introduction of GAN, that is the same mechanisms as used in GSM do apply. A GANC can be connected to only one PLMN. The available PLMNs for a mobile station are listed for GAN and GERAN/UTRAN separately. In some cases, more than one GANC-PLMN pairs is offered to the mobile station at registration. If the currently used PLMN in GERAN/UTRAN is among those, it is selected. Otherwise the selection is made based on implementation specific criteria. So the saying "the devil is in the detail" applies here as well.

---

[10] ARFCN: Absolute Radio Frequency Channel Number.
[11] BSIC: Base Station Identity Code.

15.5   Mobility Features in GAN

15.5.1   Idle Mode Mobility

Mobility between GAN and GSM in idle mode, that is when no active user transaction is ongoing is carried out by "rove-in" in the direction from GSM to GAN and "rove-out" in the opposite direction. Both almost always involve a switch between GERAN(/UTRAN) mode and GAN mode. Actually the idea of "rove in/out" is very close to the cell reselection procedure in the GSM networks. It is also worth noting that inbound mobility to GAN is only possible following successful registration to the GANC.

15.5.1.1   Rove-In

While in GERAN mode, following successful GANC registration, the mobile station is able to autonomously switch to GAN mode, that is "rove in" provided its mobility in the GSM network is not under control of the network. Once the access mode is switched to GAN mode (see Figures 15.3 and 15.6), the upper layers (mobility management) are notified that the mobile station is now served solely by GAN, that is the GSM radio protocols in the mobile no longer interact with upper layers. This means that for example possible mobile-terminated call paging is to be received from this point onwards via GAN not GSM. However, it should be noted that typically the GSM radio still needs to measure neighboring cells from time to time in order to maintain the possibility for mobility back to GSM (if needed). In GAN-only mode, this feature can be disabled and a so-called GSM radio *hibernation* can be applied to save the terminal's battery.

15.5.1.2   Rove-Out

While in GAN mode, the mobile station may monitor neighboring GERAN/UTRAN cells. When a suitable cell is known and upon loss or detachment of the IP connection, the mobile station may rove out of the GAN "cell" and restore normal GSM operation. Once the access mode is switched to GERAN mode (see Figures 15.3 and 15.6), the upper layers are notified that the mobile station is now served solely by the GSM access network.

It is worth noting that de-registration from the GANC may not occur when roving out. Indeed, the generic IP connection may be lost so suddenly that there is no time to de-register, for example when losing WLAN coverage on entering an elevator. However, in such a case, the GANC will automatically detect it has lost contact with the mobile station and then make an implicit GAN de-registration.

It should be noted as well that if the mobile station was in GAN-only mode (Section 15.4.4), it will inevitably remain in GAN-mode indicating "no service".

15.5.2   Active Mode Mobility

Mobility between GAN and GERAN/UTRAN in active mode, that is, when a transaction is ongoing between the mobile station and the network,[12] is handled by means of (network-controlled) cell reselection (for non time critical data) or handover (for time critical data e.g. voice call). In the direction from GSM to GAN, the role played by camouflaging the GAN

---

[12] The ongoing transaction may be for example a packet data session or a voice call.

"cell" as a fake GSM cell (see Section 15.4.3) is essential for it allows not only the mobile to request mobility to GAN, but also the network to order the mobile to move to GAN.

15.5.2.1 Measurement Reporting, Cellular to GAN mobility

When under network control, active mode mobility in a GSM network relies on the mobile station reporting measurements to its serving cell. These include measurements of the signal level or signal quality of neighboring cells,[13] as well that of the serving cell. The neighboring cells with the best signal are reported. Based on these measurement reports and other policies, the network is then able to select the best cell to which the mobile should go. Since GAN relies on a generic IP network, the signal level of the neighboring GAN "cell" (i.e. fake GSM cell) listed by the network cannot as such be measured and reported using the "GSM scale" results. Instead the mobile station in GERAN mode, successfully registered to the GANC, always reports a superb signal level for the GAN "cell", that is in practice the maximum possible level in the "GSM scale". This indicates a request from the mobile station to the cellular (e.g. GSM) network to move to GAN. The network may then order cell re-selection or handover to GAN as appropriate. The cleverness of this scheme is two-fold: this way the MS always gets to GAN when it has been successfully registered to it (in the background); and it is achieved with no changes to existing network elements.

15.5.2.2 Network-Controlled Cell Re-Selection

Network-controlled cell re-selection in active mode only occurs in the direction from GERAN/UTRAN to GAN and during an ongoing packet data session. Upon reception of a measurement report for the GAN "cell", the network may order the mobile station to move to the GAN "cell", that is "rove-in". The latencies of this operation are quite tolerable for most use cases such as web browsing since the GAN connection has already been established in the background and thus the needed time for the re-selection is in the range of similar plain GSM/GPRS operation.

15.5.2.3 Handover

Because handover ensures seamless mobility, it is a necessary mechanism for services that have tight delay requirements such as a voice call.

In the case of a cellular network for example GSM-to-GSM handover, all resources in the new cell and in the core network are assigned prior to the mobile station being handed over to that cell, thus guaranteeing the service gap is minimal. It is the responsibility of the network to assign and indicate these resources to the mobile station before it arrives in the cell.

In the case of GSM-to-GAN handover as dual-radio operation is used (see Section 15.4.1), there is no need to assign and indicate the (GAN) resources to the mobile station before it arrives in the GAN "cell". The connection path between the core network and the GANC must, however, be properly set up before handing over the mobile to GAN. The mobile station is able to maintain the ongoing voice call in the GSM cell until the call is fully established through the GANC.

---

[13] Only neighboring cells listed by the GSM cell may be measured and reported by the mobile station.

In the case of GAN-to-GSM handover, the same principles as in a GSM-to-GSM handover are used. All resources are assigned in the target GSM cell prior to the mobile station being handed over to that cell. Unlike the previous handover cases, however, it is the mobile that triggers the handover. The trigger may be for example a weak signal level from the WLAN access point, or an indication from the GANC that based on measurements on the RTP layer, the uplink voice quality is problematic, etc. depending on implementation decisions. Upon such a trigger occurring, the mobile station explicitly requests a handover from the GANC, after which the GANC proceeds with the handover in the same manner as a normal GSM-to-GSM handover. In its first release (Release 6), GAN-to-UMTS handover was supported in a similar manner as a GAN-to-GSM handover, and using the same principles as in a GSM-to-UMTS handover.

15.6   Voice Service over GAN

Voice telephony traffic in GAN is very similar to Voice over IP (VoIP) traffic: AMR voice frames are packetized by the Real Time Protocol (RTP) and then carried over UDP/IP packets. However, the voice traffic stream going through the IPsec tunnel voice security in GAN is higher than in average internet VoIP. The most usual voice sample transmission period in GAN is 20 ms as in GSM, but also longer periods such as 40 ms can be supported, at the expense of end-to-end latency. The benefit from the longer sample period would be an improved power saving on the air interface. As the voice traffic in GAN is packet-based, it also requires a related jitter buffer to mitigate the fact that the voice packets may arrive at the receiver's end in the wrong order as well as at different intervals. The length and characteristics of the jitter buffer are an implementation-specific topic. The GAN specifications also include an RTP redundancy feature where the reception certainty at the receiving end is increased by including with each voice packet the nth and $(n-1)^{th}$ speech frame at the senders' end. Thus if some of the packets are lost, the receiving end can still receive the lost piece of voice data from the consecutive RTP packet.

15.7   Supplementary Services and SMS over GAN

15.7.1   Supplementary Services

GSM specifications have a large amount of standardized Supplementary Services (SS) (see Chapter 1). The DTAP[14] signaling related to SS is relayed over GAN like any layer 3 level call control or other signaling. So in most cases there are no changes to the existing SS offering over GAN.

It is worth noting that in practice, the implementation of certain supplementary services such as "Call Hold" will also require following all relevant RTP related IETF RFCs,[15] where, for example, the use of Silence Insertion Descriptor (SID) frames during a discontinuous transmission (DTX) mode is specified. Another example is to address requirements which may be relevant only for some regions. For example, the US Text Telephony (TTY), which

---

[14] DTAP: Direct Transfer Application Part.
[15] IETF: Internet Engineering Task Force. RFC: Request For Comment.

can be implemented over the GAN voice channel, requires additional implementation work compared to simple DTAP signaling-based SS message relaying over GAN.

15.7.2 SMS

SMS is always conveyed over GAN when the MS is in GAN mode. Typically the implementation of the application layer for SMS service in the mobile station remains intact. Also the network side implementation is straightforward as the A interface carries the SMS traffic as usual. In principle, also "SMS over GPRS (over GAN)" could be supported but this has not been seen so far in practical implementations.

Furthermore, the Multimedia Messaging Service (MMS) can be supported as a combination of SMS and packet data support over GAN. Packet data support is further explained in the following section.

15.8 Packet Switched Data (GPRS) over GAN

Even if predominantly thought to carry voice, GAN also supports packet data (GPRS-like) connections by default. The introduction of GPRS support over GAN in the network is quite straightforward as the same GANC supports the standard Gb interface towards a 3GPP standard SGSN network element, see Section 15.3.1. On the device side, there is practically no impact on applications using GPRS normally as they can continue using any GPRS APN[16] over GAN access. The decision on whether to use the GSM or the GAN access parts is taken on the (lower layer) access mode switch as already described in Figure 15.6.

It is worth noting that since the early days of GPRS, the maximum bit rate of the Gb interface has increased not only to benefit from the evolution of the underlying access, for example higher WLAN bit rate in GAN, but also to prevent this interface from being a bottleneck. The improved Gb interface can support up to 1000 times faster traffic than initially defined. In practice, it implies that Gb can accommodate more terminals with higher bitrates. Earlier the maximum supported peak rate was 6 Mbit/s, that is supporting at most six mobile stations with 1 Mbit/s each. Now it is up to 6 Gbit/s, so virtually any speed can be achieved, that is the throughput is no longer limited by the capacity of the Gb interface.

So where are the throughput limits of the GAN PS connectivity set then? If the radio access used in GAN is, for example, IEEE 802.11g, it can accommodate an air interface throughput of around 25 Mbit/s (theoretical maximum of 54 Mbit/s) which is thus definitely not the limiting factor in this case either. With GAN, the GPRS upper layer frames are packed on top of UDP frames which then are carried through the IPsec tunnel over the 802.11g access. This multilayered protocol stack cuts some of the perceived end user performance but is typically at least on a par with the best case EGPRS or WCDMA perceived throughputs, in the order of a few hundreds of kbit/s.

A typical operator requirement for GAN today is 256–600 kbit/s. A GAN MS can easily reach 1 Mbit/s and actually even 2 Mbit/s is feasible with the current even mid-priced mobile engines. Some further throughput improvements are expected with GAN Release 8 adding UTRAN Iu-interface support (with a slightly thinner UMTS-PS protocol stack).

---

[16] APN: Access Point Name.

15.9   Emergency Call Support in GAN

There is an operator-selectable option to prefer emergency calls either over GAN or over GERAN/UTRAN access. If the access network preference is GERAN/UTRAN, the mobile station must switch to GERAN/UTRAN mode when attempting to place an emergency call. This is in order to provide also the location information data of the terminal through the GERAN/UTRAN network.

Even if the network preference for emergency call is GAN, it is recommended but not mandatory to switch to GERAN/UTRAN mode if there is not sufficient location information available via GAN. The GAN location information can be based on A-GPS[17] location information or on the location information associated with the point of IP network attachment (e.g. at the accuracy of BSSID[18] of the used access point). This latter method, however, requires an information database maintained by the GAN operator in question, for example to map the BSSID with given locations.

There are also region-specific rules such as E911 from the US authority FCC which need to be taken into account when implementing an emergency call in GAN terminal as well.

15.10   GAN in 3GPP Releases

15.10.1   GAN in Release 6

GAN was initially introduced in 3GPP Release 6 [1, 2] after adoption of UMA work to 3GPP. In practice, the first Release 6 GAN corresponds to the very last UMA Consortium release 1.0.4. Although not part of UMA Release 1.0.4, GAN Release 6 supports also Cell Broadcast.

15.10.2   GAN in Release 7

GAN was expanded in Release 7 mainly by the addition of circuit-switched video telephony and packet-switched handover. The support in GAN of the same 64 kbit/s CS video telephony bearer as used in for example UMTS is emulated over similar RTP/UDP over IP methods as used for a voice call, and as described earlier in Section 15.3.3. In practice, however, a native IP-based video calling application could, of course, be better suited for GAN which is an IP-based system.

GAN mobility procedures in active mode are also complemented by the addition of packet-switched handover. Similar to CS handover, PS handover allows seamless mobility between cellular access and GAN for packet data services with tight delay requirements.

Also, as previously mentioned, the definition of the "Gigabit Gb interface" (between a BSC or a GANC and an SGSN), see section 15.8, means that the achievable data rates over the radio access are no longer limited by the capacity of the Gb interface. It also means that more mobiles can be supported. For GAN, this means that more mobiles can use better data rates over the generic IP network for packet data transmission.

---

[17] A-GPS: Assisted GPS.
[18] BSSID: MAC address of the "access point".

15.10.3   GAN in Release 8

The major addition to GAN in Release 8 is that of the Iu interface, that is "GAN Iu", enabling the expansion of GAN within UMTS, using the same principles as when GAN was introduced in GSM. It also offers full (Iu) mobility procedures with UTRAN. With GAN Iu, GAN has also been promoted as a "femto" technology while it fulfills an essential requirement of attachment to the core network in a standardized, secure and easily scalable manner.

Beneficial to GAN network implementation, the specification of IP transport for the user-plane traffic over the A interface (complementing that of the control plane specified in Release 7), as seen in Chapter 3, means it is also better suited to carry GAN user plane AMR/UDP/IP packets.

15.11   Implementation Aspects for a GAN-enabled Device

Whereas the GAN impact in the network side can be isolated almost to one new network element (GANC), the impact on the MS can be listed as follows:

- Off-the-shelf type of features or components like a (standard) WLAN radio, TCP/IP stack, IPsec tunneling with IKEv2, etc. which in most cases can already be present but that may require some further tweaking because of the real-time and mobility aspects of the services running over GAN.
- A few fairly simple protocols and functionalities accommodating the signaling between the 3GPP core and the "IP world".
- Complementary parts for the product creation such as user-interface menus related to GAN and WLAN, new icons to distinguish GAN access from cellular access (e.g. a "house" symbol indicating all incoming and outgoing traffic goes through GAN), and also items related to out-of-the-box experiences such as provisioning for the service or any kind of tutorial when putting it into use.

15.12   Considerations for GAN Deployment

15.12.1   Systematic Approach to GAN Set-up

The system is as strong as its weakest link is a rule that applies also to telecommunication systems such as GAN. The high quality implementation of both the GANC network entity as well as its sub-entity SEGW is highly significant, given the deployment of GAN within an existing system.

Going towards the MS, there needs to be an IP network providing sufficient bandwidth for the GAN operation. In practice, the minimum throughput for the bidirectional voice over GAN service is in the range of 100 kbps, whereas for packet data over GAN the throughput of the used IP network should be at a minimum 350–400 kbps to accommodate the typically required 256 kbps user perceived data speed for GAN. Normally the throughput of the IP network is not an issue as even the lowest fixed-line broadband connections are sufficient. A more severe impact, however, can be noted in the voice quality if the IP network is badly congested. In this case there can be gaps in the voice which even a good jitter buffer implementation cannot cope with. As stated earlier, there are ways in the mobile station to detect such an abnormal situation and make a request for handover to GSM.

In most GAN implementations so far, the dominant radio access technology used between the IP broadband network (such as ADSL) and the mobile station is based on 802.11 WLAN. 802.11 WLAN is specified under the IEEE regime but in addition the Wi-Fi Association has defined additional feature sets and is also facilitating testing plug-fests for those feature sets. It is highly recommended to follow these definitions and to conduct the necessary testing for them. Indeed, there is quite a lot of variance in the performance of different WLAN chipset products which means that, if not carefully verified, the WLAN "cell" range may be shorter than planned. Also the parameterization of handovers between cellular and GAN must be done properly to exploit the WLAN range, while avoiding the increase of handover failures and call drops.

The MS supporting GAN at the end of production line is not ready for GAN service at start-up for the first time unless it is provisioned with operator-specific settings (e.g. provisioning GANC contact details including SEGW) and a certificate from a (trusted) Certificate Authority for security reasons. However, these can also be provided over-the-air at a later phase (e.g. the CA certificate can be securely located with the help of a mobile browser). Assuming that there is no additional operator-specific feature needed, this is in principle sufficient to start GAN service with the mobile.

References

[1] 3GPP TS 43.318 v6.12.0, "Generic Access Network (GAN); Stage 2", May 2008.
[2] 3GPP TS 44.318 v6.13.0, "Generic Access Network (GAN); Mobile GAN Interface Layer 3 Specification", September 2009.
[3] IETF RFC 4306, "Internet Key Exchange (IKEv2) Protocol", December 2005.

Index